The Circular Economy and their Management

The Circular Economy and their Management

Recycling, Reuse, and Waste-to-Energy Innovations

Vikas Khare
School of Technology, Management and Engineering, NMIMS, Indore
Madhya Pradesh, India

Monica Bhatia
School of Business Management, NMIMS, Indore, India

Miraj Ahmed Bhuiyan
Guangdong University of Finance and Economics, China

ELSEVIER

Elsevier
Radarweg 29, PO Box 211, 1000 AE Amsterdam, Netherlands
125 London Wall, London EC2Y 5AS, United Kingdom
50 Hampshire Street, 5th Floor, Cambridge, MA 02139, United States

For accessibility purposes, images in electronic versions of this book are accompanied by alt text descriptions provided by Elsevier. For more information, see https://www.elsevier.com/about/accessibility.

Books and Journals published by Elsevier comply with applicable product safety requirements. For any product safety concerns or queries, please contact our authorised representative, Elsevier B.V., at productsafety@elsevier.com.

Publisher's note: Elsevier takes a neutral position with respect to territorial disputes or jurisdictional claims in its published content, including in maps and institutional affiliations.

Notices

Knowledge and best practice in this field are constantly changing. As new research and experience broaden our understanding, changes in research methods, professional practices, or medical treatment may become necessary.

Practitioners and researchers must always rely on their own experience and knowledge in evaluating and using any information, methods, compounds, or experiments described herein. In using such information or methods they should be mindful of their own safety and the safety of others, including parties for whom they have a professional responsibility.

To the fullest extent of the law, neither the Publisher nor the authors, contributors, or editors, assume any liability for any injury and/or damage to persons or property as a matter of products liability, negligence or otherwise, or from any use or operation of any methods, products, instructions, or ideas contained in the material herein.

ISBN: 978-0-443-44617-7

For information on all Elsevier publications visit our website at https://www.elsevier.com/books-and-journals

Publisher: Matthew Deans
Acquisitions Editor: Glyn Jones
Editorial Project Manager: Arushi Rawat
Production Project Manager: Kiruthigadevi
Cover Designer: Bakyalakshmi S

Typeset by TNQ Tech

Dedication

This book is dedicated to the visionaries, practitioners, researchers, policymakers, and communities who believe that progress does not have to come at the cost of the planet. It is dedicated to those who understand that economic growth and environmental responsibility are not opposing forces, but interconnected dimensions of a sustainable future. We dedicate this work to the generations before us who lived closer to nature, often practicing reuse and resource efficiency not as a strategy, but as a way of life. Their intuitive understanding of balance, restraint, and stewardship laid the earliest foundations of what we now call the circular economy. Long before modern terminology emerged, their practices reflected wisdom that today's world urgently needs to rediscover. This book is also dedicated to the present generation of innovators—engineers designing advanced recycling systems, entrepreneurs transforming waste into value, urban planners reimagining cities, and scientists developing waste-to-energy technologies that convert challenges into opportunities. Your work stands at the intersection of necessity and creativity, proving that sustainable solutions can be technologically robust, economically viable, and socially inclusive. We further dedicate this book to students and young professionals who are preparing to inherit a planet shaped by today's decisions. May this work inspire critical thinking, ethical responsibility, and bold innovation. May it encourage you to question linear systems of "take–make–dispose" and replace them with regenerative models that restore, renew, and respect natural systems. To policymakers and institutional leaders, this dedication acknowledges your role in shaping frameworks that enable circular transitions. The policies you design today will influence how industries operate, how cities manage waste, and how societies measure success—not merely in economic output, but in environmental resilience and social well-being. Finally, this book is dedicated to the Earth itself—our shared home. Every material recycled, every product reused, and every unit of energy recovered from waste is a step toward honoring the finite nature of planetary resources. The circular economy is not merely a technical or managerial concept; it is a moral commitment to coexist responsibly within ecological limits. May this book serve as both a tribute and a call to think systemically, act responsibly, and innovate courageously for a future where prosperity and sustainability move forward together.

Contents

Preface

The global economy stands at a critical crossroads. Rapid industrialization, urban expansion, and consumption-driven growth have delivered unprecedented material prosperity, yet they have also resulted in resource depletion, environmental degradation, and mounting waste challenges. The traditional linear economic model, extract, produce, consume, and dispose, is increasingly unsustainable in a world of finite resources and growing populations. It is within this context that the concept of the circular economy has emerged as both a necessity and an opportunity. This book, *The Circular Economy and Their Management: Recycling, Reuse, and Waste-to-Energy Innovations*, has been conceived to provide a comprehensive, interdisciplinary understanding of circular economy principles and their practical implementation. It aims to bridge the gap between theory and practice by exploring how circular strategies can be effectively managed across industries, municipalities, and energy systems. The focus on recycling, reuse, and waste-to-energy is deliberate. These three pillars represent critical leverage points in closing material loops and reducing environmental footprints. Recycling extends material life cycles, reuse preserves embedded value, and waste-to-energy technologies recover energy from residual waste streams that cannot be otherwise utilized. Together, they form a synergistic framework for transforming waste from an environmental liability into an economic and social asset. This book is intended for a diverse audience, including students, researchers, academicians, industry professionals, policymakers, and sustainability practitioners. The content integrates managerial perspectives with technological insights, recognizing that successful circular transitions require not only innovation in materials and processes, but also effective governance, strategic planning, stakeholder engagement, and behavioral change.

Throughout the chapters, emphasis is placed on systems thinking understanding material flows, energy cycles, and value chains as interconnected systems rather than isolated components. Case studies, emerging innovations, and global best practices are woven into the discussion to illustrate how circular economy principles are being applied in real-world contexts. The intent of this book is not to present the circular economy as a perfect or universal solution, but as an evolving framework that must be adapted to local conditions, cultural contexts, and economic realities. By critically examining both opportunities and limitations, this work encourages informed decision-making and responsible innovation. It is our hope that this book contributes meaningfully to the ongoing global dialogue on sustainability and serves as a practical guide for those committed to building resilient, low-carbon, and resource-efficient economies.

Acknowledgments

The completion of this book has been made possible through the support, guidance, and contributions of many individuals and institutions, to whom we extend our sincere gratitude. First and foremost, we acknowledge the academic and research communities whose work has laid the foundation for circular economy thinking. The insights, models, and empirical studies developed by scholars across disciplines, environmental science, engineering, economics, and management, have deeply informed the perspectives presented in this book. We express our heartfelt appreciation to colleagues and peers who provided valuable feedback, constructive discussions, and intellectual encouragement throughout the development of this manuscript. Their diverse viewpoints helped refine arguments, strengthen clarity, and ensure academic rigor.

We are grateful to industry professionals, sustainability practitioners, and policy experts whose real-world experiences enriched this work. Their practical insights into recycling systems, reuse strategies, and waste-to-energy technologies ensured that the book remains grounded in operational realities rather than purely conceptual frameworks. Special thanks are extended to educational institutions and research organizations that fostered an environment conducive to interdisciplinary exploration and sustainability-focused inquiry. Institutional support, access to resources, and collaborative platforms played a vital role in shaping this work.

We also acknowledge the contributions of students—past and present—-whose curiosity, questions, and engagement with sustainability topics continue to inspire deeper inquiry. Their enthusiasm reinforces the importance of education in driving long-term systemic change. The publishing team deserves sincere appreciation for their professionalism, patience, and commitment to quality. Their guidance throughout the editorial and production process has been invaluable in bringing this book to fruition.

Finally, we extend our deepest gratitude to our families and well-wishers for their unwavering support, understanding, and encouragement. The time and effort required to complete this work would not have been possible without their belief in the purpose behind it. This book is the result of collective effort and shared commitment to a more sustainable future. Any shortcomings remain our own, but its aspirations are shaped by the many minds and hands that contributed along the way.

Chapter 1

Introduction to the circular economy and clean energy

1.1 Introduction

The contemporary world is witnessing an unprecedented convergence of environmental, economic, and social challenges. Rapid industrialization, population growth, and urban expansion have led to the overexploitation of natural resources, environmental degradation, and escalating greenhouse gas emissions. These challenges have highlighted the urgent need to rethink the conventional linear economic model, which follows a "take-make-dispose" approach, and to embrace sustainable alternatives that harmonize economic growth with environmental stewardship. Among these alternatives, the concepts of circular economy and clean energy have emerged as pivotal solutions for shaping a resilient, sustainable, and low-carbon future. The circular economy represents a transformative shift from the linear paradigm. It is a regenerative system designed to minimize waste, optimize resource use, and extend the lifecycle of products through practices such as recycling, reuse, remanufacturing, and design for longevity. Unlike the traditional linear economy, which often treats waste as an inevitable by-product, the circular economy views waste as a resource that can be reintegrated into production cycles. This philosophy not only reduces environmental impact but also creates economic opportunities by unlocking the value embedded in materials that would otherwise be discarded. The circular economy, therefore, aligns closely with the principles of sustainable development, balancing economic growth, environmental preservation, and social well-being. Central to the circular economy is the concept of resource efficiency. Efficient use of materials, energy, and water reduces dependency on finite resources and mitigates the environmental footprint of industrial activities. For example, adopting eco-design principles enables manufacturers to produce goods that are easier to disassemble, repair, or recycle. Similarly, shifting towards service-based models, such as product-as-a-service offerings, encourages manufacturers to retain ownership and responsibility for their products, fostering a system in which materials circulate continuously. The integration of digital technologies,

The Circular Economy and their Management. https://doi.org/10.1016/B978-0-443-44617-7.00008-5

including the Internet of Things (IoT), artificial intelligence (AI), and blockchain, further enhances the efficiency of circular systems by enabling real-time monitoring, traceability, and optimization of resource flows. In parallel, the transition to clean energy is indispensable for achieving a sustainable and low-carbon economy. Clean energy encompasses energy derived from renewable sources such as solar, wind, hydro, geothermal, and biomass that produce minimal environmental impact and reduce dependence on fossil fuels. Fossil fuel-based energy systems have historically been the backbone of industrial development but are also the primary contributors to greenhouse gas emissions, air pollution, and climate change. The adoption of clean energy technologies not only addresses environmental concerns but also strengthens energy security, fosters technological innovation, and supports economic diversification. In essence, clean energy serves as the foundation upon which circular economic principles can be operationalized effectively. Fig. 1.1 illustrate interaction between circular economy & clean energy.

The intersection of circular economy and clean energy presents a unique opportunity to redefine sustainability. While circular practices reduce resource consumption and waste generation, clean energy solutions provide a low-carbon power source to support these practices. For instance, the recycling of materials in a circular system often requires energy inputs, which can be sustainably supplied by renewable energy sources. Similarly, the deployment of smart grids, energy storage, and micro-grids enhances the integration of distributed renewable energy into industrial and urban systems, thereby

FIGURE 1.1 Interaction between circular economy and clean energy.

reinforcing circular operations. This symbiotic relationship underscores the importance of adopting holistic strategies that simultaneously address resource efficiency, energy transition, and environmental protection. Several global trends underscore the urgency of embracing circular economy and clean energy. Climate change, resource scarcity, and environmental degradation have emerged as pressing concerns for policymakers, businesses, and civil society. The Intergovernmental Panel on Climate Change (IPCC) emphasizes that achieving net-zero carbon emissions by mid-century is critical to limiting global warming to 1.5°C above pre-industrial levels. Similarly, the United Nations Sustainable Development Goals (SDGs) highlight the interconnected nature of environmental sustainability, economic growth, and social equity. Goals such as responsible consumption and production (SDG 12), affordable and clean energy (SDG 7), and climate action (SDG 13) directly align with the principles of circular economy and clean energy, demonstrating their global relevance and strategic importance.

The shift towards circularity and clean energy is also driven by economic imperatives. Resource-intensive industries face rising costs due to material scarcity, regulatory pressures, and fluctuating commodity prices. Businesses that adopt circular practices and integrate renewable energy into their operations can achieve cost savings, reduce supply chain vulnerabilities, and enhance competitiveness. Furthermore, the development of circular and clean energy markets stimulates innovation, creates new business models, and generates employment opportunities in sectors such as renewable energy installation, waste management, recycling technologies, and sustainable product design. Governments and financial institutions are increasingly recognizing these opportunities, offering incentives, subsidies, and regulatory frameworks to accelerate the transition towards a circular, low-carbon economy. Technological innovation plays a central role in realizing the potential of circular economy and clean energy. Advanced materials, energy-efficient manufacturing, digital platforms for resource tracking, and renewable energy technologies collectively enable the design of resilient systems that minimize environmental impact. For instance, additive manufacturing or 3D printing reduces material waste by building products layer by layer, while smart sensors monitor energy consumption and optimize operational efficiency. Similarly, advancements in solar photovoltaic panels, wind turbines, and energy storage systems make renewable energy more accessible, affordable, and reliable. By leveraging these technologies, industries and societies can create a synergistic framework in which economic growth and environmental sustainability are mutually reinforcing rather than contradictory. In addition to environmental and economic benefits, circular economy and clean energy foster social well-being and resilience. Sustainable practices contribute to cleaner air and water, reduced health risks, and improved quality of life for communities. Moreover, inclusive participation in circular and renewable energy initiatives can empower local populations, enhance skill development, and promote equitable access to resources. For example, community-based renewable energy projects not only supply clean energy but also generate local employment and strengthen energy independence. Similarly, recycling and repair programs can create jobs in informal and formal sectors, supporting livelihoods while promoting sustainable consumption patterns. Despite the compelling benefits, the transition to a circular economy and clean energy system is not without challenges. Structural barriers, such as entrenched linear production models, inadequate infrastructure, lack of awareness, and policy gaps, hinder widespread adoption. Financial constraints, technological limitations, and behavioral resistance further complicate the shift. Addressing these challenges requires coordinated efforts among governments, businesses, academia, and civil society. Policy instruments, regulatory frameworks, public-private partnerships, and educational initiatives play a crucial role in creating enabling environments for circular and clean energy solutions. Additionally, fostering a culture of innovation, collaboration, and long-term thinking is essential for sustaining momentum and achieving systemic transformation. The integration of circular economy principles and clean energy technologies represents a transformative approach to sustainable development. By reimagining production, consumption, and energy systems, societies can reduce environmental impact, enhance resource efficiency, and create resilient, low-carbon economies. The circular economy provides a framework for optimizing material flows and minimizing waste, while clean energy offers a sustainable power source to support industrial, urban, and societal activities. Together, these paradigms offer a comprehensive strategy for addressing the pressing challenges of climate change, resource scarcity, and environmental degradation. As the world navigates the complex landscape of sustainability, understanding the principles, opportunities, and challenges of circular economy and clean energy becomes essential for policymakers, business leaders, researchers, and citizens alike. This chapter aims to provide a foundational understanding of these concepts, highlighting their relevance, interconnections, and transformative potential in shaping a sustainable future.

1.2 Status of the circular economy worldwide

The circular economy has emerged as a globally recognized framework for addressing pressing challenges related to resource depletion, waste generation, and environmental degradation. Governments, international organizations, and businesses increasingly acknowledge its potential to decouple economic growth from resource consumption. However, despite growing awareness and policy momentum, the actual global transition toward circularity remains slow and uneven. Current evidence indicates a significant gap between ambition and implementation, highlighting the need for systemic transformation rather than incremental improvements. At the global level, material circularity remains alarmingly low. Recent global assessments reveal that only a small fraction of materials circulating in the world economy are reused, recycled, or recovered. The majority of economic activity continues to rely on virgin material extraction, reinforcing the dominance of the traditional linear "take–make–dispose" model. Over the past few years, global circularity has not only stagnated but has declined, primarily due to rising material consumption driven by population growth, urbanization, and increased demand for infrastructure, consumer goods, and energy. This trend suggests that improvements in recycling and reuse are being outpaced by overall growth in material throughput. While material circularity remains limited, the economic landscape of the circular economy tells a more optimistic story. The global circular economy market has expanded rapidly, driven by investments in recycling technologies, waste management, renewable materials, and circular business models. Market projections indicate substantial growth in the coming decades, with circular economy activities expected to contribute significantly to global economic output. This expansion reflects increasing interest from industries seeking cost savings, supply chain resilience, regulatory compliance, and reputational benefits. Nevertheless, the growth of the circular economy market does not automatically translate into higher material circularity, as many initiatives focus on efficiency improvements rather than fundamental system redesign. Policy and institutional frameworks play a crucial role in shaping the global status of the circular economy. Over the past decade, numerous countries have introduced national circular economy strategies, action plans, and regulatory instruments. These policies include measures such as extended producer responsibility, eco-design standards, waste reduction targets, and recycling mandates. Among global regions, Europe has emerged as a leader in circular economy policy, with comprehensive regulatory frameworks and coordinated action at the regional and national levels. However, even in regions with strong policy support, the transition remains partial, and circular practices are often limited to specific sectors rather than integrated across entire value chains. Regional disparities in circular economy adoption are pronounced. Advanced economies generally exhibit stronger regulatory frameworks, better infrastructure, and higher levels of technological innovation, enabling more effective implementation of circular strategies. Several European countries, along with selected Asian economies, have embedded circular principles into industrial planning and urban development. In contrast, many developing and emerging economies face structural constraints, including inadequate waste management infrastructure, limited access to finance, and competing development priorities. Nevertheless, these regions also present significant opportunities, as informal recycling systems, community-based initiatives, and localized circular solutions can play an important role in advancing circularity while supporting livelihoods. Despite growing global momentum, multiple challenges continue to hinder the widespread adoption of circular economy principles. Existing economic systems are deeply rooted in linear production and consumption patterns, with market incentives favoring high-volume production and short product lifecycles. Technological barriers further complicate circular transitions, particularly in sectors dealing with complex products such as electronics, composites, and mixed plastics, where material recovery is technically difficult or economically unviable. Financial constraints, lack of standardized metrics, insufficient consumer awareness, and fragmented governance structures also slow progress and limit scalability. At the same time, the global transition toward a circular economy presents significant opportunities. Increasing corporate engagement reflects a growing recognition that circularity can enhance competitiveness, reduce risks associated with resource scarcity, and support long-term sustainability goals. Circular practices such as repair, remanufacturing, refurbishment, and recycling have strong potential for job creation, particularly in local and regional economies. Moreover, international organizations, development banks, and multilateral institutions are expanding financial and technical support for circular initiatives, especially in low- and middle-income countries. In summary, the worldwide status of the circular economy is characterized by a paradoxical combination of strong conceptual acceptance and weak material outcomes. While policies, investments, and market interest are expanding rapidly, actual progress in reducing resource extraction and closing material loops remains insufficient. Achieving a truly circular global economy will require transformative changes in production systems, consumption behavior, infrastructure development, and governance mechanisms. Coordinated global action, supported by technological innovation, inclusive financing, and long-term policy commitment, is essential to bridge the gap between circular economy aspirations and real-world impact.

1.3 Framework of the circular economy

The circular economy is an economic system designed to replace the traditional linear model of production and consumption by promoting the continuous circulation of materials, products, and resources within the economy. It emphasizes the reduction of resource extraction, minimization of waste, and regeneration of natural systems through strategies such as reuse, repair, refurbishment, remanufacturing, recycling, and sustainable product design. By maintaining the value of materials and energy for as long as possible, the circular economy seeks to decouple economic growth from finite resource consumption while simultaneously reducing environmental degradation and supporting long-term economic, social, and ecological sustainability. Fig. 1.2 illustrate framework of the circular economy.

The circular economy (CE) constitutes a normative socio-technical regime whose principal objective is to decouple aggregate value creation from virgin resource throughput and environmental externalities. Conceptually, it replaces the linear "take–make–dispose" metabolic sequence with a multi-loop system that prioritizes absolute reduction, lifetime extension, reuse, remanufacturing,

FIGURE 1.2 Framework of the circular economy.

and high-quality recycling. Material loops are closed through integrative eco-design, service-based business models, and advanced recovery technologies that reincorporate end-of-life products as productive inputs, thereby minimizing both front-end resource extraction and back-end pollutant emissions.

By internalizing ecological costs and distributional equity considerations, the CE framework offers a systemic strategy for mitigating planetary pressures such as resource scarcity, ecosystem degradation, and anthropogenic climate forcing.

Operationalization of the CE is guided by the hierarchical 3R principle reduce, reuse, recycle augmented by innovation in governance architectures and cognitive paradigms. This recursive configuration disrupts the unidirectional material flows characteristic of linear industrial systems and fosters regenerative feedback cycles that synchronize economic expansion with resource conservation and ecological resilience. Consequently, the CE reframes the relationship between anthropogenic activity and the biosphere, shifting the policy focus from end-of-pipe waste management to a fundamental reconfiguration of production–consumption networks in accordance with thermodynamic and ecological boundaries.

Reduction focuses on controlling production and consumption at their source, reducing resource input and waste generation through optimized design, lightweight production, and simplified consumption. Reuse emphasizes extending product lifecycles through repair, refurbishment, and secondhand trading, allowing products to be used multiple times, as exemplified by appliance trade-ins and the widespread adoption of ride-sharing. Resource recovery targets unusable waste, transforming it into renewable resources through recycling, dismantling, and processing technologies.

The three principles work synergistically and progressively: reduction lowers resource consumption at the source, reuse extends the value of resources, and recycling achieves the complete transformation of waste, together forming a complete practical system for the circular economy. Under the "dual carbon" goals and the requirements of high-quality development, deeply practicing the 5R principles can not only alleviate resource constraints and improve the ecological environment, but also foster new industries such as green manufacturing and renewable resource utilization, injecting lasting momentum into the sustainable development of the economy and society.

The circular economy is built on a hierarchy of strategies that aim to minimize resource extraction, extend product life cycles, and eliminate waste. The 5R framework Rethink, Reduce, Reuse, Recycle, and Recover represents a progressive approach to value retention, where priority is given to strategies that prevent waste at the source before managing it at later stages.

1.3.1 Rethink

Concept: Rethink refers to fundamentally reconsidering how products, services, and systems are designed, produced, and consumed. It challenges traditional linear thinking and encourages life-cycle-based decision-making.

Role in circular economy: Rethink is the most strategic and impactful element of the 5R framework, as decisions made at the design and planning stage determine most of a product's environmental footprint. It involves reimagining products, business models, and consumption patterns to align with circular principles.

Key focus areas

- Design for durability, modularity, and disassembly
- Shifting from ownership to access (product-as-a-service)
- Selecting sustainable and recyclable materials
- Integrating life cycle assessment (LCA) in decision-making

Example: Designing electronic devices with modular components that can be easily upgraded rather than replaced.

1.3.2 Reduce

Concept: Reduce aims to minimize the use of raw materials, energy, and water while lowering waste and emissions throughout the product life cycle.

Role in circular economy: Reduction improves efficiency and directly lowers environmental impact by limiting resource input and unnecessary consumption.

Key focus areas

- Lightweight product design
- Process optimization in manufacturing
- Energy-efficient technologies
- Minimizing packaging and material use

Example: Using lightweight materials in automotive manufacturing to reduce material usage and fuel consumption.

1.3.3 Reuse

Concept: Reuse involves using products or components multiple times for the same or a new purpose without significant reprocessing.

Role in circular economy: Reuse retains the highest possible value of products by extending their functional life, making it more sustainable than recycling.

Key focus areas

- Repair, refurbishment, and remanufacturing
- Sharing and leasing models
- Second-life applications

Example: Refurbishing electronic devices and reselling them in secondary markets.

1.3.4 Recycle

Concept: Recycle refers to processing waste materials into new raw materials that can be used to manufacture new products.

Role in circular economy: Recycling closes material loops but often involves energy use and some loss of material quality. It is therefore lower in priority than rethink, reduce, and reuse.

Key focus areas

- Material separation and sorting
- Mechanical and chemical recycling technologies
- Improving recyclability through product design

Example: Recycling aluminum cans into new aluminum products.

1.3.5 Recover

Concept: Recover involves extracting value from waste that cannot be reused or recycled, usually in the form of energy or secondary resources.

Role in circular economy: Recovery is the last preferred option in the circular hierarchy. While it reduces landfill use, it does not preserve material value.

Key focus areas

- Waste-to-energy processes
- Heat and energy recovery
- Safe handling of non-recyclable waste

Example: Generating electricity from non-recyclable municipal waste through incineration.

- Hierarchy of the 5R framework in the circular economy
  1. Rethink—Prevent waste through better design and system thinking
  2. Reduce—Minimize resource use and emissions
  3. Reuse—Extend product life and retain value
  4. Recycle—Convert waste into secondary materials
  5. Recover—Extract remaining energy or value from waste
- **Importance of the 5R framework in circular economy**
  - Minimizes dependence on virgin resources
  - Reduces environmental and carbon footprints
  - Encourages sustainable consumption and production
  - Supports climate change mitigation goals
  - Promotes innovation and circular business models

The 5R framework provides a structured pathway for implementing the circular economy by prioritizing prevention, value retention, and system redesign over waste management. Among the five, **Rethink** plays the most transformative role, as it influences all subsequent actions. Together, Rethink, Reduce, Reuse, Recycle, and Recover guide economies toward regenerative, resilient, and sustainable systems.

1.4 Principles and core strategies of circular economy systems

The Circular Economy (CE) represents a transformative shift from the traditional linear economic model of *take–make dispose* toward a regenerative, restorative, and sustainable system. At its core, the circular economy aims to decouple economic growth from resource depletion by redesigning how products, materials, and energy are produced, consumed, and recovered. The effectiveness of circular economy systems rests on a set of foundational principles supported by core operational strategies that guide implementation across industries, cities, and value chains.

1.4.1 Core principles of circular economy systems

1.4.1.1 Designing out waste and pollution

The first and most fundamental principle of the circular economy is the elimination of waste at the design stage, rather than managing it after it is created. Traditional systems treat waste as an inevitable by-product, whereas circular systems recognize waste as a design flaw.

This principle emphasizes:

- Eco-design and sustainable product development
- Reduction of hazardous substances
- Minimization of material complexity
- Designing products for durability, reparability, modularity, and disassembly

By embedding circular thinking into product design, waste and pollution are prevented before they occur, significantly reducing environmental damage and resource loss. Fig. 1.3 illustrate core principles of circular economy system.

1.4.1.2 Keeping products and materials in use

A central objective of the circular economy is to extend the useful life of products, components, and materials. Instead of rapid disposal, circular systems prioritize reuse, repair, refurbishment, and remanufacturing.

This principle supports:

- Maximizing product lifespan
- Retaining the highest possible value of materials
- Creating secondary markets for refurbished goods

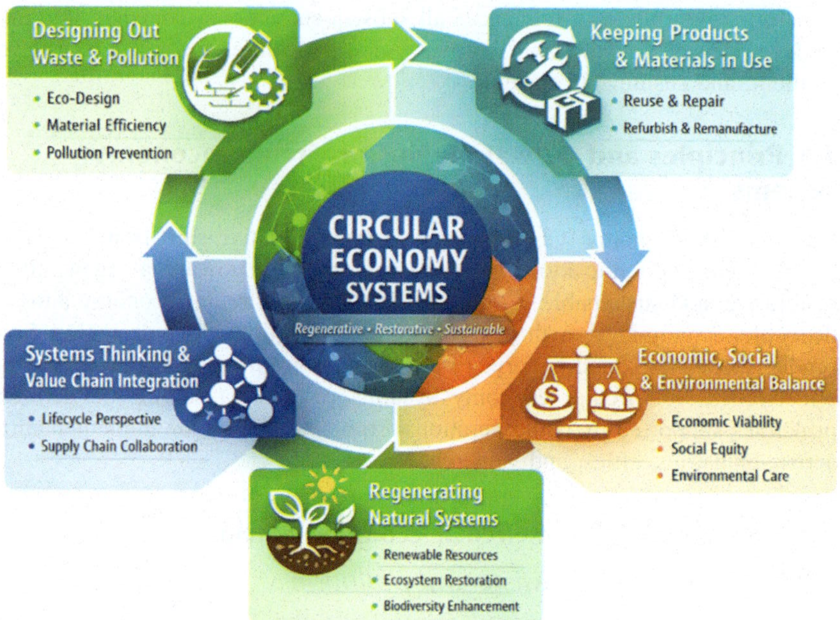

FIGURE 1.3 Core principles of circular economy system.

- Reducing demand for virgin resources

The longer products and materials remain in circulation, the lower the pressure on natural resources and the environment.

1.4.1.3 Regenerating natural systems

Unlike conventional sustainability approaches that focus mainly on minimizing harm, the circular economy actively seeks to restore and regenerate natural capital.

This principle involves:

- Returning biological nutrients safely to ecosystems
- Enhancing soil health through composting and regenerative agriculture
- Promoting biodiversity and ecosystem resilience
- Using renewable energy sources instead of fossil fuels

Circular systems thus move beyond "less bad" solutions toward net-positive environmental outcomes.

1.4.1.4 Systems thinking and value chain integration

The circular economy operates on the principle of systems thinking, recognizing that economic, environmental, and social elements are deeply interconnected. Decisions at one stage of the value chain influence outcomes at multiple other stages.

Key aspects include:

- Integration across supply chains
- Collaboration among producers, consumers, recyclers, and policymakers
- Transparency in material flows
- Lifecycle-based decision-making

This holistic perspective ensures that circular interventions generate long-term and scalable benefits.

1.4.1.5 Economic, social, and environmental balance

Circular economy systems aim to achieve triple-bottom-line sustainability, balancing:

- Economic viability
- Environmental protection
- Social well-being

This principle highlights inclusive growth, green job creation, ethical sourcing, and equitable access to resources, ensuring that circularity contributes to broader societal development.

1.4.2 Core strategies of circular economy systems

The circular economy is operationalized through a set of core strategies that translate its philosophical principles into practical, implementable actions. These strategies aim to retain the highest possible value of products, materials, and energy while minimizing waste, environmental degradation, and dependence on virgin resources. Unlike linear economic strategies, circular strategies focus on slowing, narrowing, and closing resource loops, thereby creating resilient and regenerative systems. Fig. 1.4 illustrate core strategies of circular economy system.

1.4.2.1 Resource reduction and efficiency strategy

The most impactful circular economy strategy is resource reduction at the source, which focuses on minimizing material and energy inputs during product design, manufacturing, and consumption.

Key components include:

- Lightweight and minimalist product design
- Efficient use of raw materials and energy
- Reduction of packaging and single-use materials
- Process optimization and lean manufacturing

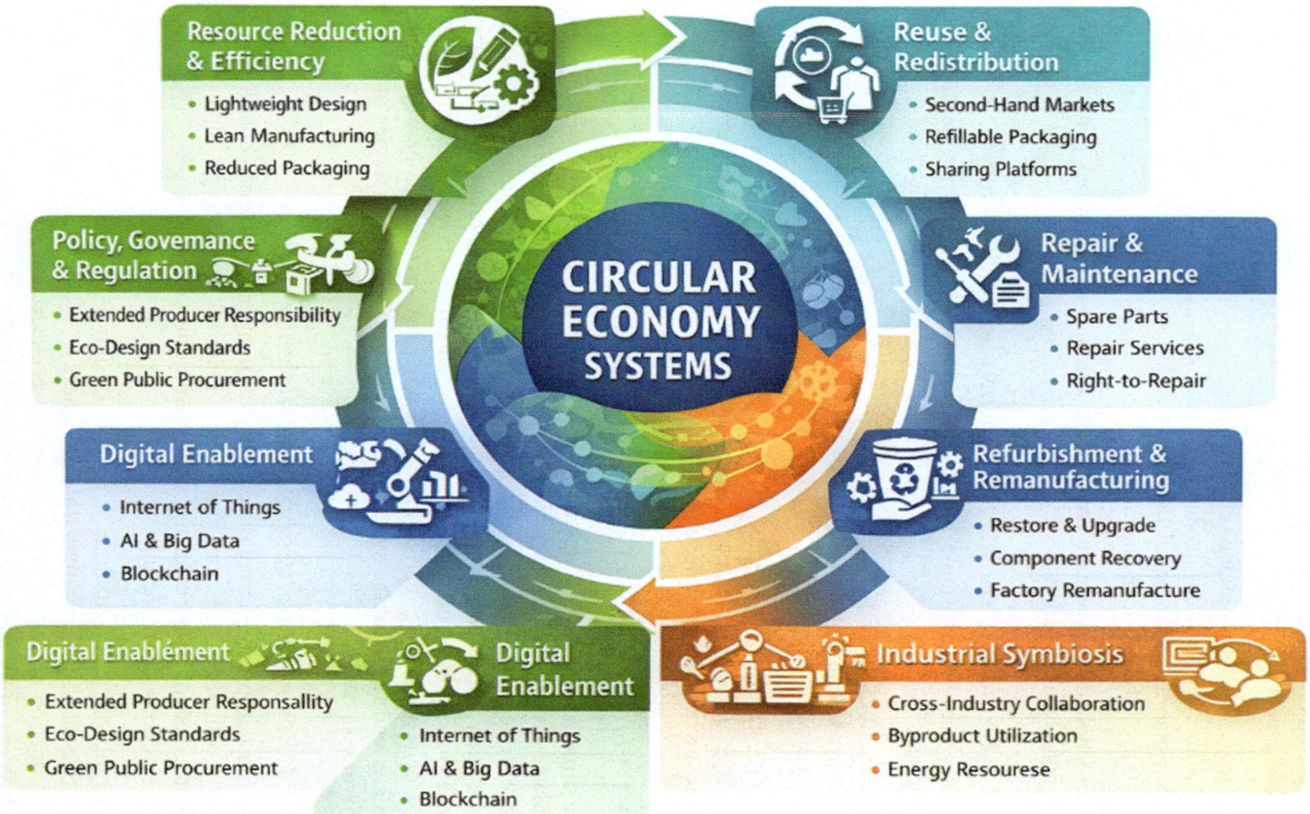

FIGURE 1.4 Core strategies of circular economy system.

By reducing resource demand upfront, organizations prevent waste generation and lower costs, making this strategy economically and environmentally advantageous.

1.4.2.2 Reuse and redistribution strategy

Reuse strategies aim to extend the functional life of products by enabling their use multiple times without significant reprocessing.

This strategy involves:

- Second-hand and resale markets
- Refillable and returnable packaging systems
- Redistribution of surplus goods
- Sharing and collaborative consumption models

Reuse maintains the product's original form and value, significantly reducing the need for new production and conserving embedded energy and resources.

1.4.2.3 Repair and maintenance strategy

Repair and maintenance strategies focus on restoring defective or worn-out products to working condition rather than replacing them.

Core elements include:

- Design for repairability and modularity
- Availability of spare parts and repair manuals
- Skilled repair services and infrastructure
- Right-to-repair regulations

By empowering users to repair products, this strategy enhances product longevity, supports local employment, and reduces electronic and mechanical waste.

1.4.2.4 Refurbishment and remanufacturing strategy

Refurbishment and remanufacturing involve restoring used products to a high functional or "like-new" condition through component replacement, upgrading, and quality testing.

Key characteristics:

- Refurbishment improves cosmetic and functional performance
- Remanufacturing rebuilds products to original specifications
- Retention of core components and materials
- Industrial-scale circular production systems

These strategies are widely used in sectors such as electronics, automotive, heavy machinery, and medical equipment, offering high economic and environmental value.

1.4.2.5 Recycling and material recovery strategy

Recycling transforms waste materials into secondary raw materials that can re-enter production cycles.

Important aspects include:

- Efficient waste segregation and collection systems
- Mechanical, chemical, and biological recycling technologies
- Quality preservation of recycled materials
- Development of markets for recycled content

Although recycling is essential, it is considered a downstream strategy because material quality and energy are often partially lost during processing.

1.4.2.6 Energy recovery strategy

When materials cannot be reused or recycled economically, energy recovery extracts remaining value through controlled processes.

This strategy includes:

- Waste-to-energy plants
- Biogas production from organic waste
- Thermal recovery systems

Energy recovery helps reduce landfill dependence while generating alternative energy, but it should only be used when higher-value circular options are no longer feasible.

1.4.2.7 Circular product design strategy

Circular product design embeds circularity at the earliest stage of the value chain.

Design strategies include:

- Modular and upgradeable product architecture
- Use of recyclable and biodegradable materials
- Design for disassembly and material separation
- Avoidance of toxic and composite materials

Designing products with end-of-life recovery in mind is critical for effective circular systems.

1.4.2.8 Circular business model strategy

Circular business models shift the economic focus from product ownership to performance and value delivery.

Common models include:

- Product-as-a-Service (leasing, subscription)
- Take-back and buy-back schemes

- Pay-per-use models
- Sharing platforms

These models align business incentives with durability, repair ability, and recyclability.

1.4.2.9 Industrial symbiosis strategy

Industrial symbiosis involves collaboration among different industries to utilize waste or by-products from one process as inputs for another.

Key features:

- Cross-industry material and energy exchange
- Shared infrastructure and utilities
- Reduced waste disposal and raw material demand

This strategy enhances regional circularity and strengthens industrial ecosystems.

1.4.2.10 Digital enablement strategy

Digital technologies play a critical role in supporting circular economy implementation.

Enabling technologies include:

- Internet of Things (IoT) for tracking assets and materials
- Artificial Intelligence for predictive maintenance and demand forecasting
- Blockchain for traceability and transparency
- Digital twins for lifecycle optimization

Digital tools improve efficiency, accountability, and scalability of circular systems.

1.4.2.11 Consumer engagement and behavioral strategy

Circular systems require active participation from consumers and users.

This strategy emphasizes:

- Awareness and education on circular consumption
- Incentives for returns, reuse, and repair
- Behavioral nudges for sustainable choices
- Transparent product information

Informed consumers act as co-creators of circular value.

1.4.2.12 Policy, governance, and regulatory strategy

Supportive policy frameworks are essential for scaling circular economy systems.

Policy instruments include:

- Extended Producer Responsibility (EPR)
- Eco-design standards and labeling
- Fiscal incentives and green procurement
- Waste reduction and recycling mandates

Governance mechanisms ensure accountability and long-term system transformation. The core strategies of circular economy systems form a multi-layered and interconnected framework that enables the transition from linear to circular production and consumption. Their effectiveness depends on co-ordinated action across design, business models, technology, policy, and consumer behavior. When strategically implemented, these approaches deliver economic resilience, environmental regeneration, and social value key pillars of sustainable development.

1.5 From linear to circular—Evolution of economic models

Economic systems have historically evolved in response to technological progress, resource availability, population dynamics, and societal priorities. For much of the industrial era, economic development has been dominated by the **linear economic model**, characterized by continuous extraction of natural resources, large-scale production, mass consumption, and disposal. While this model has delivered significant economic growth and improvements in living standards, it has also resulted in profound environmental degradation, resource depletion, and systemic inefficiencies. The transition from a linear to a circular economic model represents a fundamental shift in how value is created, preserved, and sustained over time.

The linear economy is commonly described by the sequence **"take–make–use–dispose."** Raw materials are extracted from nature, processed into products, consumed, and eventually discarded as waste. This model emerged during the Industrial Revolution, when resources appeared abundant, environmental impacts were poorly understood, and economic success was measured primarily by output growth and consumption levels. Efficiency in the linear system focused on reducing production costs and increasing throughput rather than minimizing waste or conserving resources.

A defining characteristic of the linear model is its reliance on **finite resources** and **fossil energy**, which has led to escalating environmental pressures. Waste generation is treated as an externality, often managed through landfilling or incineration rather than prevention or recovery. Products are typically designed for short lifespans, limited reparability, and planned obsolescence, reinforcing a culture of disposability. While economically effective in the short term, the linear model has proven increasingly unsustainable in the face of global challenges such as climate change, biodiversity loss, and material scarcity. The linear economy follows a one-directional flow of resources and materials, commonly described as "Take–Make–Use–Dispose."

Key characteristics

- Extraction of virgin raw materials
- Mass production of short-life products
- High energy and resource consumption
- Limited product durability and reparability
- Disposal of waste through landfilling or incineration

Limitations of the linear model

- Resource depletion: Finite natural resources are rapidly exhausted.
- Environmental pollution: Waste accumulation causes land, air, and water pollution.
- Economic inefficiency: Loss of material value at the end of product life.
- Climate impact: High greenhouse gas emissions across the value chain.
- Social inequality: Environmental burdens often fall on vulnerable communities.

The linear model assumes unlimited resources and unlimited capacity of nature to absorb waste an assumption no longer valid. Several interconnected factors have exposed the limitations of the linear economic model and accelerated the search for alternatives. Rapid population growth and urbanization have intensified demand for energy, materials, and infrastructure, placing unprecedented pressure on natural systems. At the same time, the environmental consequences of linear production such as greenhouse gas emissions, pollution, and ecosystem degradation have become more visible and scientifically documented. Economic vulnerabilities have also emerged. Volatile commodity prices, supply chain disruptions, and geopolitical dependencies on critical raw materials have revealed the fragility of linear systems. Moreover, social concerns related to inequality, public health, and environmental justice have highlighted the uneven distribution of the costs and benefits of linear growth. These challenges have prompted policymakers, businesses, and scholars to reconsider traditional economic assumptions and explore more resilient and regenerative models.

- **Early foundations of circular thinking**

The evolution toward circular economic models did not occur abruptly but was shaped by decades of intellectual and practical developments. Concepts such as **industrial ecology, cradle-to-cradle design, performance economy**, and **systems thinking** laid the groundwork for circularity. These approaches emphasized the interdependence between economic activities and natural systems, advocating for closed-loop material flows and energy efficiency. In the late twentieth century, environmental movements and sustainability discourse gained momentum, leading to the recognition that economic growth must be aligned with ecological limits. The introduction of life cycle assessment, cleaner production, and waste minimization strategies represented early efforts to improve resource efficiency within predominantly linear systems. However, these incremental improvements often failed to address the structural roots of unsustainability.

The **circular economy** emerged as a comprehensive framework that goes beyond efficiency improvements to reimagine the entire economic system. Unlike the linear model, the circular economy is designed to maintain the value of products, components, and materials for as long as possible. It seeks to close material loops through reuse, repair, refurbishment, remanufacturing, and recycling, while minimizing waste and environmental harm. This model redefines economic success by emphasizing **value retention** rather than throughput. Products are designed for durability, modularity, and recoverability, enabling multiple life cycles. Waste is reframed as a resource, and economic activities are organized to mimic natural systems, where outputs from one process serve as inputs for another. Clean and renewable energy sources play a critical role in powering circular systems, reducing reliance on fossil fuels and lowering carbon emissions. The transition from linear to circular economics has led to significant changes in business strategies and production models. Traditional ownership-based models are increasingly complemented or replaced by **product-as-a-service** and **sharing economy** approaches, where access and functionality matter more than possession. These models incentivize manufacturers to design high-quality, long-lasting products and to take responsibility for maintenance and end-of-life recovery. Production systems are also evolving through practices such as industrial symbiosis, where waste streams from one industry become inputs for another. Advances in digital technologies, including artificial intelligence, the Internet of Things, and data analytics, enable real-time monitoring, predictive maintenance, and efficient resource management. These innovations support the operationalization of circular principles at scale (Table 1.1).

Governments and international organizations have played a critical role in accelerating the shift from linear to circular models. Regulatory instruments such as extended producer responsibility, eco-design standards, waste reduction targets, and green public procurement encourage circular practices across value chains. International frameworks, including the Sustainable Development Goals and climate agreements, further reinforce the integration of circular economy principles into broader development agendas. Policy evolution reflects a growing recognition that market forces alone are insufficient to drive systemic change. Coordinated governance, long-term planning, and cross-sector collaboration are essential for overcoming structural barriers and aligning economic incentives with sustainability objectives.

The evolution from linear to circular economic models represents more than a technical or operational change; it signifies a shift in economic philosophy. While the linear economy prioritizes extraction and consumption, the

TABLE 1.1 Core differences—Linear v/s circular economy.

Aspect	Linear economy	Circular economy
Resource use	Extract–consume–discard	Reduce–reuse–regenerate
Waste	Inevitable outcome	Design flaw
Product design	Short life, planned obsolescence	Durable, modular, repairable
Energy source	Predominantly fossil fuels	Renewable and clean energy
Economic focus	Volume-driven growth	Value-driven, sustainable growth
Environmental impact	High pollution and emissions	Reduced footprint, regeneration
Role of consumers	End-users	Co-creators of value

circular economy emphasizes regeneration, resilience, and shared value creation. This transition aligns economic activity with planetary boundaries and societal well-being, offering a pathway toward sustainable prosperity.

The movement from linear to circular economic models reflects an adaptive response to the environmental, economic, and social challenges of the modern world. By redesigning systems to eliminate waste, retain value, and regenerate natural resources, the circular economy offers a viable alternative to the limitations of linear growth. As global pressures intensify, the evolution toward circularity is not merely an option but a necessity for ensuring long-term sustainability and resilience.

1.6 Clean energy

Clean energy, also known as renewable or green energy, refers to forms of energy that are produced with minimal environmental impact, primarily low or zero greenhouse gas (GHG) emissions, and that rely on naturally replenishing resources. Unlike fossil fuels, which are finite and polluting, clean energy aims to reduce the ecological footprint while meeting the growing energy demand of modern economies.

Key principles of clean energy:

1. Sustainability—relies on resources that naturally replenish.
2. Low Carbon Emission—reduces carbon dioxide and other pollutants.
3. Environmental Protection—minimal disruption to ecosystems.
4. Energy Security—diversification of energy sources to reduce dependency on fossil fuels.

Clean energy comes from naturally replenishing resources that produce minimal environmental impact. Solar energy harnesses sunlight through photovoltaic (PV) panels or concentrated solar power (CSP) systems to generate electricity and heat. Solar PV converts sunlight directly into electricity, while CSP uses mirrors or lenses to concentrate sunlight to produce thermal energy, which then drives turbines. Solar energy is abundant and scalable, suitable for residential, commercial, and industrial applications, though its intermittency due to day-night cycles and weather conditions poses a challenge that can be mitigated with energy storage solutions. Fig. 1.5 illustrate different types of clean energy system.

Wind energy captures kinetic energy from moving air using wind turbines. Onshore and offshore wind farms convert wind flow into mechanical energy, which is then transformed into electricity. Wind energy is cost-effective on a large scale and produces no direct emissions, though its performance is weather-dependent, and installation may face spatial and esthetic constraints.

Hydropower generates electricity from the movement of water, usually through dams or run-of-river systems. It is a reliable, controllable, and large-scale source of renewable energy, often providing base-load power. Beyond energy generation, hydropower projects can support irrigation and flood control. However, they can have significant environmental and social impacts, including ecosystem disruption and displacement of local communities.

Geothermal energy taps heat from beneath the Earth's surface, either for electricity generation or direct heating. Geothermal power plants and ground-source heat pumps exploit the Earth's thermal energy, offering a stable and low-emission energy source. Geothermal energy is limited by geographic availability and requires significant upfront investment, but it provides consistent, base-load power unlike intermittent sources such as solar or wind.

Biomass energy is produced from organic materials, including agricultural residues, wood, and waste. Through processes like combustion, anaerobic digestion, and biofuel production, biomass can provide electricity, heat, and transportation fuels. Biomass is particularly beneficial because it converts waste into energy, supports carbon neutrality if managed sustainably, and reduces landfill use. However, it can compete with food production and, if improperly managed, may contribute to emissions.

Finally, ocean and tidal energy utilize the mechanical power of tides, waves, and currents to produce electricity. This energy source is highly predictable and sustainable, particularly for coastal regions, and technologies such as tidal turbines and wave energy converters are emerging to harness this potential. The main challenges include high installation costs and possible environmental impacts on marine life.

Together, these clean energy sources provide diverse, low-carbon, and sustainable alternatives to fossil fuels, forming the backbone of a renewable energy-driven future. By integrating these technologies, societies can reduce

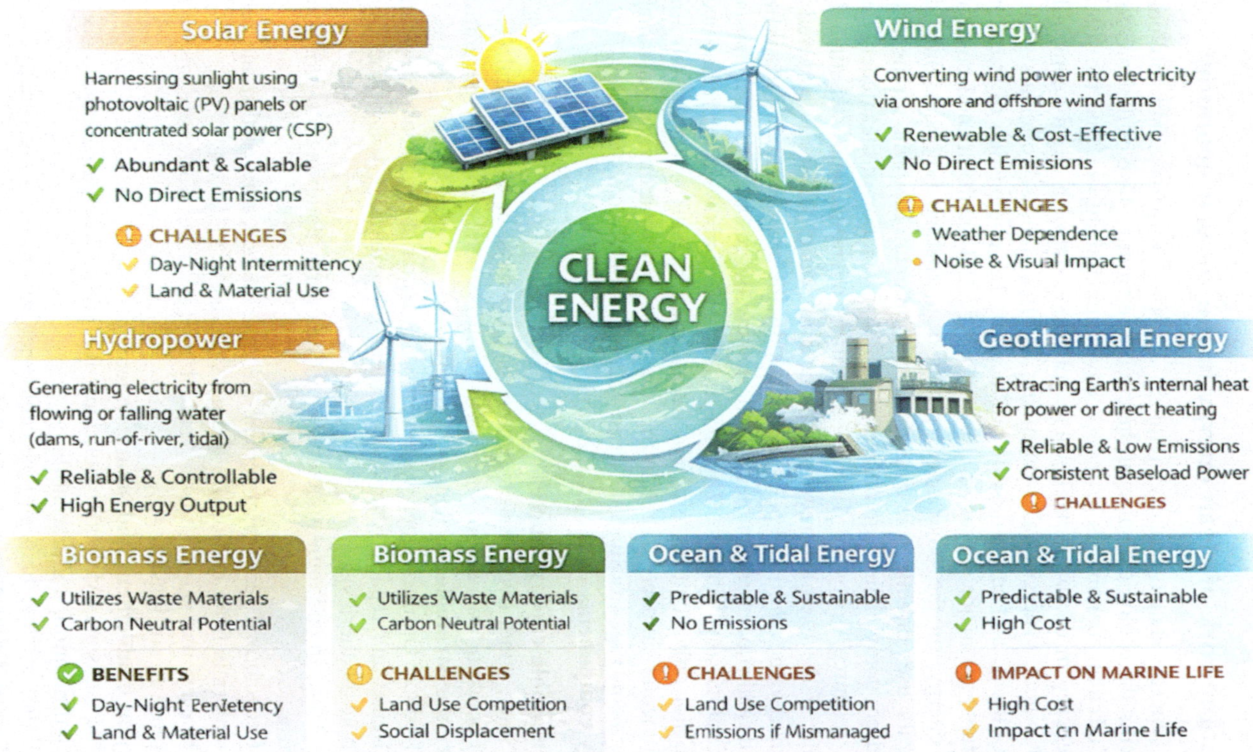

FIGURE 1.5 Different types of clean energy system.

greenhouse gas emissions, mitigate climate change, and support the principles of the circular economy through sustainable and regenerative energy systems.

1.7 Framework of clean energy in the circular economy

Clean energy synergistically reduces carbon emissions and pollution, enhancing the ecological value of the circular economy. One of the core goals of the circular economy is to reduce environmental impact, and the role of clean energy in carbon reduction and pollution control is highly compatible with this goal, synergistically enhancing its effectiveness. Pollutants such as carbon dioxide and sulfur dioxide produced by the combustion of traditional fossil fuels are major causes of climate change and environmental pollution. Clean energy, on the other hand, produces almost no pollutant emissions during its use, directly reducing the environmental footprint of the entire circular economy chain. Simultaneously, the combination of clean energy and waste resource recovery further amplifies environmental benefits. For example, if the chemical recycling process of waste plastics is powered by solar energy, the additional pollution caused by the consumption of fossil fuels in traditional processes can be avoided, making the resource recovery process more ecologically valuable. This synergistic effect allows the circular economy to not only achieve "efficient resource utilization" but also "minimize environmental impact" highlighting the core connotations of sustainable development. Fig. 1.6 illustrate clean energy in the circular economy.

Key components of the framework

1. **Renewable energy sources**
   - Solar, wind, hydropower, geothermal, biomass, ocean and tidal.
   - Focus on low-carbon, sustainable, and regenerative energy generation.
2. **Circular strategies in energy systems**
   - Reduce: Optimize energy efficiency and material use in energy systems.
   - Reuse and repair: Extend equipment lifespan (e.g., solar panels, turbines, batteries).
   - Refurbish and remanufacture: Upgrade energy equipment for continued use.
   - Recycle and recover materials: Recover metals, rare earths, and other materials from end-of-life energy systems.
   - Energy recovery: Utilize residual energy from waste streams when reuse is not feasible.
3. **Digital enablement and smart systems**
   - IoT sensors, AI, and blockchain for predictive maintenance, energy optimization, and traceability.
   - Smart grids to manage renewable energy distribution efficiently.
4. **Policy and governance enablers**

FIGURE 1.6 Clean energy in the circular economy.

- Regulations promoting renewable energy adoption, EPR (Extended Producer Responsibility) for equipment, incentives for circular practices.
- Alignment with global sustainability goals (e.g., SDG 7—Affordable and Clean Energy).

5. **Socio-economic integration**
   - Creating local employment in repair, recycling, and renewable energy installation.
   - Promoting energy access, affordability, and social equity.

Clean energy is giving rise to new business models and expanding the boundaries of circular economy development. In the transportation sector, the combination of new energy vehicles and renewable energy has created a circular chain of "electricity-transportation-recycling." In the construction sector, technologies such as solar heating and geothermal cooling are combined with the recycling of green building materials to form a building circular system of "building material recycling-clean energy." Furthermore, the synergistic development of energy storage technology and clean energy has solved the intermittency problem of renewable energy, providing a stable energy guarantee for the circular economy while giving rise to new business models such as energy storage battery manufacturing and energy management services. This promotes the integration of the circular economy into multiple fields such as energy, industry, construction, and transportation, forming a comprehensive circular development pattern. Clean energy in the circular economy integrates renewable energy technologies with circular strategies to minimize waste, maximize resource efficiency, and restore natural systems. It ensures energy production and consumption are sustainable, regenerative, and aligned with economic, environmental, and social goals.

Clean energy in the circular economy represents a transformative pathway toward sustainable development by harmonizing renewable energy systems with circular principles of resource efficiency, regeneration, and long-term value creation. By shifting away from the traditional linear model of extract–consume–dispose, this integrated approach ensures that energy production not only minimizes environmental degradation but actively contributes to ecological restoration and economic resilience. Renewable energy sources such as solar, wind, hydropower, geothermal, biomass, and ocean energy form the backbone of this transition, while circular strategies—reduce, reuse, repair, refurbish, recycle, recover, and regenerate—extend the life cycle of energy infrastructure and critical materials. The incorporation of digital technologies, including smart grids, artificial intelligence, IoT, and blockchain, further enhances system efficiency, transparency, and adaptability, enabling optimized energy flows and predictive maintenance across the value chain. Equally important is the role of policy, governance, and regulatory frameworks in incentivizing innovation, ensuring responsible resource management, and fostering stakeholder collaboration. When aligned with socio-economic objectives such as job creation, energy accessibility, and community engagement, clean energy within a circular economy becomes a powerful driver of inclusive growth and climate action. Ultimately, this integrated model not only addresses the urgent challenges of climate change and resource scarcity but also establishes a resilient, low-carbon, and regenerative energy system capable of supporting present and future generations.

1.8 Importance of transitioning to a circular economy for sustainability

The transition to a circular economy is a critical requirement for achieving long-term environmental, economic, and social sustainability in an era marked by climate change, resource depletion, and growing socio-economic inequalities. The prevailing linear economic model based on "take, make, use, and dispose" has resulted in excessive extraction of finite resources, large-scale waste generation, environmental pollution, and ecological degradation. In contrast, the circular economy offers a regenerative and restorative framework that redefines growth by decoupling economic development from resource consumption and environmental harm.

1.8.1 Environmental sustainability and resource conservation

One of the most significant reasons for transitioning to a circular economy is its ability to conserve natural resources and protect ecosystems. Circular systems emphasize reducing material extraction, extending product life cycles, and recovering valuable materials through reuse, repair, remanufacturing, and recycling. By minimizing reliance on virgin raw materials, the circular economy helps preserve forests, minerals, water bodies, and biodiversity. Moreover, designing out waste and pollution reduces landfill use, marine litter, and soil contamination, thereby restoring ecological balance and supporting planetary health.

1.8.2 Climate change mitigation and low-carbon development

The circular economy plays a crucial role in reducing greenhouse gas emissions across the entire value chain. Emissions associated with raw material extraction, manufacturing, transportation, and waste disposal are significantly lowered when materials are kept in circulation for longer periods. When combined with clean and renewable energy, circular practices further accelerate the shift toward a low-carbon economy, supporting global climate goals such as the Paris Agreement. This makes circularity a key strategy for climate action and carbon neutrality.

1.8.3 Economic resilience and sustainable growth

Transitioning to a circular economy enhances economic resilience by reducing dependency on volatile global supply chains and scarce resources. By promoting secondary material markets, local repair services, and remanufacturing industries, circular systems create new revenue streams and business opportunities. This shift fosters innovation, improves productivity, and enables value creation without proportional increases in resource consumption. As a result, the circular economy supports sustainable and inclusive economic growth rather than growth at the cost of environmental degradation.

1.8.4 Waste reduction and efficient resource management

Waste in a circular economy is viewed as a resource rather than a burden. Materials that would otherwise be discarded are reintegrated into production cycles, minimizing landfill use and environmental pollution. Efficient resource management ensures that materials maintain their highest possible value throughout their life cycle. This approach is particularly critical in addressing growing waste challenges such as plastic waste, electronic waste, and industrial by-products, which pose severe environmental and health risks under linear systems.

1.8.5 Social sustainability and quality of life

The circular economy contributes significantly to social sustainability by creating employment opportunities in sectors such as repair, recycling, remanufacturing, renewable energy, and waste management. These jobs are often localized, supporting regional development and community resilience. Furthermore, reduced pollution leads to improved public health, while access-based business models make products and services more affordable and accessible. In this way, circularity enhances overall quality of life and promotes social equity.

1.8.6 Innovation and technological advancement

Transitioning to a circular economy stimulates technological innovation in product design, materials science, digital technologies, and industrial processes. Advanced technologies such as artificial intelligence, IoT, blockchain, and digital product passports enable traceability, predictive maintenance, and efficient resource utilization. These innovations not only support circular practices but also drive competitiveness and long-term sustainability in industries.

1.8.7 Alignment with global sustainability goals

The circular economy directly supports several United Nations Sustainable Development Goals (SDGs), including responsible consumption and production (SDG 12), climate action (SDG 13), affordable and clean energy (SDG 7), and sustainable cities and communities (SDG 11). By integrating economic, environmental, and social objectives, circular systems provide a comprehensive pathway to achieving global sustainability targets.

The importance of transitioning to a circular economy lies in its ability to transform sustainability from a reactive approach into a proactive, regenerative, and systemic solution. Rather than merely reducing harm, the circular economy actively restores natural systems, strengthens economic resilience, and enhances social well-being. As resource pressures and environmental challenges intensify, embracing circularity is no longer optional it is essential for ensuring a sustainable, equitable, and resilient future for present and future generations.

1.9 Synergy between renewable energy and circular practices

The synergy between renewable energy and circular economy practices represents a holistic sustainability paradigm in which clean energy systems are designed, deployed, operated, and retired in ways that maximize resource efficiency, minimize waste, and regenerate natural systems. While renewable energy focuses on replacing fossil fuels with low-carbon energy sources, circular practices ensure that the materials, technologies, and value chains supporting renewable energy remain sustainable over their entire life cycle. Together, they form a mutually reinforcing system that accelerates the transition toward climate-resilient and resource-secure economies.

1.9.1 Circular design of renewable energy technologies

Circular practices begin at the design stage of renewable energy systems such as solar panels, wind turbines, batteries, and energy storage units. Products are designed to be durable, modular, repairable, and upgradable, allowing components to be replaced or improved without discarding entire systems. For example, modular wind turbine blades and easily replaceable battery cells reduce material waste and extend operational lifetimes. This design philosophy reduces dependence on virgin raw materials, particularly critical minerals like lithium, cobalt, and rare earth elements.

1.9.2 Lifecycle extension through reuse, repair, and refurbishment

Renewable energy infrastructure benefits significantly from reuse, repair, refurbishment, and remanufacturing strategies. Solar panels that no longer meet high-efficiency standards can be reused in secondary markets or refurbished for off-grid and rural applications. Wind turbine components, power electronics, and inverters can be remanufactured to meet performance standards at lower economic and environmental cost. These practices retain embedded energy and material value, reducing both waste generation and the carbon footprint of energy systems.

1.9.3 Material recovery and recycling in energy systems

At the end of life of renewable energy equipment, circular material recovery becomes crucial. Advanced recycling technologies allow recovery of valuable metals such as aluminum, copper, silver, lithium, and rare earths from solar panels, batteries, and turbines. Recycling reduces landfill pressure, prevents environmental contamination, and strengthens supply chain resilience by reintroducing secondary materials into production cycles. Renewable energy thus becomes part of a closed-loop material system, aligning clean energy goals with circular resource flows.

1.9.4 Renewable energy as an enabler of circular operations

Renewable energy actively powers circular economy processes, such as recycling plants, remanufacturing facilities, wastewater treatment, and material recovery operations. Using clean energy to operate these processes ensures that circular practices themselves are low-carbon and environmentally sustainable. For instance, solar-powered recycling units and wind-powered industrial symbiosis parks significantly reduce emissions across the circular value chain.

1.9.5 Waste-to-energy and bioenergy integration

Circular practices transform organic waste and agricultural residues into biomass energy, biogas, and biofuels, creating a strong link between renewable energy and waste valorization. Anaerobic digestion of food waste, manure, and sewage sludge produces biogas while simultaneously generating nutrient-rich digestate for agriculture. This integration closes biological loops, reduces landfill methane emissions, and contributes to decentralized renewable energy generation.

1.9.6 Digital technologies strengthening the synergy

Digital tools such as IoT, artificial intelligence, digital twins, and blockchain strengthen the synergy between renewable energy and circular practices by enabling predictive maintenance, lifecycle tracking, and transparent material flows. Smart monitoring systems extend asset lifetimes, optimize energy generation, and ensure responsible end-of-life management. Digital product passports, for example, record material composition and usage history, simplifying repair, reuse, and recycling of renewable energy equipment.

1.9.7 Economic and industrial symbiosis benefits

The integration of renewable energy with circular practices creates new business models and industrial symbiosis opportunities. Energy-intensive industries can use renewable power while exchanging waste heat, by-products, or materials with nearby firms. Circular business models such as energy-as-a-service, leasing of solar panels, and shared energy infrastructure reduce upfront costs and encourage responsible resource use. This synergy enhances economic resilience while supporting sustainable industrial growth.

1.9.8 Environmental and climate synergies

From an environmental perspective, the synergy between renewable energy and circular practices delivers compounded climate benefits. Renewable energy reduces greenhouse gas emissions from fossil fuels, while circular practices minimize emissions associated with material extraction, processing, and disposal. Together, they lower overall environmental footprints, protect ecosystems, and support biodiversity regeneration. This combined approach aligns strongly with global climate targets and sustainable development goals.

1.9.9 Social and policy dimensions

The renewable–circular synergy promotes inclusive and decentralized energy systems, improving access to clean energy in remote and underserved regions. It also generates employment in repair, recycling, remanufacturing, and renewable energy sectors. Policy frameworks such as Extended Producer Responsibility (EPR), renewable energy mandates, and circular economy regulations play a critical role in enabling this synergy by encouraging sustainable design, responsible material management, and long-term system thinking.

The synergy between renewable energy and circular practices moves sustainability beyond energy transition alone toward a regenerative economic model. Renewable energy provides the clean power needed to decarbonize societies, while circular practices ensure that the technologies enabling this transition are themselves sustainable, efficient, and resilient. Together, they create a powerful, integrated system that supports climate action, resource security, economic innovation, and long-term environmental stewardship.

1.10 How clean energy supports circularity in industries

Clean energy plays a foundational and enabling role in advancing circularity within industrial systems by decarbonizing production processes, improving resource efficiency, and supporting closed-loop material flows. While circularity focuses on keeping products, components, and materials in use for as long as possible, clean energy ensures that these circular processes themselves operate with minimal environmental impact. Together, they transform industries from resource-intensive, linear systems into resilient, low-carbon, and regenerative value chains.

1.10.1 Decarbonizing circular industrial processes

Many circular industrial activities such as recycling, remanufacturing, material recovery, and waste treatment—are energy-intensive. When powered by fossil fuels, these processes can undermine sustainability goals. Clean energy sources such as solar, wind, biomass, and green hydrogen enable industries to carry out circular operations with significantly lower carbon emissions. For example, using renewable electricity in metal recycling or plastic reprocessing drastically reduces the carbon footprint compared to primary material production.

1.10.2 Enabling resource-efficient manufacturing

Clean energy supports resource-efficient and cleaner production systems by reducing energy losses and promoting high-efficiency technologies. Electrification of industrial processes using renewable energy improves process control, reduces heat losses, and enables precise material usage. Energy-efficient machinery powered by clean energy aligns with circular principles by minimizing both material and energy waste across manufacturing stages.

1.10.3 Supporting industrial symbiosis

In circular industrial ecosystems, waste or by-products from one industry become resources for another. Clean energy strengthens this industrial symbiosis by enabling shared renewable energy infrastructure, waste heat recovery, and decentralized energy systems. For instance, excess renewable electricity or waste heat from one plant can power nearby facilities, creating interconnected industrial networks that optimize resource use and reduce overall emissions.

1.10.4 Powering recycling, remanufacturing, and material recovery

Recycling and remanufacturing processes central pillars of circularity require stable and affordable energy supply. Clean energy ensures these operations are environmentally sustainable and economically viable. Renewable-powered recycling facilities for metals, electronics, paper, and plastics reduce dependency on fossil fuels while improving the overall environmental performance of secondary material production.

1.10.5 Facilitating circular product design and lifecycle management

Clean energy enables industries to adopt circular product design and lifecycle thinking. Renewable-powered digital technologies such as IoT, AI, and digital twins monitor product performance, predict maintenance needs, and extend product lifespans. These technologies help industries design products for durability, modularity, and easy disassembly key requirements for reuse, refurbishment, and recycling.

1.10.6 Waste-to-energy and bioenergy integration

Industries generate significant amounts of organic and process waste. Clean energy technologies such as biogas, biomass energy, and waste-to-energy systems convert these waste streams into usable energy, closing both material and energy loops. For example, agro-industrial residues can be converted into biogas for heat and power, reducing waste disposal while providing renewable energy for industrial operations.

1.10.7 Reducing dependency on virgin resources

By supporting the use of recycled and secondary materials, clean energy indirectly reduces reliance on virgin resource extraction. Producing materials such as aluminum, steel, and plastics from recycled sources using renewable energy significantly lowers energy consumption and emissions. This synergy enhances supply chain resilience and supports circular material flows across industries.

1.10.8 Enabling circular business models

Clean energy supports innovative circular business models such as product-as-a-service, leasing, and shared industrial assets. Renewable-powered operations reduce operating costs and emissions, making these models more attractive and scalable. For example, leasing solar-powered industrial equipment encourages manufacturers to design long-lasting, repairable products while retaining ownership and responsibility for end-of-life management.

1.10.9 Improving compliance and sustainability performance

Industries face increasing regulatory pressure to reduce emissions and improve resource efficiency. Clean energy helps industries meet environmental standards, carbon reduction targets, and circular economy regulations. This alignment enhances corporate sustainability performance, improves ESG ratings, and strengthens stakeholder trust.

Clean energy acts as the energy backbone of circular industries, ensuring that material loops, resource recovery, and product life extension occur within a low-carbon and environmentally responsible framework. By powering circular processes with renewable energy, industries can achieve deeper sustainability outcomes—reducing emissions, conserving resources, improving economic resilience, and fostering innovation. Ultimately, clean energy does not merely support circularity; it amplifies its impact, enabling industries to transition toward truly sustainable and regenerative systems.

1.11 Resource scarcity, climate change, and waste management crises

The global economy is currently confronting three deeply interconnected and escalating challenges resource scarcity, climate change, and waste management crises which together pose serious threats to environmental stability, economic resilience, and social well-being. These challenges are largely driven by the prevailing linear economic model of "take–make–use–dispose," which relies on continuous extraction of finite resources, intensive energy consumption, and the generation of massive amounts of waste. Addressing these crises requires a systemic shift toward sustainable and circular approaches.

1.11.1 Resource scarcity

Resource scarcity has emerged as a critical global concern due to rapid industrialization, population growth, urbanization, and rising consumption patterns. Many natural resources such as fossil fuels, minerals, freshwater, fertile soil, and rare earth elements are finite and unevenly distributed. Overexploitation of these resources has led to depletion, supply chain disruptions, price volatility, and geopolitical tensions. Industries that depend heavily on virgin raw materials face increasing risks related to availability and cost, while ecosystems suffer from habitat destruction, biodiversity loss, and land degradation. Resource scarcity highlights the urgent need for resource efficiency, material circularity, and regenerative practices that decouple economic growth from resource consumption.

1.11.2 Climate change

Climate change represents one of the most profound environmental challenges of the modern era. It is primarily driven by greenhouse gas emissions from fossil fuel combustion, industrial processes, deforestation, and unsustainable land use. Rising global temperatures, extreme weather events, sea-level rise, and changing precipitation patterns are already impacting ecosystems, agriculture, infrastructure, and human health. Climate change not only exacerbates resource scarcity by affecting water availability, crop yields, and energy systems but also increases vulnerability across societies, particularly in developing regions. Mitigating climate change requires a rapid transition to clean energy, low-carbon technologies, and sustainable production systems, alongside adaptation strategies to build resilience.

1.11.3 Waste management crises

The global waste crisis has intensified due to increased consumption, short product lifecycles, and inadequate waste management infrastructure. Massive quantities of solid waste, plastic waste, electronic waste, and industrial byproducts are generated each year, overwhelming landfills and polluting land, air, and water bodies. Improper waste disposal contributes to soil contamination, marine pollution, and public health risks, while valuable materials are lost instead of being recovered and reused. The waste crisis reflects systemic inefficiencies in product design, consumption patterns, and end-of-life management, underscoring the need for waste reduction, recycling, recovery, and circular design.

1.11.4 Interconnected nature of the crises

Resource scarcity, climate change, and waste management crises are not isolated problems; they reinforce one another. Extracting virgin resources increases energy use and emissions, accelerating climate change. Climate impacts disrupt resource availability and waste management systems. Growing waste streams lead to additional emissions and environmental degradation. This interconnectedness calls for integrated solutions that address all three challenges simultaneously rather than in isolation.

1.11.5 The way forward

Overcoming these crises requires a fundamental transformation of economic systems through the adoption of circular economy principles and clean energy transitions. By keeping resources in use for longer, minimizing waste, recovering materials, and powering systems with renewable energy, societies can reduce environmental pressures while fostering sustainable growth. Policy interventions, technological innovation, responsible consumption, and collaborative governance are essential to addressing these global challenges and ensuring a resilient, sustainable future for generations to come.

1.12 Opportunities for innovation and economic growth

The transition toward sustainable development driven by the circular economy and clean energy systems creates significant opportunities for innovation and economic growth across industries, regions, and societies. Rather than limiting growth, sustainability-oriented transitions unlock new markets, business models, technologies, and employment pathways while enhancing resilience and long-term competitiveness. By rethinking how resources, energy, and value are created and retained, economies can move from extractive growth to regenerative and inclusive growth.

1.12.1 Innovation in products and materials

One of the most powerful opportunities lies in product and material innovation. Circular economy principles encourage the design of products that are durable, modular, repairable, and recyclable. This drives innovation in advanced materials, biodegradable alternatives, lightweight composites, and high-performance recycled materials. Industries that invest in eco-design and material science gain competitive advantages by reducing input costs, improving product longevity, and meeting rising consumer and regulatory demands for sustainability.

1.12.2 Emergence of new business models

Sustainable transitions stimulate innovative circular business models such as product-as-a-service, leasing, sharing platforms, take-back schemes, and extended producer responsibility systems. These models shift value creation from volume-based sales to performance-based services, enabling companies to generate recurring revenues, strengthen customer relationships, and reduce resource dependency. Clean energy integration further lowers operating costs, enhancing profitability and scalability.

1.12.3 Clean energy–driven industrial growth

The expansion of renewable energy technologies solar, wind, bioenergy, green hydrogen, and energy storage creates vast opportunities for industrial innovation and economic expansion. Investments in clean energy infrastructure stimulate manufacturing, installation, operation, and maintenance sectors, while reducing energy costs and exposure to fossil fuel price volatility. Clean energy also enables energy-intensive industries to decarbonize, remain globally competitive, and attract sustainable investment.

1.12.4 Digital innovation and smart systems

Digital technologies such as Artificial Intelligence (AI), Internet of Things (IoT), blockchain, and data analytics play a critical role in unlocking circular and clean energy innovation. These technologies enable real-time monitoring of resource flows, predictive maintenance, energy optimization, and transparent supply chains. Digital enablement enhances operational efficiency, reduces waste, and supports the development of smart grids, digital product passports, and intelligent recycling systems creating new digital service markets.

1.12.5 Growth of green jobs and skills development

The shift toward circular and clean energy systems generate new employment opportunities across multiple skill levels. Jobs emerge in renewable energy deployment, recycling and remanufacturing, sustainable product design, environmental data analytics, and circular supply chain management. This transition encourages upskilling and reskilling, fostering a future-ready workforce while supporting inclusive economic growth and regional development.

1.12.6 Strengthening industrial competitiveness and resilience

Innovation driven by sustainability enhances industrial resilience by reducing dependency on imported raw materials and fossil fuels. Circular material loops and local renewable energy systems improve supply chain security and reduce exposure to global disruptions. Companies that adopt sustainable innovations are better positioned to comply with evolving regulations, access green finance, and maintain long-term competitiveness in global markets.

1.12.7 Attracting investment and financing opportunities

Sustainability transitions attract increasing flows of green finance, impact investment, and ESG-focused capital. Governments, financial institutions, and investors prioritize projects aligned with clean energy, circular economy, and climate goals. Innovative enterprises benefit from grants, incentives, carbon markets, and sustainability-linked financing, accelerating commercialization and market expansion.

1.12.8 Regional and community-level economic development

Circular and clean energy initiatives promote localized economic growth by encouraging decentralized energy systems, local recycling hubs, and community-based enterprises. These initiatives create value within regions, reduce waste transportation costs, and support small and medium enterprises (SMEs). Community participation further strengthens social cohesion and economic inclusion.

Opportunities for innovation and economic growth lie at the heart of the transition to a circular economy powered by clean energy. By fostering technological advancement, business model transformation, workforce development, and investment mobilization, sustainable systems do not constrain economic progress they redefine and expand it. Economies that embrace circularity and clean energy today will lead tomorrow's markets, create resilient industries, and deliver long-term prosperity while safeguarding environmental and social well-being.

1.13 Conclusion

The transition toward a circular economy powered by clean energy represents a fundamental shift in how societies produce, consume, and manage resources. As demonstrated throughout this chapter, the traditional linear economic model is no longer viable in the face of mounting pressures from resource scarcity, climate change, and waste accumulation. Circular economy principles—focused on reducing resource extraction, extending product life cycles, recovering materials, and regenerating natural systems offer a sustainable alternative that aligns economic development with environmental stewardship.

Clean energy acts as a critical enabler of circular systems by reducing greenhouse gas emissions, enhancing energy security, and supporting low-carbon industrial processes. The integration of renewable energy sources into circular value chains not only minimizes environmental impacts but also improves operational efficiency and long-term cost effectiveness. When combined with digital technologies such as smart grids, data analytics, and intelligent resource tracking, clean energy further strengthens circularity by optimizing material and energy flows across industries.

The chapter also highlights that the circular economy is not solely an environmental solution but a powerful driver of innovation and economic growth. Circular business models, sustainable product design, and clean energy infrastructure create new market opportunities, stimulate investment, and generate employment across diverse sectors. Moreover, supportive policy frameworks, regulatory mechanisms, and collaborative governance are essential to accelerate adoption and ensure that the benefits of circular and clean energy transitions are equitably distributed.

In conclusion, the synergy between clean energy and the circular economy provides a comprehensive pathway toward sustainable development. By addressing economic, environmental, and social dimensions simultaneously, this integrated approach fosters resilience, competitiveness, and long-term prosperity. The successful implementation of circular and clean energy systems requires coordinated efforts from governments, industries, academia, and consumers alike. As global sustainability challenges intensify, embracing clean energy within circular economy frameworks is no longer optional it is imperative for achieving a sustainable, inclusive, and regenerative future.

1.14 Case study

1.14.1 Case study: Integrating clean energy and circular economy in the Indian manufacturing sector

1.14.1.1 Background

India's rapid industrial growth has significantly increased energy demand, material consumption, and waste generation. Traditionally, the manufacturing sector has operated under a linear economic model, relying heavily on fossil fuels and virgin raw materials while generating large volumes of industrial waste. These practices have contributed to resource scarcity, rising carbon emissions, and waste management challenges, prompting regulatory pressure and growing demand for sustainable industrial transformation. ABC Manufacturing Ltd. (pseudonym), a mid-sized electrical equipment manufacturer based in Maharashtra, produces industrial motors, control panels, and energy-efficient appliances for domestic and export markets. Despite strong market demand, the company faced rising operational costs, energy price volatility, and sustainability compliance challenges.

1.14.1.2 Problem statement

ABC Manufacturing Ltd. encountered the following critical challenges:

- High dependency on grid electricity generated from fossil fuels
- Increasing costs of raw materials such as copper, aluminum, and steel
- Growing industrial waste, including metal scrap and electronic waste
- Pressure to meet environmental regulations and ESG standards
- Limited resilience to supply chain disruptions

Management realized that incremental efficiency improvements were insufficient and decided to adopt a circular economy framework supported by clean energy integration.

1.14.1.3 Strategic intervention

The company designed a Clean Energy Circular Economy Transformation Plan with the following components:

1. **Clean energy integration**
    - Installed rooftop solar photovoltaic systems to supply 40% of electricity demand
    - Adopted biomass-based boilers for process heating
    - Implemented energy storage systems for load balancing
2. **Circular resource management**
    - Introduced design-for-disassembly for electrical equipment
    - Established in-house repair and refurbishment units
    - Implemented metal scrap recovery and recycling loops
    - Partnered with authorized e-waste recyclers

3. **Digital enablement**
    - Used IoT-based energy monitoring systems
    - Deployed AI-driven predictive maintenance to reduce equipment downtime
    - Implemented digital tracking for material flows
4. **Circular business model innovation**
    - Launched a product-as-a-service model for industrial motors
    - Introduced take-back schemes for end-of-life products
    - Offered refurbished equipment to cost-sensitive markets

Implementation challenges

During implementation, the firm faced:

- High initial capital investment requirements
- Workforce resistance to new operational practices
- Need for skill upgradation in digital and circular systems
- Supplier alignment issues for circular procurement

These challenges were addressed through phased investments, training programs, and long-term supplier contracts.

Outcomes and impact

After 3 years of implementation, ABC Manufacturing Ltd. achieved:

- 35% reduction in energy-related carbon emissions
- 25% reduction in raw material procurement costs
- Near-zero industrial waste to landfill
- Improved operational resilience and cost stability
- Creation of new jobs in repair, refurbishment, and energy management
- Enhanced ESG ratings and improved brand reputation

1.15 Exercise questions

1. Explain the transition from a linear to a circular economic model with examples.
2. Discuss the role of clean energy in enabling circular economy practices.
3. Analyse the interconnection between resource scarcity, climate change, and waste management.
4. Describe the core principles of circular economy systems.
5. Explain how renewable energy supports industrial circularity.
6. Evaluate the importance of circular product design in sustainability.
7. Discuss the role of digital technologies in circular energy systems.
8. Examine circular business models and their economic benefits.
9. Explain the concept of industrial symbiosis with suitable examples.

10. Analyse policy and regulatory frameworks supporting clean energy and circular economy.
11. Discuss waste-to-energy technologies in a circular economy context.
12. Explain how circular economy contributes to climate change mitigation.
13. Evaluate challenges in implementing circular economy in developing countries.
14. Discuss the social benefits of circular economy and clean energy integration.
15. Propose strategies for industries to adopt clean energy within circular systems.

1.16 Objective type questions

1. The linear economy follows which model?
   a) Reduce–reuse–recycle
   b) Take–make–use–dispose
   c) Design–regenerate–restore
   d) Share–lease–recover

2. Which principle aims to eliminate waste at the design stage?
   a) Recycling
   b) Resource extraction
   c) Designing out waste and pollution
   d) Energy recovery

3. Clean energy primarily helps circular economy by:
   a) Increasing waste
   b) Reducing emissions
   c) Increasing raw material use
   d) Promoting landfills

4. Which renewable energy source uses organic waste?
   a) Solar
   b) Wind
   c) Biomass
   d) Geothermal

5. Industrial symbiosis refers to:
   a) Competition between industries
   b) Sharing of waste and resources
   c) Disposal of industrial waste
   d) Linear production systems

6. Digital Product Passports support circularity by enabling:
a) Increased consumption
b) Product traceability
c) Waste dumping
d) Energy loss

7. Which is NOT a circular strategy?
a) Repair
b) Remanufacture
c) Dispose
d) Reuse

8. Circular economy aims to decouple:
a) Energy from waste
b) Growth from resource consumption
c) Technology from society
d) Policy from industry

9. Waste-to-energy supports circularity by:
a) Increasing landfill
b) Recovering energy from waste
c) Polluting ecosystems
d) Reducing recycling

10. Which sector benefits most from circular job creation?
a) Repair and remanufacturing
b) Mining
c) Fossil fuel extraction
d) Landfilling

11. Clean energy systems are typically:
a) Fossil-based
b) High-emission
c) Renewable and low-carbon
d) Non-renewable

12. Which technology supports predictive maintenance?
a) Manual inspection
b) IoT and AI
c) Landfilling
d) Incineration

13. Circular business models emphasize:
a) Ownership
b) Performance and access
c) Disposal
d) Planned obsolescence

14. The triple bottom line includes:
  a) Profit only
  b) People, planet, profit
  c) Energy, waste, cost
  d) Technology, data, AI

15. Circular economy primarily promotes:
  a) Short product life
  b) Resource regeneration
  c) Increased extraction
  d) Linear growth

Policy-oriented definition: *The circular economy refers to a systemic approach to economic development that minimizes waste, reduces reliance on virgin resources, and promotes sustainable production and consumption through closed-loop material and energy flows.*

Business-focused definition: *The circular economy is a business and industrial model that creates value by extending product lifecycles, optimizing resource efficiency, and transforming waste into valuable inputs, thereby enhancing competitiveness and sustainability.*

Chapter 2

Foundations of waste management in the circular economy

2.1 Introduction

The rapid pace of urbanization, industrialization, and consumerism has dramatically increased the volume and complexity of waste generated across the globe. From household garbage and industrial refuse to packaging materials and electronic scrap, the management of this ever-growing volume of waste has become one of the most pressing environmental, social, and economic challenges of the 21st century. Traditional linear models of production and consumption based on the "take-make-dispose" paradigm have proven to be unsustainable, leading to resource depletion, environmental pollution, and mounting pressure on landfill capacities. In this context, the principles of waste management are evolving, and a transition toward more sustainable and regenerative systems is not only desirable but imperative. This book chapter, titled "*Waste Management Foundations with Circular Economy Practices*," explores into the foundational concepts of waste management while placing them within the framework of circular economy principles, offering a comprehensive understanding of how the two can and must work in tandem to build a sustainable future. Fig. 2.1 illustrate circular economy: a paradigm shifts in waste management.

At the heart of modern waste management lies the need for a systematic and scientifically informed approach to the collection, treatment, and disposal of waste in a manner that minimizes harm to the environment and human health. Foundational waste management encompasses several key elements: source reduction, segregation, collection, transportation, recycling, composting, energy recovery, and safe final disposal. However, these elements must now be reinterpreted through the lens of the circular economy a paradigm that emphasizes keeping resources in use for as long as possible, extracting the maximum value from them while in use, and then recovering and regenerating products and materials at the end of their service life. Unlike traditional waste

The Circular Economy and their Management. https://doi.org/10.1016/B978-0-443-44617-7.00003-6

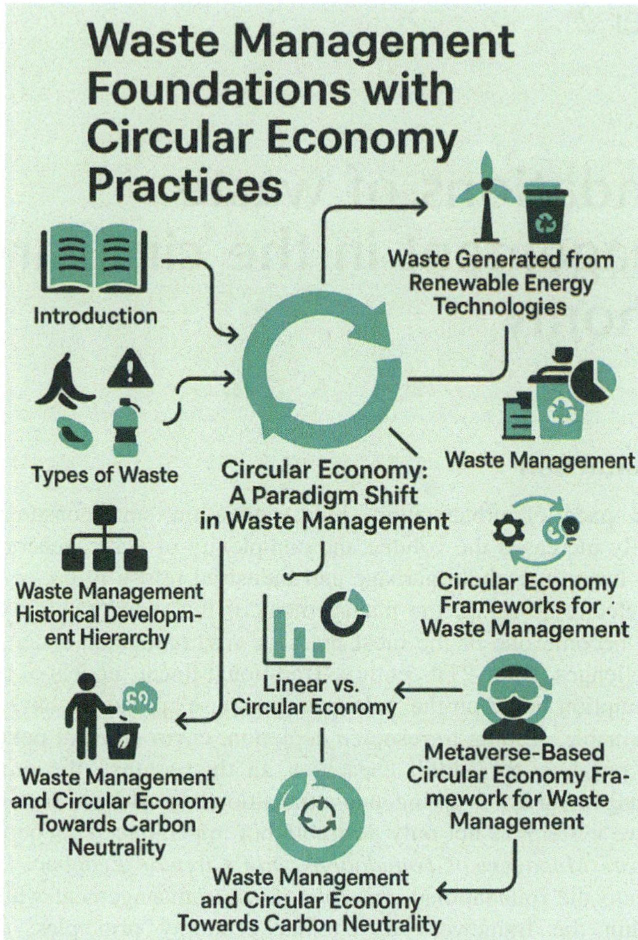

FIGURE 2.1 Circular economy: A paradigm shift in waste management.

management strategies, which often focus primarily on end-of-pipe solutions, circular economy practices aim to intervene at every stage of a product's life cycle from design to disposal, promoting eco-efficiency, closed-loop systems, and cradle-to-cradle thinking.

This chapter begins by tracing the evolution of waste management practices and their limitations in the context of growing global waste volumes, particularly in emerging economies and urban centers. It then explores the core principles of the circular economy reduce, reuse, recycle, redesign, and recover and illustrates how these principles can be integrated into waste management systems. The circular economy calls for systemic changes that go beyond waste diversion and landfill reduction, focusing instead on designing out waste altogether through innovations in product design, business models, material recovery, and resource optimization. For example, designing products for disassembly, promoting the use of biodegradable materials, implementing extended producer responsibility (EPR), and developing reverse logistics chains are strategies that exemplify this alignment.

Furthermore, this chapter highlights the role of policy frameworks, public awareness, technology adoption, and inter-sectoral collaboration in enabling this alignment. Policy tools such as mandatory source segregation, landfill taxes, material bans, and recycling targets provide the regulatory push needed to transform waste management systems. At the same time, the growing accessibility of technologies such as smart bins, waste-to-energy solutions, and AI-based sorting systems facilitates more efficient and data-driven waste management. Local communities, educational institutions, non-governmental organizations (NGOs), and private-sector stakeholders also play a vital role in fostering a culture of circularity, particularly through grassroots movements, decentralized composting initiatives, and zero-waste campaigns.

Aligning waste management foundations with circular economy practices offers a pathway to sustainable development that is economically viable, environmentally sound, and socially inclusive. It transforms the way society views waste from an unwanted burden to a valuable resource and encourages a mind-set that prioritizes regeneration over extraction, collaboration over isolation, and long-term resilience over short-term convenience. By embracing these integrated approaches, nations, cities, and communities can move closer to achieving their environmental targets, advancing the Sustainable Development Goals (SDGs), and ensuring the health and wellbeing of both people and the planet.

2.2 Types of waste

Waste can be categorized into several types based on its source, composition, and potential impact on the environment and human health. Municipal solid waste (MSW) includes everyday items discarded by households, such as food scraps, packaging, paper, plastics, and glass. Industrial waste originates from manufacturing and industrial activities, often consisting of chemicals, metal scraps, and hazardous by-products. Hazardous waste is particularly dangerous and requires special handling, as it contains toxic, corrosive, or flammable materials, commonly found in industries, hospitals, or laboratories. Biomedical waste is generated from healthcare facilities and includes items like syringes, bandages, and body fluids that can pose infection risks. E-waste, or electronic waste, comes from discarded electronic devices like computers, phones, and TVs, and contains both valuable and toxic components. Agricultural waste results from farming activities, including crop residues, manure, and pesticides. Lastly, construction and demolition waste includes materials like concrete, wood, and bricks from building activities. Each type of waste requires specific management practices to minimize environmental damage and promote sustainable development. Fig. 2.2 illustrate types of wastes.

2.2.1 Municipal solid waste (MSW)

Municipal solid waste refers to the everyday waste generated by households, offices, schools, and small businesses. This includes items like food scraps, packaging materials, paper, cardboard, plastic containers, glass bottles, textiles, and broken household items. Often called "trash" or "garbage," MSW is a major concern in urban areas due to population growth and increased consumerism. Improper management of MSW can lead to unsanitary conditions, clogged drains, pest infestations, and harmful environmental effects. Common disposal methods include landfilling, incineration, composting, and recycling. Effective segregation at source and community awareness are crucial for sustainable waste management. Recycling and composting play a major role in reducing the load on landfills and promoting circular economy

FIGURE 2.2 Types of waste.

principles. MSW is collected primarily by local municipal bodies or contracted agencies through door-to-door services. Waste is segregated at the source into biodegradable and non-biodegradable categories, using color-coded bins. Collection is typically carried out daily using manual carts or mechanized vehicles. Public participation and awareness play a significant role in the effectiveness of MSW collection.

2.2.2 Industrial waste

Industrial waste is produced by manufacturing or industrial processes such as factories, mines, and mills. This type of waste can be solid, liquid, or gaseous and includes materials like scrap metal, chemical solvents, dyes, paints, ash, and sludge. Depending on the industry, industrial waste may be hazardous or non-hazardous. For instance, textile industries generate large amounts of dyes and chemicals, while thermal power plants produce fly ash. If not treated properly, industrial waste can contaminate soil, air, and water bodies, posing serious health and ecological risks. Industries are legally obligated to follow waste disposal regulations, including treatment at Effluent Treatment Plants (ETPs) or Common Hazardous Waste Treatment, Storage, and Disposal Facilities (TSDFs). Waste minimization, process optimization, and cleaner production techniques are increasingly being adopted to reduce environmental footprints. Industries are responsible for collecting their waste from production lines and storage areas. This waste is stored in designated containers and managed through authorized waste contractors. Segregation into hazardous and non-hazardous waste is performed on-site, ensuring compliance with environmental regulations.

2.2.3 Hazardous waste

Hazardous waste is any waste material that poses substantial or potential threats to public health or the environment. It is typically generated from chemical manufacturing, metal processing, petroleum refining, and medical or electronic industries. Hazardous waste includes toxic, corrosive, flammable, or reactive substances such as batteries, asbestos, pesticides, acids, and solvents. Because of its dangerous nature, hazardous waste must be handled, stored, transported, and disposed of with utmost care and in compliance with regulatory standards. Improper disposal can lead to contamination of groundwater, air pollution, and harmful exposure to humans and wildlife. Regulations such as the Resource Conservation and Recovery Act (RCRA) in the U.S. and Hazardous Waste Management Rules in India govern its management. Technologies like incineration, chemical treatment, and secure landfilling are commonly used for safe disposal. Hazardous waste is collected in leak-proof, chemically resistant containers with appropriate hazard labels. Certified handlers transport this waste in specialized vehicles. Safety protocols such as the use of PPE, secondary containment, and manifest tracking are critical during the collection process.

2.2.4 Biomedical waste

Biomedical waste, also known as clinical or hospital waste, originates from healthcare facilities like hospitals, clinics, laboratories, and research centers. It includes used syringes, bandages, surgical gloves, body tissues, blood samples, and pharmaceutical products. This waste is potentially infectious and can transmit diseases such as HIV, hepatitis B and C, and other viral and bacterial infections if not handled properly. Biomedical waste is classified into different categories infectious, pathological, sharps, chemical, and pharmaceutical and must be treated accordingly. The Biomedical Waste Management Rules prescribe strict guidelines for segregation, collection, storage, transportation, and disposal. Incineration, autoclaving, and microwaving are commonly used treatment methods. Color-coded bins and staff training are essential to ensure safe handling and prevent public health hazards. Healthcare facilities segregate biomedical waste at the point of generation using a standardized color-coded bin system. Authorized biomedical waste treatment facilities collect the waste daily in sealed containers. Barcoding systems and trained personnel ensure compliance with the Biomedical Waste Management Rules.

2.2.5 Electronic waste (E-waste)

Electronic waste, or e-waste, refers to discarded electrical and electronic devices such as computers, televisions, smartphones, refrigerators, and other appliances. With rapid technological advancements and shorter product lifespans, the generation of e-waste has increased significantly in recent years. E-waste contains valuable materials like gold, silver, and copper, as well as toxic substances like lead, mercury, and cadmium. Improper disposal of e-waste especially in developing countries where informal recycling is common can result in severe environmental pollution and health issues for workers and nearby communities. Safe management involves collection, dismantling, material recovery, and recycling. Extended Producer Responsibility (EPR) policies require manufacturers to take back and recycle used electronic products. Public awareness, safe recycling infrastructure, and policy enforcement are key to managing e-waste responsibly. E-waste is collected through designated drop-off centers, take-back programs, or e-waste drives. Consumers, manufacturers, and recyclers collaborate under Extended Producer Responsibility (EPR) mandates to ensure safe and accountable collection of electronic devices.

2.2.6 Agricultural waste

Agricultural waste is generated from farming and livestock activities and includes crop residues, animal manure, spoiled grains, pesticide containers, and organic matter. While much of this waste is biodegradable and can be repurposed, improper management can lead to serious environmental problems such as water pollution, soil degradation, and greenhouse gas emissions. Crop residues, when burned in the fields, contribute significantly to air pollution and smog, as seen in regions of northern India. Animal waste can be a valuable resource for compost and biogas generation if handled properly. Techniques such as vermicomposting, anaerobic digestion, and mulching help convert agricultural waste into useful by-products. Promoting sustainable farming practices and educating farmers on waste utilization can significantly improve waste management in agriculture. Farmers manually or mechanically collect crop residues, manure, and other organic waste. This waste is either composted, used in biogas production, or stored for other agricultural applications. Organized farms often use designated collection pits or bins. Fig. 2.3 illustrate % distribution of the different types of waste.

2.2.7 Construction and demolition (C&D) waste

Construction and demolition waste arises from the construction, renovation, and demolition of buildings, roads, and other structures. It includes materials

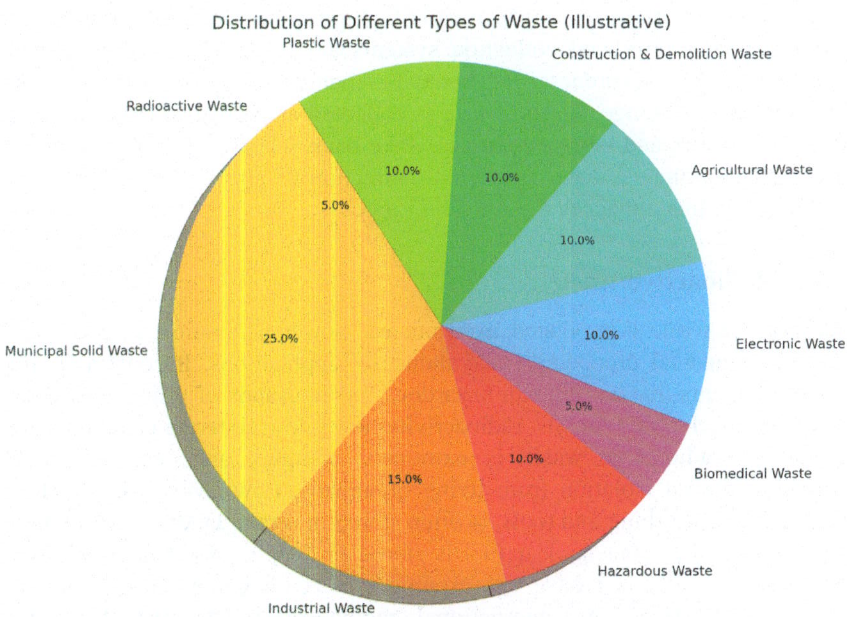

FIGURE 2.3 % Distribution of the different types of waste.

such as concrete, bricks, wood, steel, glass, asphalt, and insulation. C&D waste is one of the heaviest and most voluminous waste types generated globally. If not managed properly, it occupies vast landfills and may cause environmental degradation. However, most C&D waste is recyclable. For example, crushed concrete can be reused as road base, and salvaged wood can be repurposed for furniture. Effective management involves segregation at source, on-site recycling, and use of recycled materials in new construction projects. Green building codes and regulations increasingly promote the reuse of construction materials to minimize environmental impact and reduce carbon emissions. C&D waste is collected during site operations using trucks and dumpers. On-site segregation into categories such as wood, metal, and concrete is common practice. Large construction firms employ reuse strategies and send recyclable material to specialized facilities.

2.2.8 Plastic waste

Plastic waste is a growing global concern due to its non-biodegradable nature and widespread use. Common sources include packaging, single-use items (bottles, bags, straws), household goods, and industrial applications. While plastic is lightweight, durable, and inexpensive, its durability becomes a problem when discarded irresponsibly. Plastic waste accumulates in landfills and oceans, harming marine life and entering the food chain. Micro plastics are particularly concerning as they are difficult to remove from the environment. Recycling of plastic is possible, but the rate of recycling remains low in many countries. Bans on single-use plastics, promotion of biodegradable alternatives, and improved collection systems are essential to reduce plastic pollution. Public awareness and behavior change also play a vital role in tackling this challenge. Plastic waste is collected as part of MSW or through informal sectors and private agencies. Take-back schemes, buy-back centers, and EPR compliance drive the formal collection of plastic waste. Sorting at collection points enhances the recycling process.

2.2.9 Radioactive waste

Radioactive waste is produced from nuclear power generation, research laboratories, medical procedures, and industrial applications. It contains radioactive elements that can remain hazardous for thousands of years. This waste is categorized based on its radioactivity level low, intermediate, or high. Managing radioactive waste is extremely complex and requires secure containment and isolation from living organisms. Low-level waste, such as contaminated clothing and tools, is often stored in secure landfills, while high-level waste like spent nuclear fuel is stored in deep geological repositories. Radiation leakage can have catastrophic effects on human health and the environment. Therefore, international regulatory bodies like the IAEA

(International Atomic Energy Agency) monitor and guide safe practices for its disposal. Radioactive waste is collected in shielded containers by specially trained teams using robotic or remote handling systems. Transport to storage or disposal sites is strictly regulated and monitored, ensuring full compliance with international safety standards. Table 2.1 illustrate comparison of different types of waste. Table 2.2 illustrate mathematical interpretation of different types of waste.

Figs. 2.4–2.6 a clear comparative analysis of waste categories based on risk level, treatment availability, and environmental impact. The Risk Level graph shows that radioactive, hazardous, and biomedical wastes pose the highest danger due to their toxic, reactive, or long-lasting nature, while MSW and C&D waste fall in the lower-risk range. The Treatment Options graph highlights that most waste types have around three major treatment methods available, with municipal solid waste having the highest variety, whereas radioactive waste has the fewest due to its complex and specialized handling requirements. Finally, the Environmental Impact graph demonstrates that radioactive, hazardous, and biomedical wastes cause the most severe environmental harm if improperly managed, while agricultural, C&D, and MSW have moderate impacts. Together, these graphs provide an integrated understanding of how different wastes vary in risk, manageability, and potential damage to the ecosystem.

Each type of waste whether solid, liquid, hazardous, biodegradable, or non-biodegradable has unique characteristics, sources, and environmental impacts. Effective waste management requires a multi-pronged approach involving segregation at source, appropriate treatment technologies, recycling and reuse, strict regulations, and public participation. With rising global waste generation, there is an urgent need to shift towards a circular economy model that emphasizes waste reduction, resource efficiency, and environmental sustainability. Awareness, innovation, and policy integration are key to ensuring a cleaner and healthier future for all. The collection of waste must be tailored to the type and risk profile of the waste stream. Adhering to standardized procedures, regulatory compliance, and promoting stakeholder awareness are key elements of an effective collection strategy. When done correctly, proper waste collection not only protects human health and the environment but also supports recycling and resource recovery initiatives.

2.3 Waste generated from renewable energy technologies

While renewable energy sources are cleaner alternatives to fossil fuels, they are not entirely free from environmental impact. One key concern is the waste generated at different stages manufacturing, installation, operation, and especially end-of-life disposal. Each technology solar, wind, hydro, biomass, and geothermal produces distinct forms of waste.

TABLE 2.1 Comparison of different types of waste.

Type of waste	Source	Composition	Nature	Risk level	Treatment options	Environmental impact
Municipal solid waste (MSW)	Households, offices, commercial places	Organic waste, plastic, paper, glass, metal	Biodegradable & non-biodegradable mix	Low to moderate	Recycling, composting, landfilling, incineration	Land pollution, foul odor, vector breeding
Industrial waste	Factories, mines, refineries	Chemicals, solvents, scrap metal, sludge	Hazardous or non-hazardous	High	ETPs, secured landfills, recycling	Soil, water, and air pollution if untreated
Hazardous waste	Chemical plants, labs, refineries	Toxic chemicals, acids, flammable materials	Toxic, corrosive, reactive, flammable	Very high	Incineration, chemical treatment, encapsulation	Severe environmental and health threats if mishandled
Biomedical waste	Hospitals, clinics, labs	Infected bandages, syringes, tissues, medicines	Infectious, pathological	Very high	Incineration, autoclaving, microwaving	Spread of diseases, water and soil contamination
Electronic waste (E-waste)	Discarded electronics	Circuit boards, plastics, metals, batteries	Contains toxic and valuable materials	High	Dismantling, material recovery, formal recycling	Soil/water pollution, health hazards from heavy metals
Agricultural waste	Farms, livestock operations	Crop residue, manure, pesticide containers	Mostly biodegradable	Moderate	Composting, biogas, vermiculture	Water pollution, methane emissions, stubble burning
Construction & demolition	Construction and renovation sites	Concrete, bricks, wood, glass, tiles	Bulky, inert, mostly recyclable	Low to moderate	On-site recycling, reuse, landfilling	Land degradation, dust, large landfill usage
Plastic waste	Packaging, consumer goods, industrial use	PET, HDPE, PVC, etc.	Non-biodegradable, persistent	High	Recycling, pyrolysis, upcycling	Marine pollution, microplastic contamination
Radioactive waste	Nuclear plants, hospitals, research labs	Radioisotopes, spent fuel, contaminated tools	Radioactive, long-lasting	Extremely high	Encapsulation, deep geological disposal	Radiation risk, irreversible environmental damage

TABLE 2.2 Mathematical interpretation of different types of waste.

Type of waste	Key variables/ parameters	Mathematical expression or model	Purpose
Municipal solid waste (MSW)	Q_t: Total waste generated per day, *P*: Population W_p: Waste per capita (kg/day)	$Q_t = P \times W_p$	To estimate total daily MSW for city planning
Industrial waste	Q_i: Industrial waste per unit, N: No. of industries *f*: Frequency of disposal	$Q_{total} = \sum_{i=1}^{N} Q_i \times f$	Estimate cumulative industrial waste generation
Hazardous waste	H_r: Hazard index Q_h: Quantity of hazardous waste *S*: Safety compliance factor	$R_{risk} = H_r \times Q_h \div S$	Risk quantification and compliance level estimation
Biomedical waste	*B*: Total biomedical waste generated daily *bp*: Avg. biomedical waste per patient P_h: Patients/day	$B = b_p \times P_h$	For planning biomedical waste collection infrastructure
Electronic waste (E-waste)	E_i: No. of devices L_i: Avg. Lifespan	$D_e = \sum_{i=1}^{n} E_i \div L_i$	E-waste forecasting and reverse logistics design
Agricultural waste	*Y*: Crop yield *r*: Residue ratio (waste/yield)	$W_{agri} = Y \times r$	To estimate waste biomass for composting or bioenergy
C&D waste	*A*: Area under construction *d*: Avg. Debris per sq. meter	$Q_{cnd} = A \times d$	Planning for transportation and recycling of debris
Plastic waste	u_i: plastic usage r_i: Recycling rate	$p_t = \sum_{i=1}^{n} u_i \times r_i$	Plastic pollution estimation and recycling gap analysis
Radioactive waste	A_0: Initial activity λ: Decay constant	$A(t) = A_0 e^{-\lambda t}$	Predicting storage safety levels and long-term disposal planning

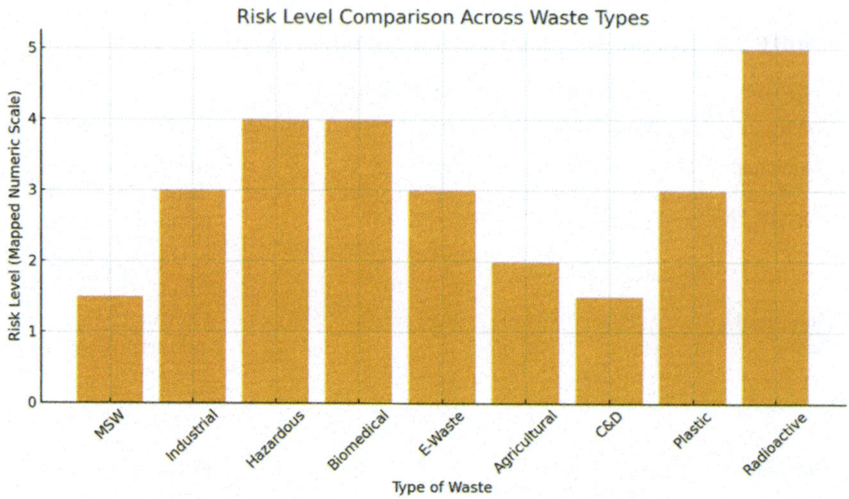

FIGURE 2.4 Risk level comparison across waste types.

2.3.1 Solar power plant

Solar photovoltaic (PV) systems, while emission-free during operation, generate substantial waste when panels reach the end of their 25–30-year lifespan. Waste includes glass, plastic, silicon, aluminum frames, copper wiring, and toxic metals like cadmium, lead, and selenium (in thin-film panels). Improper disposal can lead to heavy metal leaching into soil and

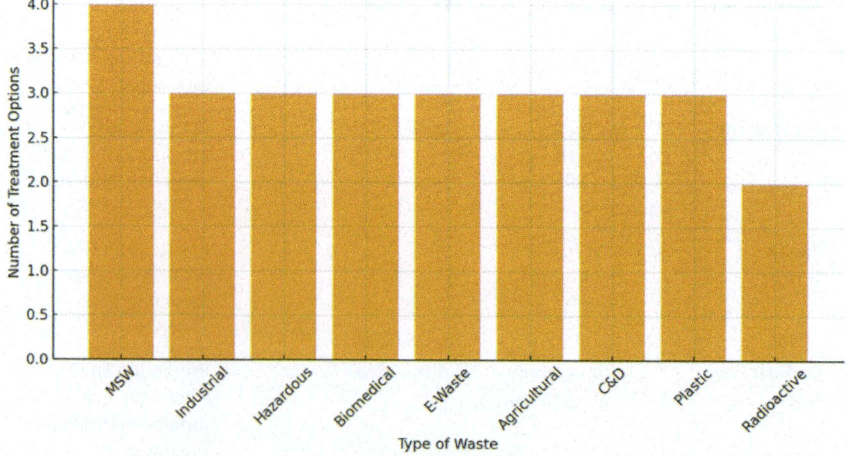

FIGURE 2.5 Treatment level comparison across waste types.

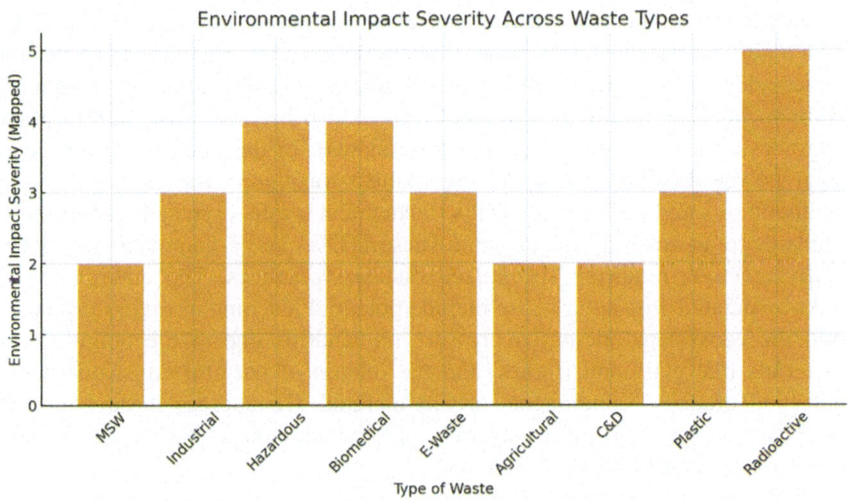

FIGURE 2.6 Environmental impact severity across waste types.

groundwater. Additionally, the production process emits chemical by-products such as silicon tetrachloride. Without robust recycling infrastructure, solar waste is expected to reach millions of tons globally by 2050. While solar energy is widely celebrated as a clean and renewable alternative to fossil fuels, it is essential to recognize the growing concern of waste generated from solar power plants across their lifecycle from manufacturing and operation to decommissioning. The most substantial form of waste arises from photovoltaic (PV) panels, which generally have a lifespan of 25 to 30 years. These panels are composed of materials such as silicon, aluminum, tempered glass, and trace amounts of hazardous elements like cadmium, selenium, and lead. As panels degrade, break during transportation, or become obsolete due to technological upgrades, they contribute significantly to solar waste volumes. According to the International Renewable Energy Agency (IRENA), by 2050, the global PV waste could exceed 78 million tonnes. Alongside panels, another significant waste stream originates from batteries used in solar systems, particularly in off-grid and hybrid applications. These batteries most commonly lead-acid or lithium-ion have shorter lifespans (5–10 years) and often contain toxic components including lead, lithium, cobalt, and sulfuric acid. Improper disposal or spillage can lead to contamination of soil and groundwater and pose serious health hazards. Equally important is the electronic and electrical waste (e-waste) generated from auxiliary components such as inverters, charge controllers, sensors, and monitoring systems. These elements, containing circuit boards, plastic casings, and metallic wiring, add to the complexity of solar waste management due to their classification under e-waste protocols.

In addition to these, packaging waste such as plastic films, wooden pallets, and foam inserts generated during transportation and installation, while not toxic, adds to municipal solid waste if not responsibly handled. Structural components like aluminum and steel frames also become waste during plant upgrades or when systems are decommissioned. In the case of concentrated solar power (CSP) systems, the use of heat transfer fluids (HTFs) such as synthetic oils and molten salts leads to chemical waste, especially when these fluids degrade or need replacement. These HTFs can be flammable or chemically reactive, requiring specialized disposal practices. The environmental risks associated with solar waste include potential leaching of toxic substances into the ecosystem, increased pressure on landfills due to the bulkiness of materials like glass and frames, and the release of harmful emissions when non-biodegradable components are incinerated. To address these concerns effectively, a multi-pronged approach must be adopted by both policymakers and industry stakeholders.

Strategic actions should include:

- **Adoption of extended producer responsibility (EPR)**: Mandating that manufacturers take responsibility for end-of-life collection and recycling of solar components can promote circular economy practices and encourage better product design.
- **Investment in recycling infrastructure**: Establishing solar panel-specific recycling facilities that can separate glass, silicon, and metals efficiently will be critical, especially as waste volumes grow in the coming decades.
- **Standardized waste handling protocols**: Guidelines must be established for safe storage, transport, and processing of hazardous waste from solar batteries and CSP fluids.
- **Data-driven policy planning**: Regular auditing, tracking, and mapping of waste flows from solar installations will allow targeted interventions and resource allocation.
- **Promotion of reuse markets**: Creating pathways for the reuse of functional but outdated panels in low-energy applications such as rural lighting or small-scale irrigation can reduce the waste burden.

Ultimately, the transition to clean energy must be holistic, ensuring that solar technology not only reduces carbon emissions but also incorporates sustainability in waste management practices. As solar installations expand globally, the cumulative waste from panels, batteries, electronic systems, and associated components poses a looming environmental challenge. By proactively addressing these issues with strategic planning, investment in recycling, and robust regulatory frameworks, the solar industry can maintain its promise of being a truly green energy solution. A shift toward circular economy principles and sustainable waste practices will be essential in realizing the long-term environmental benefits of solar energy.

2.3.2 Wind power plant

Wind turbines have long operational lifespans, but their composite blades made of fiberglass and resin are difficult to recycle. These blades often end up in landfills due to their size and non-biodegradable nature. Other components like gearboxes and generators may contain rare earth elements, oils, and lubricants that require careful handling during decommissioning. Metal parts like steel towers are recyclable, but the large volume and logistical challenges pose environmental risks if not managed properly. Wind power is globally recognized as one of the cleanest and most sustainable forms of renewable energy, playing a pivotal role in the reduction of greenhouse gas emissions and reliance on fossil fuels. However, an often-overlooked challenge lies in the lifecycle waste generated by wind power plants. From manufacturing and construction to operation, maintenance, and eventual decommissioning, wind turbines and their associated infrastructure produce a variety of waste streams that require planned management strategies. One of the primary concerns revolves around the wind turbine blades, which are typically made from composite materials such as fiberglass, epoxy resin, carbon fiber, and sometimes balsa wood. These composites are incredibly strong and lightweight but are difficult to recycle using conventional methods. Wind turbine blades have an average lifespan of 20 to 25 years, and when retired, they are often disposed of in landfills due to the lack of scalable recycling solutions. The growing number of aging wind farms worldwide, especially in Europe, the United States, and India, means that blade waste is expected to surge in the coming years some estimates suggest that by 2050, nearly 43 million tonnes of blade waste could be generated globally.

Another critical waste stream comes from the electromechanical components of wind turbines, including gearboxes, generators, bearings, and electronic control systems. These components contain valuable but also hazardous materials such as copper, aluminum, lubricating oils, and rare earth magnets. Over time, wear and tear, corrosion, or overheating can render these parts unusable. Although many of the metallic parts can be recycled, lubricants and oils used in gearboxes must be treated as hazardous waste due to the potential for soil and water contamination. Additionally, electronic waste (e-waste) from control panels, sensors, and SCADA (Supervisory Control and Data Acquisition) systems accumulates, especially as turbines are upgraded or replaced. This e-waste contains circuit boards, wiring, and heavy metals, posing similar risks as in solar power systems. Concrete and steel waste from tower foundations is another large contributor during the dismantling of old turbines. The foundations, often made of reinforced concrete, can weigh hundreds of tons and are energy-intensive to remove and process. Steel from towers and nacelles is highly recyclable but requires dismantling infrastructure and proper logistics.

Furthermore, packaging and transport waste—such as wood pallets, plastic coverings, and strapping—produced during the transportation of turbine components to remote wind farm sites contributes to the overall environmental footprint. With offshore wind farms, additional marine waste can be generated from the use of anti-fouling paints, corrosion-prevention systems, and subsea cables. While this waste is less voluminous, it has a higher potential for marine pollution if not managed carefully. Given these diverse and sometimes hazardous waste streams, strategic and forward-looking waste management is critical for maintaining the sustainability of wind energy.

To address the challenges of wind turbine waste, several strategic interventions can be employed:

- **Blade recycling innovations:** Research and investment in new recycling technologies such as pyrolysis, solvolysis, and mechanical shredding can help recover fibers and resins from composite blades. Initiatives such as using old blades in construction materials or for urban furniture are also gaining momentum.
- **Circular design approaches:** Encouraging manufacturers to design wind turbines with modular, recyclable, and bio-based components can reduce end-of-life waste. Blades made from thermoplastics instead of thermoset resins may be easier to remold and reuse.
- **Lifecycle assessment and extended producer responsibility (EPR):** Manufacturers and developers should be required to track the environmental impact of turbines from cradle to grave and take responsibility for collection and disposal post-decommissioning.
- **Recycling of metallic components:** Steel, copper, and aluminum from turbines should be directed to certified recycling facilities. Establishing decentralized dismantling units near wind farms can optimize logistics and reduce environmental cost.
- **Safe disposal of hazardous waste:** Used gear oils, coolants, and electronic components must be collected and processed under hazardous waste rules to prevent leaching and pollution.
- **Foundation reuse and deconstruction:** Instead of complete demolition, foundations can sometimes be reused for repowering projects or adapted for new installations, reducing raw material consumption.
- **Regulatory frameworks and waste audits:** Governments should implement mandatory waste audits and decommissioning plans for all large wind farms, especially those nearing the end of operational life.

Despite the waste-related challenges, wind power remains far less polluting over its lifecycle compared to fossil-based generation. However, the industry must recognize that true sustainability goes beyond clean energy production and includes responsible material use, waste recovery, and recycling. With the global expansion of wind energy, especially in offshore regions, managing turbine waste will become an integral part of energy transition strategies. By embracing a circular economy mind-set, the wind energy sector can ensure that its ecological footprint remains as light as the air it harnesses to generate power.

2.3.3 Hydro power plant

Hydropower systems generally have minimal operational waste. However, the construction and maintenance of dams lead to concrete, metal, and sediment waste. Siltation behind dams can reduce efficiency and impact aquatic ecosystems, necessitating dredging and waste disposal. Additionally, aging infrastructure can result in the release of hazardous materials like lead-based paints and asbestos, particularly in older plants. There's also the ecological "waste" in terms of habitat loss and biodiversity disruption. Hydropower, long considered the backbone of renewable electricity generation, is often perceived as an environmentally benign technology due to its low operational emissions and ability to store energy through pumped hydro mechanisms. However, like all infrastructure-intensive energy sources, hydropower plants also produce a range of waste materials throughout their construction, operation, maintenance, and decommissioning phases. The most prominent waste streams associated with hydropower plants begin at the construction stage, where massive amounts of civil waste are generated. This includes surplus concrete, steel reinforcements, wooden scaffolding, excavation debris, and packing materials. Large-scale dam construction also leads to the generation of geological waste, particularly during tunneling and excavation activities where excess rock, soil, and sediment are removed and often left in open disposal sites. While much of this construction waste is inert and can be reused in road construction or landscaping, poor management may result in soil erosion, sediment runoff, and ecological disturbance in riverine systems.

Another significant environmental issue stemming from hydropower installations is the organic waste that accumulates in the reservoirs behind dams. When forests, vegetation, or agricultural land are submerged during reservoir formation, they begin to decompose anaerobically, leading to the release of methane, a potent greenhouse gas. In addition, floating biomass, tree trunks, leaves, and algae accumulate at the water surface or clog turbine inlets, requiring regular removal and disposal. Over time, these residues are either incinerated, composted, or dumped in nearby landfills, contributing to biodegradable waste loads. The issue becomes more pronounced in tropical and subtropical regions, where vegetation grows and decays rapidly, creating a constant stream of plant-based organic waste in reservoirs.

During regular operation and maintenance of hydropower plants, another category of waste arises from the mechanical and electrical systems. These include lubricating oils, greases, coolants, hydraulic fluids, and metallic wear particles from turbines, gearboxes, and generators. These fluids, while essential for machinery health, must be replaced periodically and are classified as hazardous waste due to their chemical composition. Improper disposal can lead to contamination of soil and aquatic systems, especially since hydropower plants are located adjacent to rivers. Used filters, rags, oil drums, and chemical containers also contribute to the operational waste stream. Moreover, e-waste is generated from control rooms, automation systems, monitoring equipment, and SCADA interfaces. These components include circuit boards, cables, and outdated sensors, all of which require disposal in compliance with e-waste management rules.

Hydropower also necessitates the construction of access roads, transmission lines, staff quarters, and maintenance facilities, which leads to ongoing municipal solid waste generation such as plastic packaging, paper waste, construction debris, food waste, and sanitary waste. These wastes, if not properly segregated and managed, burden local landfills or pose risks of open dumping near water bodies. Additionally, sediment waste becomes an operational issue in river-fed hydropower plants. Over time, silt and sand accumulate in reservoirs, reducing storage capacity and affecting efficiency. Dredging operations, often needed every few years, generate sediment waste that must be treated for contaminants (especially in industrial regions) and disposed of safely.

At the end of a hydropower plant's lifecycle or during major renovations, decommissioning waste is produced. This includes massive metallic structures, turbines, steel gates, pipes, valves, electrical cables, concrete blocks, and scrap machinery. While most of these materials are recyclable, dismantling and logistics require heavy-duty planning and specialized handling to minimize environmental impact.

To address and minimize the waste generated from hydropower projects, several strategic interventions are essential:

- **Reservoir biomass management**: Pre-clearing vegetation before reservoir filling and regular removal of floating biomass can reduce methane emissions and minimize organic waste accumulation.
- **Oil and chemical waste control**: Installing oil separators, bunded storage, and conducting regular audits for hazardous fluid management helps prevent water contamination.
- **Construction waste recycling**: Civil and demolition waste should be reused in embankment strengthening, road repair, or foundation works to minimize disposal volumes.
- **Sediment reuse planning**: Dredged sediment can be evaluated for use in agriculture, brick making, or land restoration, provided it's free from pollutants.
- **Establishing waste management zones**: Dedicated waste handling areas with segregation, composting, and recycling infrastructure should be developed at large hydropower sites.

- **Lifecycle waste audits**: From project design to decommissioning, lifecycle-based waste assessments can guide better material choices and ensure closed-loop systems.
- **E-waste and municipal waste integration**: Partnering with local municipalities and authorized recyclers can improve safe disposal and recovery from staff quarters and admin blocks.

Hydropower offers a low-carbon source of renewable electricity, its environmental performance is incomplete without addressing the multiple layers of waste it produces across its lifecycle. By adopting proactive waste minimization strategies and investing in eco-friendly operational practices, the hydropower sector can significantly enhance its sustainability credentials. Planning for waste at the design stage, integrating circular economy models, and ensuring compliance with evolving environmental standards will be crucial as the global hydropower capacity continues to grow.

2.3.4 Biomass power plant

Biomass power plants burn organic materials such as crop residues, wood, and animal waste to produce energy. The combustion process results in fly ash, bottom ash, and particulate emissions. If biomass feedstocks are not clean or are contaminated (e.g., treated wood), the resulting ash may contain harmful substances like heavy metals and dioxins. Moreover, logistics around sourcing and storing biomass can generate waste in the form of expired or degraded materials. Biomass power plants, which convert organic materials like agricultural residues, wood chips, animal waste, and industrial organic by-products into energy, are often seen as a sustainable solution to both energy generation and waste management. However, like any other power generation system, biomass power plants also produce several types of waste throughout their operational lifecycle. These wastes can vary in type, volume, and environmental impact depending on the feedstock, conversion technology used (such as combustion, gasification, pyrolysis, or anaerobic digestion), and local management practices. One of the most significant by-products is ash waste, particularly bottom ash and fly ash generated during combustion processes. Bottom ash accumulates at the base of the combustion chamber and consists of unburned carbon, soil, silica, and trace metals, while fly ash is a finer particulate matter carried with flue gases and captured using electrostatic precipitators or bag filters. Although in small quantities these ashes can be used in brick manufacturing, soil stabilization, or as cement additives, they often contain heavy metals and polycyclic aromatic hydrocarbons (PAHs), which, if not managed properly, can lead to soil contamination and groundwater pollution.

In addition to ash, air pollution control residues such as filter dust, spent lime, and used sorbents from flue gas cleaning systems also form a part of hazardous waste in biomass power plants. These materials can be corrosive or toxic and require proper handling, storage, and disposal according to environmental regulations. Another substantial category of waste includes unutilized feedstock. Biomass feedstock is often stored in open or semi-covered areas, where it may decompose, get infested with pests, or become wet and unusable. This rotted biomass emits methane and foul odors and needs to be composted or sent to landfill, contributing to greenhouse gas emissions if unmanaged.

Moreover, biomass plants often require pre-processing of raw material—such as chipping, shredding, drying, or sieving which produces significant quantities of dust, bark, husk, and other particulates. These are sometimes reused, but more often disposed of as low-grade waste, particularly when contaminated or non-uniform. In biochemical conversion processes like anaerobic digestion, which are used in biogas plants, the main waste product is digestate a nutrient-rich slurry left after gas extraction. Although digestate can be used as a fertilizer, its effectiveness depends on the input feedstock and pathogen content. If derived from municipal or mixed organic waste, it may contain plastic residues, pathogens, heavy metals, or drug residues, which limit its usability and call for controlled application or further treatment.

Wastewater is another environmental issue linked to biomass power plants, especially when wet biomass, fermentation, or chemical treatment is involved. Washing of biomass, cooling processes, and scrubbers generate wastewater that may be high in biochemical oxygen demand (BOD), chemical oxygen demand (COD), and organic matter which, if discharged untreated, can pollute local water bodies. Routine maintenance and operations also produce waste in the form of used oils, greases, cleaning agents, filters, and safety gear, which are categorized under hazardous operational waste and require segregation and disposal through authorized handlers.

Furthermore, metallic and electrical waste arises from the periodic replacement of boilers, pipelines, valves, sensors, and electrical systems. This scrap metal and e-waste may be recyclable but needs systematic collection. Packaging waste, including plastic wrappings, metal drums, and wooden pallets used in feedstock transport and machinery supply, adds to the solid waste stream, especially if not segregated at the source. At the end of a biomass plant's life or during major overhauls, decommissioning waste such as structural steel, insulation material, refractory bricks, and damaged concrete is generated and must be processed responsibly. To mitigate and manage the waste generated from biomass power plants, the following strategic actions can be employed:

- **Ash valorization programs**: Explore certified uses of bottom and fly ash in construction, paving, or agriculture, while ensuring heavy metal testing and environmental safety.

- **Controlled feedstock management**: Implement improved feedstock storage systems with moisture protection and pest control to prevent degradation and minimize waste.
- **Digestate quality standards**: Develop and enforce digestate quality protocols to promote safe reuse as bio-fertilizer while preventing the spread of contaminants.
- **Wastewater treatment units**: Install on-site wastewater treatment systems to handle high-BOD effluents from biomass pre-treatment and combustion processes.
- **Waste segregation protocols**: Train staff and install facilities to sort hazardous, recyclable, and biodegradable waste streams at the plant level.
- **Recycling and reuse of metals and plastics**: Establish tie-ups with recyclers for consistent removal and processing of scrap metals, packaging, and e-waste.
- **Lifecycle waste planning**: Include comprehensive waste audits and material flow analysis in the project design phase to prepare for waste generated during operational and decommissioning stages.

In conclusion, while biomass power plants serve a dual function of renewable energy generation and organic waste utilization, they are not free from environmental externalities. The waste generated from ash and digestate to feedstock residues and hazardous fluids must be proactively managed to ensure the overall sustainability of the biomass energy system. By adopting circular economy strategies, enforcing environmental safeguards, and investing in waste reduction technologies, biomass plants can significantly improve their ecological performance and contribute to a greener energy future.

2.3.5 Ocean power plant

Ocean power also known as marine or blue energy refers to renewable energy harnessed from ocean resources, including tidal power, wave energy, ocean thermal energy conversion (OTEC), and salinity gradient power. While ocean energy is celebrated for its immense potential and zero-emission operation, it is not entirely free from environmental and material waste throughout its lifecycle. The construction, operation, maintenance, and decommissioning of ocean power plants generate a variety of physical, chemical, and biological wastes, many of which are unique due to the harsh marine environment and complexity of offshore installations. One of the primary categories of waste comes from structural components used in turbines, pontoons, buoys, and submerged platforms. These components are typically made of high-grade steel, reinforced composites, anti-corrosive coatings, and synthetic polymers to withstand saltwater corrosion and biofouling. Over time, due to constant exposure to moisture, salt, and pressure, parts of these structures corrode, degrade, or fracture. The replacement of such components generates metallic and polymer waste, some of which may contain toxins from paints or coatings such as zinc, chromium, or copper-based biocides.

Another significant concern in ocean power systems is the waste associated with anti-biofouling and corrosion prevention techniques. Marine organisms such as algae, barnacles, and mussels rapidly colonize underwater equipment, leading to inefficiency and mechanical strain. To prevent this, operators apply anti-fouling paints and coatings, often containing hazardous substances like tributyltin (TBT) or copper compounds. Over time, these coatings slough off into the marine ecosystem, contributing to chemical waste and marine pollution. Additionally, periodic re-coating of submerged surfaces generates hazardous paint waste, solvent waste, and contaminated application tools that must be disposed of according to marine pollution control protocols. The use of sacrificial anodes (typically made of zinc or aluminum) for cathodic protection leads to metal leaching into seawater, resulting in long-term chemical accumulation and potential ecological disturbance.

In addition to physical and chemical waste, ocean power plants can create biological waste during installation and maintenance. Seabed disturbance while anchoring tidal turbines or laying underwater cables often displaces benthic organisms and generates sediment plumes, which can contain decomposed organic matter, contaminants, or microplastics. These sediments, if dredged or removed, need safe offshore relocation or land-based containment. Furthermore, during maintenance shutdowns, the cleaning of bio-fouled equipment using high-pressure water jets or chemical baths produces sludge that must be carefully processed to avoid introducing invasive species or pathogens into new environments.

On the operational side, lubricants, hydraulic fluids, and coolants used in underwater turbines and converters are crucial for functionality but pose an environmental hazard if leaked. These fluids are considered marine hazardous waste and must be replaced periodically, generating used oil, grease, and chemical waste. Special precautions, such as sealed transmission systems and remotely operated maintenance equipment, are required to minimize leakage into ocean ecosystems. Moreover, electronic waste (e-waste) is another by-product of marine energy systems. The control panels, sensors, telemetry units, and communication equipment degrade over time due to salt exposure and must be replaced. These components contain circuit boards, batteries, and metallic wiring, which need specialized disposal and recycling facilities.

Transportation and logistics in marine environments also contribute to the overall waste footprint. Packing materials, ropes, plastic barriers, and unused hardware are often left at sea or discarded improperly, creating risks of marine litter and ghost gear entanglement for aquatic species. At the decommissioning phase, waste generated from subsea foundations, mooring systems, electrical cables, and damaged floatation devices is massive and must be retrieved, sorted, and processed often under challenging sea conditions. To manage and mitigate waste from ocean power systems, the following strategic actions are essential:

- **Eco-friendly anti-fouling technologies**: Shift to biodegradable or non-toxic coatings that reduce marine pollution while controlling biofouling.
- **Closed-loop hydraulic systems**: Design systems that minimize fluid loss and leakage, with automatic alarms and maintenance alerts.
- **Marine-safe coatings and materials**: Use non-toxic, corrosion-resistant alternatives to harmful heavy metals or organotins in paints and structures.
- **Sediment and sludge management plans**: Monitor and control seabed disruption and treat sludge from bio-cleaning and dredging responsibly.
- **Recyclable design philosophy**: Encourage modular and recyclable marine components that can be reused or processed after decommissioning.
- **Lifecycle waste audits**: Conduct waste assessments at design, operation, and decommissioning phases for better tracking and compliance.
- **Marine debris recovery initiatives**: Collaborate with marine conservation groups to recover and recycle waste during routine offshore inspections.

In conclusion, while ocean energy holds vast promise as a sustainable and inexhaustible resource, it also brings with it a unique set of waste management challenges that must be addressed through proactive design, sustainable materials, rigorous maintenance, and responsible decommissioning practices. Incorporating circular economy principles and environmental compliance can ensure that marine power technologies remain as clean in their waste footprint as they are in their energy generation potential. Table 2.3 illustrate comparison of waste from renewable energy technologies. Fig. 2.7 presents waste to resource mapping.

2.4 Waste management

Waste management is a critical aspect of environmental sustainability, encompassing the collection, transportation, processing, recycling, and disposal of waste materials. It plays a pivotal role in minimizing the adverse impacts of waste on human health, natural resources, and the environment. With the rapid pace of urbanization, industrialization, and population growth, the generation of waste has escalated significantly, making effective waste management systems essential for maintaining ecological balance and public health.

The concept of waste management is not limited to merely discarding waste but also involves the strategic reduction of waste at its source, resource recovery through recycling and composting, and the safe disposal of residual waste. The waste hierarchy, which prioritizes waste prevention, minimization, reuse, recycling, and recovery before disposal, serves as a guiding framework for sustainable waste management practices.

TABLE 2.3 Comparison of waste from renewable energy technologies.

Technology	Type of waste	Source/Stage	Environmental impact	Recyclability	Key challenges
Solar energy	Glass, plastics, silicon, aluminum, toxic metals (cadmium, lead, selenium)	Manufacturing, end-of-life (panel disposal)	Heavy metal leaching into soil/water, landfill space	Moderate (requires specialized processes)	Lack of global recycling infrastructure, hazardous material handling
Wind energy	Fiberglass blades, oils, lubricants, metals, rare earth elements	Decommissioning of turbines	Blade disposal in landfills, chemical leakage from gear oils	Blades: Low; metals: High	Blade recycling is complex, large component transport & handling
Hydropower	Sediment, concrete debris, aging infrastructure materials (e.g., asbestos, lead)	Construction, dredging, maintenance	Aquatic ecosystem disruption, siltation issues	High (concrete and metals); some hazardous materials	Sediment handling, hazardous waste in old structures
Biomass energy	Ash (fly ash, bottom ash), particulates, heavy metals, dioxins	Combustion, feedstock processing	Air and soil pollution, toxic residue	Depends on feedstock and contamination level	Safe disposal of ash, emissions from contaminated biomass
Geothermal energy	Mineral-laden brine, arsenic, boron, mercury, scale, drilling debris	Extraction, wastewater discharge, maintenance	Water and soil contamination from untreated brine	Brine: Low; metals: Some	Proper reinjection or treatment of wastewater, corrosion of infrastructure

FIGURE 2.7 Waste to resource mapping.

Waste

W—Waste Reduction: Minimizing the amount of waste generated at the source.

A—Awareness: Educating people about responsible consumption and disposal.

S—Segregation: Sorting waste into categories (organic, recyclable, hazardous, etc.).

T—Treatment: Processing waste to reduce harm (composting, recycling, incineration, etc.).

E—Environment: Protecting natural resources through sustainable waste practices.

Management

M—Monitoring: Keeping track of waste types and volumes for better control.

A—Action: Implementing practical strategies and policies for waste handling.

N—Neutralization: Rendering hazardous waste harmless before disposal.

A—Accountability: Ensuring all stakeholders take responsibility for waste.

G—Governance: Establishing rules, regulations, and enforcement mechanisms.

E—Efficiency: Using resources smartly to minimize waste and costs.

M—Minimization: Continuously working to reduce waste output.

E—Engagement: Involving communities, industries, and governments in solutions.

N—Network: Building systems of collaboration across sectors.

T—Technology: Leveraging innovations like AI, IoT, and automation for smart waste management.

The principles of waste management are grounded in reducing waste generation, maximizing recycling and resource recovery, and minimizing environmental impact. Integrated waste management systems incorporate multiple waste treatment methods, such as landfilling, incineration, composting, and anaerobic digestion, to handle different waste streams effectively. While landfilling remains a common disposal method, it poses significant environmental challenges, including greenhouse gas emissions and groundwater contamination. Incineration, though effective in reducing waste volume, can release harmful pollutants if not properly managed. In recent years, the circular economy model has gained prominence in waste management, advocating for a shift from the traditional linear model of "take-make-dispose" to a closed-loop system where resources are continuously reused and recycled. This approach emphasizes product design for longevity, reparability, and recyclability, thus minimizing waste generation and conserving natural resources. Fig. 2.8 illustrate process of waste management.

Government policies, regulations, and international agreements play a crucial role in promoting effective waste management practices. For instance, the Basel Convention regulates the transboundary movement of hazardous waste to prevent environmental dumping. National governments implement waste management policies, such as extended producer responsibility (EPR), which holds manufacturers accountable for the end-of-life management of their products, encouraging sustainable product design and waste reduction.

Public awareness and community participation are vital for the success of waste management initiatives. Educational campaigns, waste segregation programs, and incentives for recycling can foster a culture of responsible waste disposal. Additionally, technological advancements in waste treatment, such as waste-to-energy (WtE) systems, smart waste bins, and sensor-based

FIGURE 2.8 Process of waste management.

monitoring, are transforming traditional waste management practices, making them more efficient and environmental friendly. However, despite significant progress, numerous challenges persist in the field of waste management. Developing countries face issues related to inadequate infrastructure, lack of regulatory enforcement, and insufficient public awareness, leading to uncontrolled dumping and open burning of waste. On the other hand, high-income countries grapple with managing complex waste streams, such as electronic and hazardous waste, which require advanced treatment technologies and stringent regulations. Addressing these challenges requires a multifaceted approach involving policy frameworks, technological innovations, stakeholder engagement, and international cooperation. As global waste generation continues to rise, the need for sustainable waste management systems becomes increasingly urgent. By adopting a holistic and inclusive approach, societies can effectively mitigate the environmental and health impacts of waste, contributing to a cleaner, greener, and more sustainable future.

2.4.1 Historical development of waste management

The historical development of waste management has evolved significantly from ancient to modern times. In early civilizations, waste was often discarded in open spaces or buried, with little concern for hygiene or environmental consequences. The first organized waste management system appeared in ancient Greece and Rome, where municipal cleaning services and public washrooms were introduced. During the Middle Ages, waste accumulation in urban areas led to serious health issues, including the spread of diseases like the plague. The Industrial Revolution in the 18th and 19th centuries saw a dramatic increase in waste generation, prompting cities to develop formal waste collection and disposal systems. The 20th century introduced landfill engineering, incineration, and early recycling efforts. In recent decades, the focus has shifted to sustainable practices, waste minimization, and circular economy models, driven by increasing environmental awareness, technological advancements, and regulatory frameworks.

Following are key themes of the historical development of waste management:

Ancient waste management practices: The earliest forms of waste management can be traced back to ancient civilizations such as Mesopotamia, Egypt, Greece, and Rome, where waste disposal was rudimentary and primarily involved open dumping. In Mesopotamia, waste was disposed of outside city walls, while the ancient Egyptians created basic landfills to bury waste. The Greeks were among the first to introduce structured waste disposal practices, with Athens establishing the first municipal landfill in 500 BCE, requiring waste to be dumped at least one mile from the city. The Romans advanced waste management by constructing elaborate sewage systems, such as the Cloaca Maxima, to transport waste away from populated areas. Despite these early efforts, waste disposal remained largely unregulated, and public health was significantly compromised by unsanitary conditions.

Medieval period and public health concerns: During the medieval period, waste management systems regressed significantly, leading to widespread public health crises. Urban populations increased, but waste disposal practices remained unorganized and unsanitary. Streets were often used as dumping grounds for household and industrial waste, attracting rodents and leading to the proliferation of diseases. The Bubonic Plague of the 14th century, which killed millions across Europe, was exacerbated by poor waste management and lack of sanitation infrastructure. In response, cities like London and Paris issued announcements to regulate waste disposal and street cleaning, mandating that waste be removed and disposed of outside city limits. However, enforcement was inconsistent, and waste accumulation continued to be a major public health concern. Fig. 2.9 illustrate Historical Development of Waste Management.

Industrial revolution and the rise of modern waste management: The Industrial Revolution in the 18th and 19th centuries brought about unprecedented industrial growth, resulting in massive increases in waste production. Factories generated large quantities of solid, liquid, and gaseous waste, much of which was discharged untreated into rivers, lakes, and open land. Public awareness of the consequences of poor waste management grew, leading to

FIGURE 2.9 Historical development of waste management.

significant reforms. In 1842, Edwin Chadwick, a British social reformer, published a landmark report that linked waste accumulation to disease outbreaks, advocating for organized waste collection and sewage treatment. This report laid the groundwork for modern waste management practices. Shortly after, the first waste incinerator was constructed in Nottingham, England, in 1874, marking the beginning of waste-to-energy practices. Table 2.4 illustrate historical development of waste management.

Municipal solid waste management (Late 19th to early 20th century): The late 19th and early 20th centuries saw the establishment of formal municipal waste management systems in major cities worldwide. In the United States, the city of New York introduced structured waste collection services, employing horse-drawn carts to collect refuse and transport it to designated landfills. The concept of sanitary landfills, in which waste is compacted and covered with soil to prevent pest infestation and foul odors, emerged during this period. Incineration also gained popularity as a method to reduce waste volume, but concerns over air pollution limited its widespread adoption. Regulatory frameworks began to take shape, establishing guidelines for waste disposal to mitigate public health risks and environmental degradation.

Environmental movement and regulatory frameworks (1960s-1980s): The 1960s and 1970s marked a turning point in waste management, driven by growing environmental awareness and public pressure to address pollution. Landmark events such as the 1969 Cuyahoga River fire, caused by industrial waste dumping, and the Love Canal disaster in 1978, which exposed the dangers of hazardous waste disposal, prompted governments to take action. In the United States, the Solid Waste Disposal Act of 1965 was the first federal law to address waste management, followed by the Resource Conservation and Recovery Act (RCRA) of 1976, which introduced regulations for hazardous waste management and established a framework for recycling and waste reduction. Similar initiatives were implemented in Europe, focusing on reducing waste generation, promoting recycling, and enforcing stricter waste disposal regulations.

Shift toward recycling and resource recovery (1980s-1990s): The 1980s and 1990s witnessed a global shift toward recycling and resource recovery, driven by concerns over landfill capacity and resource scarcity. Countries like Germany and Japan implemented comprehensive recycling programs, requiring citizens to separate waste into recyclable, compostable, and landfill-bound categories. Municipal recycling facilities were established to process paper, plastic, glass, and metals, reducing the volume of waste sent to landfills. In the United States, the Bottle Bill legislation incentivized recycling by offering monetary refunds for returning glass and plastic bottles. Recycling became an integral part of municipal waste management, supported by public awareness campaigns and educational programs promoting the importance of waste reduction and material recovery.

TABLE 2.4 Historical development of waste management.

Era/Time period	Waste generation pattern	Waste disposal methods	Policy/Legislation development	Technological advancements	Public awareness & societal attitude
Ancient civilizations (3000 BCE–500 AD)	Minimal waste, mostly organic	Dumping in open pits or rivers	None or local customary practices	Manual collection, basic tools	Low—Waste was part of natural cycle
Medieval period (500–1500 AD)	Rise in urban waste, poor sanitation	Open dumping, animal scavenging	Local ordinances in some cities	None – Mostly manual labor	Very low—Waste seen as unavoidable
Industrial revolution (1700s–1800s)	Rapid increase due to mass production	Dumping, incineration begins	Initial public health acts in Europe	Introduction of incinerators, sewers	Growing—Linked to health & hygiene
Early 20th century (1900–1950s)	Increase in industrial & urban waste	Landfilling, burning	First national waste laws (e.g., UK, USA)	Mechanized collection, compactors	Moderate—Awareness through reformers
Late 20th century (1960s–1990s)	Rise in plastic & hazardous waste	Recycling, regulated landfilling	Environmental protection acts (EPA, EU directives)	Recycling facilities, WtE plants	High—Rise of environmental movements
21st century–Present	Explosion of e-waste, packaging waste	Integrated waste management (IWM)	Global agreements (Basel, Paris Accord, SDGs)	Smart bins, AI sorting, bio-tech	Very high—Zero-waste & sustainability

Sustainable waste management and circular economy (2000s--Present): In the 21st century, waste management evolved to encompass sustainability and circular economy principles, emphasizing the reduction, reuse, and recycling of materials to minimize environmental impact. Advanced waste treatment technologies, such as anaerobic digestion, pyrolysis, and gasification, were developed to convert waste into valuable byproducts like biogas, synthetic fuels, and fertilizers. Countries adopted Zero Waste strategies, aiming to reduce landfill-bound waste to near-zero levels. The rise of the circular economy further transformed waste management by encouraging product design for durability, recyclability, and modularity. Initiatives like the European Green Deal and the United Nations Sustainable Development Goals (SDGs) promote sustainable waste management as a key component of climate action and resource conservation.

Future trends and innovations in waste management: The future of waste management is increasingly driven by technological advancements and data-driven approaches. Innovations in smart waste management systems employ sensors, AI, and blockchain to optimize waste collection, track recycling rates, and monitor landfill emissions. Waste-to-energy plants are becoming more efficient, incorporating carbon capture technologies to mitigate greenhouse gas emissions. Additionally, the rise of Extended Producer Responsibility (EPR) frameworks holds manufacturers accountable for the entire lifecycle of their products, from production to disposal. Efforts to promote a circular economy continue to gain momentum, with initiatives aimed at reducing resource extraction, extending product lifecycles, and closing the loop on waste. The integration of green technologies, waste valorization, and digital monitoring systems is expected to reshape the waste management landscape in the coming decades.

Global waste management: a growing challenge and collective responsibility: Global waste management is one of the most pressing environmental and social challenges of the 21st century. As population growth, urbanization, and industrialization accelerate, the volume of waste generated globally continues to rise at an alarming rate. According to the World Bank, the world generates over 2.01 billion tonnes of municipal solid waste each year, with at least 33% of that not managed in an environmentally safe manner. This figure is expected to increase significantly in the coming decades unless there is a unified global response. Waste management is no longer a local issue; it has evolved into a complex, cross-border concern requiring collaborative international efforts to ensure the sustainable health of our planet.

At its core, global waste management encompasses the collection, transportation, treatment, recycling, and final disposal of waste materials in ways that minimize adverse impacts on health, the environment, and the economy. It includes various waste types such as household trash, industrial waste, hazardous materials, electronic waste (e-waste), construction debris, and more. A critical component of global waste management is the **hierarchy of waste**

treatment, often summarized by the "3Rs" Reduce, Reuse, and Recycle. This model encourages minimizing waste generation at the source, promoting the reuse of products, and recycling materials before considering disposal options like landfilling or incineration.

The disparities in waste management practices between developed and developing countries highlight the complexity of global waste issues. High-income countries, while generating more waste per capita, often have access to advanced technologies, effective regulatory frameworks, and infrastructure to manage waste sustainably. In contrast, many low- and middle-income countries struggle with inadequate waste collection systems, limited recycling facilities, and open dumping practices, which pose significant health and environmental risks. For example, in parts of Asia and Africa, open burning of waste is a common practice, releasing toxic fumes into the atmosphere and contributing to climate change and public health crises.

A particularly troubling aspect of global waste management is the international trade of waste. Developed countries often export waste, particularly hazardous and electronic waste, to developing nations with less stringent environmental regulations. While this may appear cost-effective for exporters, it often leads to severe consequences for recipient countries, including soil and water contamination, toxic exposure, and the exploitation of informal labor in dangerous recycling operations. International agreements like the Basel Convention aim to regulate such waste movements and ensure that hazardous waste is managed in an environmentally sound manner, but enforcement and compliance remain inconsistent.

Recycling plays a vital role in addressing the global waste crisis. However, recycling rates vary significantly across regions. In Europe, for instance, some countries recycle over 50% of municipal waste, while in other parts of the world, particularly in underdeveloped regions, recycling is minimal or nonexistent. A major barrier is the lack of infrastructure and public awareness. There is also a growing realization that recycling alone cannot solve the problem. A shift toward a **circular economy** where products are designed for durability, reuse, and recyclability from the outset is crucial. Such a transformation requires changes at every level, from product design and manufacturing to consumer behavior and governmental policies.

The role of technology and innovation in global waste management cannot be overstated. Smart waste bins, data-driven logistics for waste collection, AI-based sorting systems, and bio-engineered solutions for decomposing waste are revolutionizing how waste is handled. Moreover, waste-to-energy technologies are being deployed to convert non-recyclable waste into electricity or fuel, although concerns about emissions and sustainability remain.

Global waste management is also deeply linked to climate change. Landfills are among the largest sources of methane, a potent greenhouse gas. Mismanaged waste, especially plastics, contributes to ocean pollution, threatening marine ecosystems and human livelihoods. Reducing waste and managing it sustainably is, therefore, a key strategy in combating climate change and preserving biodiversity.

To address these issues, coordinated global action is essential. Governments, industries, and individuals must all play their part. Policymakers must enact and enforce robust environmental regulations. Businesses need to adopt sustainable production and packaging practices. Consumers must be educated and motivated to reduce their waste footprint through responsible consumption and participation in recycling programs. International cooperation and knowledge sharing are crucial to ensure that best practices in waste management are adopted universally.

In conclusion, global waste management is more than just a logistical necessity it is a reflection of how humanity values its environment, health, and future. By adopting sustainable practices, embracing innovation, and fostering international collaboration, the global community can transform waste from a mounting problem into an opportunity for environmental restoration, economic development, and social well-being. Table 2.5 shows top 10 countries in waste management along with their key initiatives.

2.4.2 Waste management hierarchy

The waste management hierarchy is a framework that ranks waste management strategies according to their environmental impact from most to least preferred. It is designed to encourage the most sustainable practices in managing waste. The hierarchy helps guide governments, organizations, and individuals toward practices that reduce the environmental impact of waste. By prioritizing prevention and reuse over disposal, it supports the principles of the circular economy, where materials are kept in use for as long as possible. The hierarchy is usually represented as an inverted pyramid with the following levels (from most to least preferred):

2.4.2.1 Prevention and minimization

Prevention and minimization form the foundation of a sustainable waste management strategy. This stage involves reducing waste at its source, effectively preventing waste from being generated in the first place. This can be achieved through the adoption of lean manufacturing techniques, efficient product design, and sustainable consumption practices. For instance, manufacturers can opt for materials that have a longer lifespan or are more easily recyclable. Businesses can conduct waste audits to identify potential areas for waste reduction and implement strategies such as bulk purchasing, digital communication to reduce paper use, and product redesign for minimal packaging. The goal is to conserve resources, reduce environmental impact, and minimize disposal costs.

TABLE 2.5 Top 10 countries in waste management along with their key initiatives.

Rank	Country	Recycling rate (%)	Waste-to-energy usage	Plastic waste management	Policy & legislation	Innovation & technology	Public awareness & participation
1	**Germany**	66%	High	Very effective	Strong circular economy laws	Advanced AI-based sorting	Very high
2	**Sweden**	50%	Very high (almost 99% non-landfill)	Highly regulated	Waste import for energy generation	Waste-to-energy tech leader	High
3	**South Korea**	59%	Moderate	Strict regulations	Volume-based waste fee system	RFID tracking systems	Very high
4	**Japan**	56%	Moderate	Highly organized	"Sound material-cycle society" framework	Smart waste bins, robotics	High
5	**Austria**	55%	High	Effective systems	EU-aligned waste laws	Efficient separation tech	High
6	**Switzerland**	53%	High	Well-controlled	"Polluter pays" principle	Waste heat recovery systems	Very high
7	**Netherlands**	52%	High	Actively managed	"Lansink's ladder" hierarchy	Circular economy innovation	High
8	**Norway**	49%	High	Managed via eco-funding	Extended producer responsibility (EPR)	Thermal treatment advances	Medium-high
9	**Singapore**	47%	Moderate	Strict monitoring	Zero waste masterplan	Smart bins, data analytics	High
10	**Canada**	27% (but improving)	Moderate	Varies by province	Federal waste reduction initiatives	Emerging technologies	Growing

A prime example of waste prevention is the initiative taken by **IKEA**. The company focuses on sustainable product design by using renewable and recycled materials, reducing packaging, and optimizing logistics to minimize waste. For instance, IKEA's flat-pack furniture design not only reduces the volume of materials used but also cuts down on transportation emissions. Additionally, IKEA encourages customers to purchase long-lasting products and offers furniture repair services to extend the life of its products.

2.4.2.2 Reuse

Reuse emphasizes extending the life of products and materials before they become waste. This stage focuses on identifying items that can be repurposed, repaired, or refurbished, thus preventing them from entering the waste stream prematurely. Examples include donating old furniture, repairing electronic devices, and repurposing containers and packaging materials. Establishing community reuse centers or second-hand stores can further promote reuse practices, encouraging people to donate and purchase used goods. Reuse not only conserves resources but also fosters economic savings for both consumers and businesses while reducing the overall environmental footprint associated with the production of new products.

The Patagonia Worn Wear Program is a notable example of a company prioritizing reuse. Patagonia encourages customers to trade in their old gear in exchange for store credit. The returned items are repaired, refurbished, and resold as part of the Worn Wear collection. This program keeps functional products in circulation for longer, reducing the demand for new resources and diverting waste from landfills. Similarly, many cities have established community reuse centers where residents can donate and purchase used furniture, appliances, and building materials.

2.4.2.3 Recycling and composting

Recycling involves the collection, processing, and transformation of waste materials into new products, thus closing the resource loop. This stage requires proper segregation of recyclables such as paper, plastic, metals, and glass to facilitate effective recycling. Additionally, composting organic waste, such as food scraps and yard debris, diverts waste from landfills while producing nutrient-rich compost that can be used to enrich soil. Recycling and composting reduce the demand for raw materials, conserve energy, and lower greenhouse gas emissions. Implementing effective recycling programs and educational campaigns can significantly enhance the success of recycling efforts in communities and organizations.

San Francisco has one of the most successful recycling and composting programs in the world. The city has implemented strict waste separation policies, mandating that residents and businesses separate compostable, recyclable, and landfill waste. Organic waste, including food scraps and yard debris, is collected and processed into compost, which is then sold to local farms and vineyards. The city has achieved a waste diversion rate of over 80%, making it a model for urban waste management. Another example is Terra-Cycle, which partners with brands to recycle hard-to-recycle materials such as cigarette butts, beauty product packaging, and electronics.

2.4.2.4 Energy recovery

Energy recovery comes into play when waste cannot be practically reused or recycled. This stage involves converting waste into usable energy through methods such as WtE incineration, anaerobic digestion, and pyrolysis. These processes extract energy in the form of heat, electricity, or fuel from waste materials. For example, incineration with energy recovery can reduce waste volume while generating electricity, whereas anaerobic digestion can convert organic waste into biogas and nutrient-rich digestate. However, energy recovery should be conducted with strict environmental controls to mitigate potential emissions and ensure that it does not undermine waste prevention and recycling efforts.

In Copenhagen, Denmark, the Amager Bakke Waste-to-Energy Plant (also known as CopenHill) is an advanced waste-to-energy facility that converts non-recyclable waste into electricity and district heating for thousands of households. The plant features state-of-the-art emissions control systems, making it one of the cleanest waste-to-energy plants globally. It also doubles as a recreational facility with a ski slope and climbing wall, showcasing how infrastructure can be both functional and community-centric. This facility demonstrates how energy recovery can be integrated into urban planning to create multipurpose public spaces. Fig. 2.10 illustrate waste management hierarchy.

FIGURE 2.10 Waste management hierarchy.

2.4.2.5 Treatment

Treatment is a critical step that involves processing waste to reduce its volume, toxicity, or hazardous nature before disposal. Various treatment methods include chemical, physical, and biological processes, such as neutralization, stabilization, and disinfection. Hazardous waste, for instance, may be treated through incineration, chemical neutralization, or encapsulation to minimize its environmental impact. Advanced treatment technologies like pyrolysis and gasification further break down waste into useful by-products, such as syngas and biochar. Effective treatment reduces the risks associated with waste disposal and mitigates potential harm to human health and the environment.

Singapore employs an innovative treatment method through its NEWater Initiative. Wastewater is treated using advanced membrane filtration, reverse osmosis, and ultraviolet disinfection to produce high-quality reclaimed water that meets stringent drinking water standards. The treated water is then reused for industrial processes, reducing the demand for freshwater and ensuring a sustainable water supply. This initiative exemplifies how treatment technologies can mitigate resource scarcity and minimize environmental impact.

2.4.2.6 Disposal

Disposal is the final and least preferred stage in the waste management hierarchy. It involves safely discarding residual waste that cannot be prevented, reused, recycled, or treated. Common disposal methods include landfilling and secure incineration. Modern landfills are engineered with protective liners and leachate collection systems to minimize environmental contamination. Additionally, landfill gas recovery systems can capture methane emissions for energy generation. However, due to the environmental and health risks associated with landfills, minimizing the volume of waste directed to disposal is crucial. Disposal should always be viewed as a last resort, with a strong emphasis on waste reduction, recycling, and treatment to minimize its necessity.

Japan has implemented a highly controlled and regulated approach to waste disposal through its Eco-Town Program. In Kawasaki City, industrial waste is processed through state-of-the-art incineration facilities equipped with pollution control systems. The heat generated is captured for energy production, and the ash is recycled into construction materials. Additionally, the country enforces strict waste sorting regulations to minimize the volume of waste directed to landfills. This systematic approach has helped Japan achieve a landfill diversion rate of over 80%. Table 2.6 illustrate waste management hierarchy attributes comparison.

Let's take an example of plastic water bottle:

- **Prevention:** Carry a reusable water bottle to avoid buying plastic bottles.
- **Reuse:** Refill the plastic bottle and use it multiple times.

TABLE 2.6 Waste management hierarchy attributes comparison.

Attribute	Prevention & minimization	Reuse	Recycling & composting	Energy recovery	Treatment	Disposal
Definition	Avoiding or reducing waste generation at the source	Using items again for same or new purpose	Converting waste into reusable materials or compost	Converting waste into usable energy (electricity, heat, fuel)	Physico-chemical/ biological processes to neutralize waste	Final placement of waste into landfills or incinerators
Environmental impact	Very low (most sustainable)	Low	Moderate (depends on process efficiency)	Moderate to high (some emissions)	Depends on method (moderate to high)	High (pollution, land use, GHG emissions)
Cost	Low (in long term)	Very low	Moderate (collection + processing)	High setup, medium operation	High treatment cost	Medium setup, high long-term maintenance
Efficiency	Highly efficient when implemented systemically	Efficient for select materials	High for materials like paper, metals, organics	Energy conversion efficiency ~20–40%	Varies by technology (incineration, autoclaving, etc.)	Inefficient use of materials and space
Common methods	Eco-design, efficient processes, behavior change	Refilling, repairing, donation	Composting, material recovery, MRFs	Incineration with energy recovery, anaerobic digestion	Incineration, chemical neutralization, autoclaving	Sanitary landfilling, open dumping (unsafe)
Examples	Digital bills, minimal packaging, product redesign	Reusable bottles, furniture reuse	Recycling plastics, composting food waste	Burning waste to produce electricity	Treating sewage sludge, neutralizing acids	Dumping in landfill, burning without energy recovery
Regulatory support	Strong in zero-waste/circular economy policies	Encouraged via incentives	Widely supported by law (solid waste Mgmt. Rules, 2016 India)	Supported under waste-to-energy initiatives	Required for hazardous & biomedical waste	Only as last resort (per basel convention)
Public participation	Highest (consumer decisions matter most)	High (individual actions crucial)	Moderate (segregation at source)	Low direct public involvement	Minimal involvement	Lowest (handled by authorities)

- **Recycling:** Send the plastic bottle to a recycling facility to make new plastic products.
- **Recovery:** If not recyclable, it may be burned in a waste-to-energy plant.
- **Disposal:** If all else fails, it ends up in a landfill or is incinerated without energy recovery.

2.5 Data analysis of urban waste management

Urban waste management is a critical component of sustainable urban development, particularly in fast-growing cities. Efficient handling, segregation, recycling, and disposal of waste play a key role in reducing environmental pollution, conserving resources, and enhancing public health. Data analysis in urban waste management involves evaluating the types, quantities, and processing methods of waste generated in different zones. It helps in identifying trends, inefficiencies, and opportunities for improving waste segregation, recycling rates, and landfill reduction. The provided dataset captures 30 entries from various urban zones (Zone A to Zone J) and includes key variables such as Waste Type, Daily Waste Generation (kg), Segregation Rate (%), Recyclable (%), Compostable (%), and Landfill (%). Table 2.7 shows the sample of urban waste management dataset. Analysis of this dataset allows stakeholders to:

- **Understand spatial waste patterns**: For instance, organic waste is significantly generated in Zones A, F, D, and J, with high compostable content (e.g., 79% in Zone D).
- **Assess segregation efficiency**: Paper and metal waste types show high segregation rates (up to 95%), indicating effective source separation.
- **Compare recyclable potential**: Metal and paper have the highest recyclable percentages (up to 91%), while organic and plastic waste vary more widely.
- **Monitor landfill dependency**: Plastic and e-waste show higher landfill percentages (up to 40%), signaling the need for improved recycling or treatment facilities.
- **Prioritize areas for intervention**: Zones with lower segregation (like Zone I for e-waste, 40%) and higher landfill rates can be targeted for awareness, infrastructure, and policy interventions.

In summary, such a dataset provides valuable insights into zone-wise performance, waste stream characteristics, and overall efficiency of the urban waste management system, aiding evidence-based decision-making and policy design for sustainable urban living.

The Urban Waste Management dataset presents a detailed analysis of 30 entries across 10 city zones, showcasing various waste types and associated metrics such as daily waste generation, segregation rates, and recyclability. The bar chart titled "Average Daily Waste Generation by Waste Type" reveals that organic waste dominates the daily generation rate, averaging over 100 kg/

TABLE 2.7 Urban waste management data sample.

ID	Location	Waste type	Daily waste (kg)	Segregation rate (%)	Recyclable (%)	Compostable (%)	Landfill (%)
1	Zone A	Organic	120	85	10	75	15
2	Zone B	Plastic	80	65	60	5	35
3	Zone C	Metal	45	90	88	2	10
4	Zone D	E-waste	30	50	70	0	30
5	Zone E	Paper	60	95	90	5	5
6	Zone F	Organic	110	80	12	68	20
7	Zone G	Plastic	70	55	58	3	39
8	Zone H	Glass	50	88	85	0	15
9	Zone I	E-waste	25	40	65	0	35
10	Zone J	Organic	100	78	10	70	20
11	Zone A	Plastic	90	60	55	5	40
12	Zone B	Paper	55	92	88	4	8
13	Zone C	Glass	40	85	80	0	20
14	Zone D	Organic	130	88	9	79	12
15	Zone E	Plastic	75	63	59	4	37
16	Zone F	E-waste	35	55	72	0	28

17	Zone G	Paper	70	93	89	3	8
18	Zone H	Organic	95	77	11	67	22
19	Zone I	Plastic	85	62	60	2	38
20	Zone J	Glass	60	86	83	0	17
21	Zone A	Metal	48	91	87	1	12
22	Zone B	Organic	115	82	12	70	18
23	Zone C	Plastic	78	67	62	3	35
24	Zone D	E-waste	32	52	68	0	32
25	Zone E	Paper	62	94	91	2	7
26	Zone F	Glass	45	84	81	0	19
27	Zone G	Organic	125	86	8	74	18
28	Zone H	Plastic	82	61	57	4	39
29	Zone I	E-waste	28	43	67	0	33
30	Zone J	Paper	58	90	87	3	10

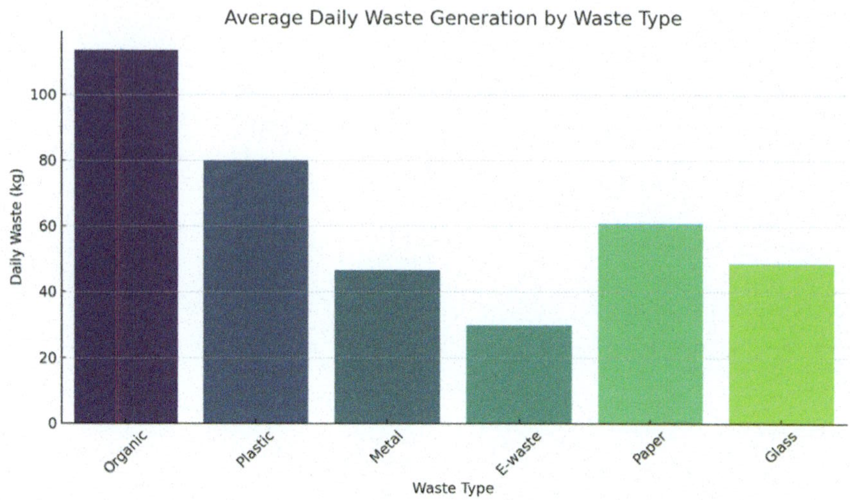

FIGURE 2.11 Average daily waste generation by waste type.

day, followed by plastic waste at approximately 85 kg/day. In contrast, e-waste and metal waste contribute the least, with daily outputs of around 30 and 45 kg respectively. This highlights the predominance of biodegradable and plastic waste in urban settings, necessitating targeted waste reduction and composting strategies. Fig. 2.11 illustrate average daily waste generation by waste type.

Fig. 2.12 illustrate average segregation rate by location. The segregation performance across different urban zones is illustrated in the chart titled "Average Segregation Rate by Location." Zones A, B, C, E, and J exhibit high

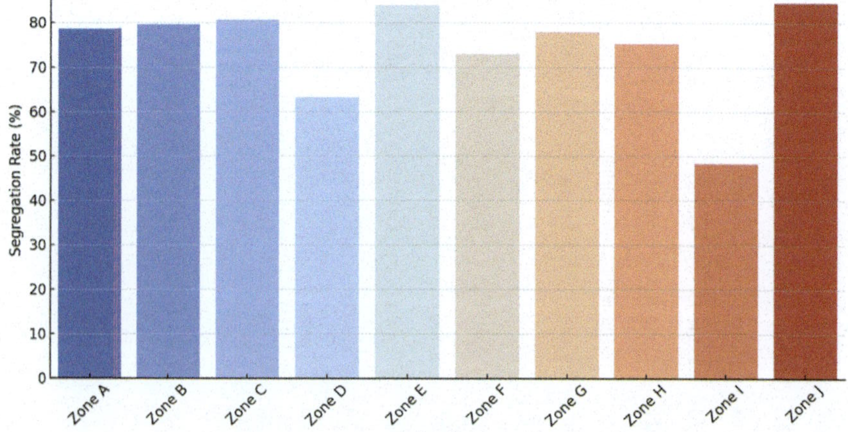

FIGURE 2.12 Average segregation rate by location.

segregation rates above 80%, suggesting robust awareness and infrastructure for source-level sorting. In contrast, Zone I lags significantly with a segregation rate below 50%, indicating potential gaps in civic engagement or municipal services. The presence of high daily waste quantities in Zone A, particularly organic and plastic waste, further underlines the importance of segregation to streamline recycling and composting processes. Fig. 2.13 illustrate Correlation Matrix for Waste Management Metrics. The correlation heatmap titled "Correlation Matrix for Waste Management Metrics" reveals significant relationships between various waste metrics. Notably, segregation rate shows a strong negative correlation with landfill percentage (−0.89), meaning better segregation practices effectively reduce landfill dependency. Furthermore, recyclable content correlates positively with segregation rate (0.66) and negatively with landfill usage, reinforcing the importance of upstream sorting in enhancing recyclability. Interestingly, daily waste correlates strongly with compostable content (0.84) but negatively with recyclable percentage (−0.82), implying that bulkier waste types like organic matter are more compostable than recyclable.

Fig. 2.14 illustrate distribution of waste types in dataset. The pie chart, which shows the distribution of waste types in the dataset, confirms that organic and plastic waste each account for 23.3% of the total waste entries. These are followed by e-waste and paper, each representing 16.7%, with glass at 13.3%, and metal being the least frequent (6.7%). This reflects the consumption patterns of urban households and the pressing need for circular waste solutions for plastics and organics. The high representation of plastic waste

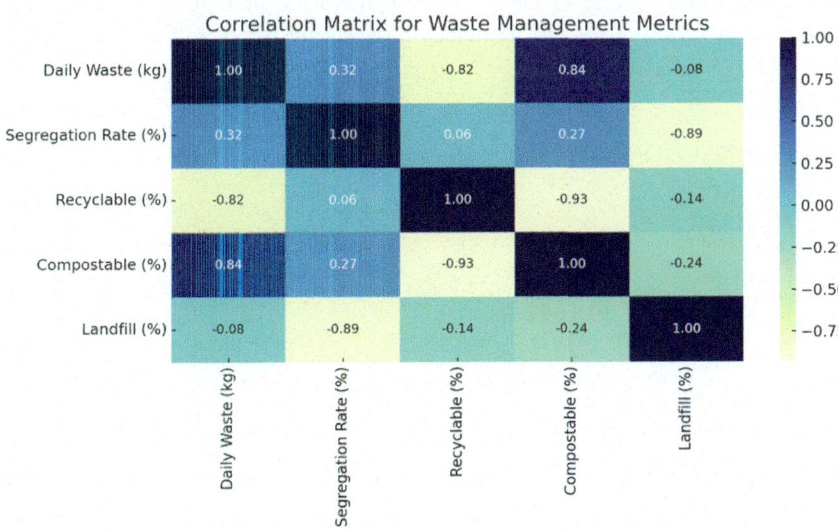

FIGURE 2.13 Correlation matrix for waste management metrics.

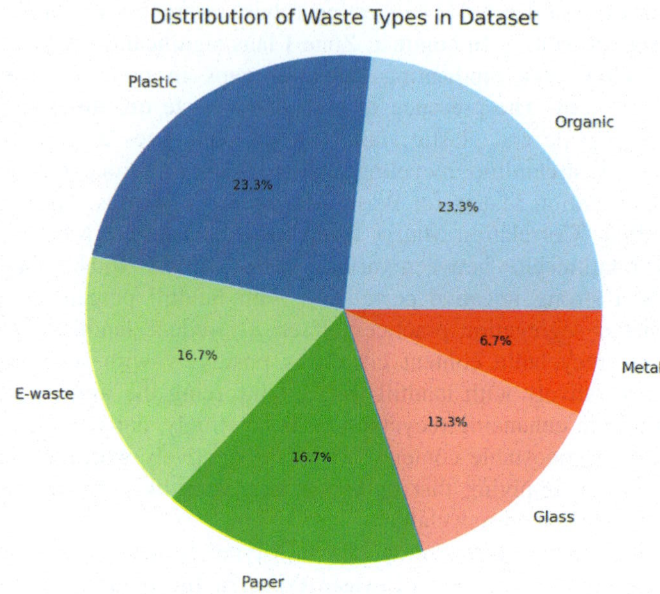

FIGURE 2.14 Distribution of waste types in dataset.

also aligns with global concerns regarding plastic pollution and the need for effective Extended Producer Responsibility (EPR) systems.

Fig. 2.15: illustrate trend of daily waste generation in zone A. Lastly, the line graph titled "Trend of Daily Waste Generation in Zone A" tracks waste data over several entries and indicates a declining trend, falling from 120 to

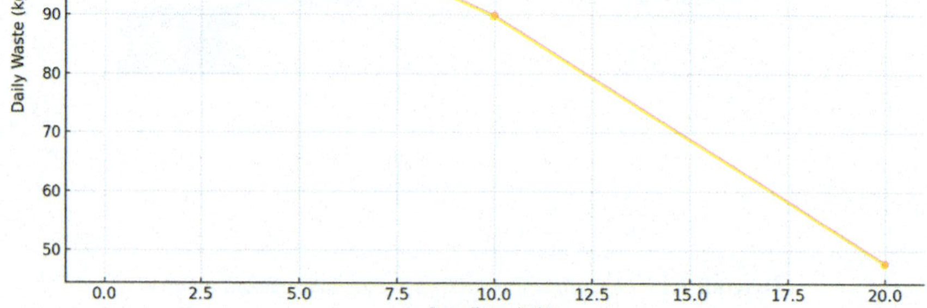

FIGURE 2.15 Trend of daily waste generation in Zone A

about 50 kg. This downward trajectory could be attributed to successful interventions such as increased public awareness, better waste collection systems, or seasonal fluctuations. As Zone A contributes substantially to the total waste volume, this trend, if sustained, could significantly ease the burden on waste management infrastructure.

The combined table and graphical analysis offer a holistic view of urban waste dynamics. The findings emphasize the critical importance of enhancing segregation efforts, particularly in lower-performing zones like Zone I, and scaling up composting and recycling initiatives for organic and plastic waste, respectively. Strong correlations between segregation and waste recovery metrics suggest that investments in community engagement and infrastructure at the source level can dramatically improve overall waste management efficiency. Additionally, trends in specific zones provide valuable insights for policymakers to identify successful practices and replicate them in lagging areas. Collectively, this integrated approach of data visualization and analysis fosters a data-driven roadmap for sustainable urban waste governance.

2.6 Circular economy: A paradigm shift in waste management

The circular economy represents a transformative approach to economic development, aiming to decouple economic growth from resource consumption and environmental degradation. Unlike the traditional linear economy, which follows a take-make-dispose model, the circular economy emphasizes the continuous circulation of resources through closed-loop systems.

Circular

C—Conserve—Preserve natural resources and energy.
I—Innovate—Design sustainable products and processes.
R—Reduce—Minimize waste and resource use.
C—Collaborate—Work together across sectors and communities.
U—Utilize—Maximize the value and use of resources.
L—Loop—Keep materials in continuous use.
A—Adapt—Evolve business models for sustainability.
R—Recover—Reclaim energy and materials from waste.

Economy

E—Engage—Involve stakeholders at every level.
C—Create—Generate value from circular innovation.
O—Optimize—Improve efficiency and productivity.
N—Nurture—Support ecosystems and communities.
O—Offset—Compensate environmental impacts responsibly.
M—Maintain—Extend the life of products and infrastructure.
Y—Yield—Achieve economic and environmental returns.

This model seeks to maximize the value of products, materials, and resources while minimizing waste generation, pollution, and environmental impact. By designing products for durability, reuse, and recyclability, the circular economy fosters sustainable consumption patterns and resource efficiency. It also encourages the regeneration of natural systems, reducing the reliance on finite resources and mitigating the adverse effects of resource extraction and waste disposal. Following are the five innovative principles of the Circular Economy, going beyond the traditional "reduce, reuse, recycle" to integrate modern technology, systems thinking, and sustainability:

2.6.1 Regenerative by design

Principle: Design systems and products that restore, renew, or regenerate natural ecosystems and social well-being.

Innovation Angle: Use biomimicry, regenerative agriculture, and cradle-to-cradle product design to ensure that every output becomes an input for another process.

2.6.2 Intelligent material stewardship

Principle: Track, trace, and manage materials digitally throughout their lifecycle.

Innovation Angle: Apply blockchain, IoT, and AI for real-time monitoring of materials, enabling efficient reuse, recycling, and zero-waste manufacturing.

2.6.3 Circular as a service (CaaS)

Principle: Shift from product ownership to product-as-a-service models.

Innovation Angle: Businesses retain ownership and responsibility for products, using subscription, leasing, or performance-based models, ensuring long-term durability, maintenance, and recycling.

2.6.4 Symbiotic industrial networks

Principle: Develop interconnected industries where waste or by-products of one become resources for another.

Innovation Angle: Create eco-industrial parks where companies exchange energy, water, and materials in closed-loop systems.

2.6.5 Circular intelligence and learning

Principle: Embed continuous learning and circular thinking in decision-making processes.

Innovation Angle: Utilize machine learning and big data analytics to forecast demand, reduce surplus, optimize logistics, and enhance circular supply chain performance. Fig. 2.16 illustrate principles of circular economy.

Based on the above principles, circular economy categorized into following three principle core components: "*Design Out Waste and Pollution*," "*Keep Products and Materials in Use*," and "*Regenerate Natural Systems*."

2.6.5.1 Design out waste and pollution

This principle emphasizes designing products and processes that prevent waste generation and minimize pollution from the outset. By incorporating sustainable design principles, manufacturers can create products that use fewer raw materials, generate less waste during production, and have longer lifecycles. For example, adopting modular designs allows for easy repair and disassembly, enabling components to be replaced or upgraded rather than discarded. Additionally, the use of biodegradable or recyclable materials reduces the environmental impact of products at the end of their life. Companies like Patagonia and IKEA have embraced this principle by incorporating recycled and renewable materials into their product lines, thus minimizing their ecological footprint. Fair phone, a company known for its modular smartphones, exemplifies this principle. Unlike traditional smartphones, which are designed with limited reparability and use of non-recyclable materials, Fair phone creates products that are modular and easy to repair. Users can replace components such as the battery, screen, or camera, thus extending the phone's lifespan. The company also sources conflict-free materials and promotes the use of recycled materials in the phone's production, significantly reducing its environmental footprint. This design approach ensures that fewer phones end

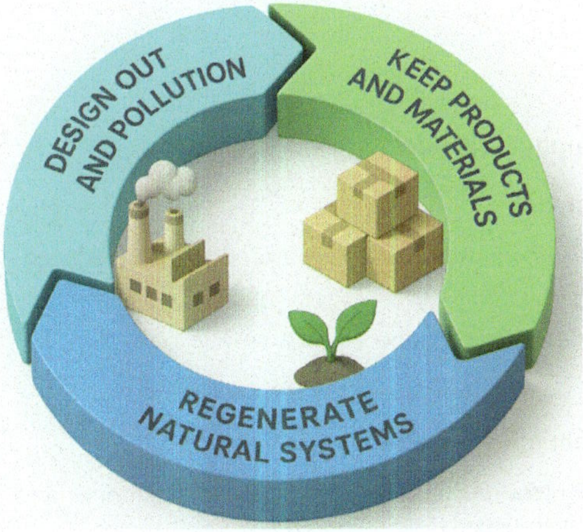

FIGURE 2.16 Principles of circular economy.

up in landfills and minimizes the pollution typically associated with the manufacturing and disposal of electronics.

- **Example: Refillable packaging system for personal care products**

Concept: A personal care company redesigns its product delivery model to eliminate single-use plastic packaging by introducing a refillable, closed-loop system for shampoos, conditioners, and body washes.

How it designs out waste and pollution:

- **Product redesign:** Products are sold in durable, reusable aluminium or glass bottles instead of single-use plastic. Labels are printed with soy-based ink, and the bottles are made with modular parts for easy repair or replacement.
- **Refill stations & reverse logistics:** Refill stations are installed at retail stores, universities, gyms, and malls. Alternatively, customers can opt for a home-refill subscription where used bottles are collected, sanitized, and refilled.
- **Supply chain innovation:** The supply chain uses clean energy and bulk distribution methods, drastically reducing emissions and transportation waste. Sensors track inventory and usage, reducing overproduction and ensuring on-demand manufacturing.
- **Customer engagement & behavior design:** Incentives (discounts, loyalty points) are provided for every returned bottle. The design incorporates aesthetics and functionality to encourage users to keep and reuse the bottles (Table 2.8).

2.6.5.2 Keep products and materials in use

The second principle focuses on extending the lifecycle of products and materials through strategies such as reuse, refurbishment, remanufacturing, and

TABLE 2.8 Circular economy impact with design out waste.

Principle	Impact
Design out waste	No single-use packaging waste, drastically reduced landfill load
Keep products in use	Bottles reused hundreds of times
Regenerate natural systems	Reduced plastic production and ocean pollution
Circular business model	Shift from product to product-as-a-service
Innovation & collaboration	Involves retailers, consumers, logistics partners, and design experts

recycling. By keeping products and components in circulation for as long as possible, the circular economy reduces the demand for virgin materials and prevents waste from ending up in landfills. Business models such as product-as-a-service (PaaS) and sharing platforms promote resource optimization by enabling consumers to access products without owning them outright. For instance, Zipcar and Rent the Runway provide temporary access to vehicles and fashion items, respectively, reducing the need for new products and encouraging a shift toward a sharing economy. Recycling infrastructure also plays a crucial role in capturing valuable materials and reintroducing them into production cycles, further closing the loop.

Patagonia, the outdoor clothing company, is a leader in this area. They have implemented the Worn Wear program, which encourages customers to buy used Patagonia products and repair their old gear rather than buying new. Through this initiative, Patagonia refurbishes old products and resells them, providing customers with an affordable, sustainable alternative to buying new items. Additionally, the company offers repair services to extend the life of their products. This model not only reduces the demand for new resources but also lowers the carbon footprint of the brand's products. Another great example is Interface, a global leader in modular carpet tiles. Interface has adopted a "Take Back" program where they collect used tiles from customers and recycle them into new products. The company also uses recycled nylon from old fishing nets to make the carpet tiles, diverting waste from oceans and landfills. By doing this, they are able to keep materials in circulation and significantly reduce the environmental impact of their products.

Example: Circular footwear with modular sneaker design

Concept: A footwear company creates modular, repairable sneakers designed to be easily disassembled and remanufactured or upgraded, keeping materials and components in use for as long as possible.

How it keeps products and materials in use:

- **Modular product design:** The sneakers are made of interchangeable components (sole, upper, laces, insoles) that snap together without glue. Customers can replace only worn-out parts instead of discarding the entire shoe.
- **Material selection:** Uses recycled rubber for soles and bio-based or recycled fabrics for uppers. Materials are selected for durability and recyclability, with easy separation at end-of-life.
- **Take-back and remanufacturing program:** The company offers a take-back program where customers return used parts. Returned parts are cleaned, repaired, or recycled into new shoes or other products.
- **Digital tracking with QR code:** Each shoe has a QR code linking to a digital passport showing material composition, repair options, and circular journey history. Helps track use, encourage maintenance, and ensure proper end-of-life recovery (Table 2.9).

TABLE 2.9 Circular economy impact with keep product and material use.

Principle	Impact
Keep products in use	Sneakers can last 2–3× longer via repair and part replacement
Reduce resource extraction	Less raw material usage through reuse and remanufacturing
Enable circular flows	Materials looped back into production rather than entering landfill
Empower consumers	Transparency and ease of repair encourage responsible ownership

2.6.5.3 Regenerate natural systems

The third principle emphasizes the importance of restoring and regenerating natural ecosystems. Instead of extracting finite resources and causing environmental degradation, the circular economy advocates for practices that enhance natural capital and maintain the balance of ecosystems. This includes using renewable energy, replenishing soil health through composting organic waste, and implementing regenerative agriculture practices. Additionally, companies are increasingly adopting biomimicry and nature-inspired design to create sustainable products and processes that align with natural systems. For example, the Ellen MacArthur Foundation has championed initiatives that integrate regenerative design into urban planning, agriculture, and manufacturing, aiming to foster biodiversity and combat climate change.

Implementing a circular economy requires a systemic shift involving multiple stakeholders, including governments, businesses, and consumers. It also necessitates changes in policy frameworks, financial models, and consumer behavior. Despite the challenges, the circular economy presents a compelling vision for a sustainable future, where economic prosperity is achieved without compromising environmental integrity. By closing the loop on resource use and fostering circular practices, it offers a pathway to mitigate climate change, conserve natural resources, and transition to a low-carbon economy.

Loop Industries, a company focusing on plastics recycling, offers a fantastic example of this principle in action. They use a proprietary technology to chemically recycle PET (Polyethylene Terephthalate) plastic, breaking it down to its base building blocks so it can be reused indefinitely. This process helps to reduce the need for virgin plastic, thereby conserving resources and minimizing pollution. By integrating sustainable practices into their recycling process, they not only reduce plastic waste but also contribute to preserving ecosystems and preventing further damage caused by plastic pollution.

Another excellent example is Kellogg's, which has committed to sustainable agricultural practices through their Sustainable Agriculture Program. They work directly with farmers to improve soil health, reduce water usage, and adopt agroforestry methods. This helps regenerate the soil by encouraging practices that restore nutrients and biodiversity, creating a regenerative cycle for the ecosystem. By adopting regenerative practices, they ensure long-term food security while supporting the natural environment.

Example: Bio-based packaging from agricultural waste

Concept: A packaging company replaces petroleum-based plastics with biodegradable packaging made from agricultural waste such as wheat husks, sugarcane bagasse, or mushroom mycelium. The packaging not only serves its function but enriches the soil after use.

How it regenerates natural systems:

- **Use of agricultural waste:** The company sources waste from farms (like rice husks or corn stalks), creating value from residue that would otherwise be burned or discarded, thus reducing air and soil pollution.
- **Compostable end-of-life:** After use, the packaging biodegrades naturally and becomes compost within 60–90 days under the right conditions. When composted, it returns nutrients to the soil, improving fertility and supporting future plant growth.
- **Decentralized composting:** Partnerships with local municipalities and communities promote urban composting programs, where consumers drop off used packaging for community gardens or greenhouses.
- **Zero chemical inputs:** The production process is chemical-free, and the packaging is safe to degrade in home compost bins, helping avoid micro plastic pollution in soil and water bodies (Table 2.10).

TABLE 2.10 Circular economy impact with regenerate natural system.

Principle	Impact
Regenerate natural systems	Returns nutrients to soil, enhances biodiversity, and prevents pollution
Design out waste	Eliminates non-recyclable plastics and reduces landfill burden
Use renewable inputs	Leverages natural, renewable agricultural by-products
Support local economies	Farmers gain new revenue from crop waste; local jobs in composting

- **Key stakeholders in a circular economy**

A circular economy involves a systemic shift in how products are designed, manufactured, consumed, and disposed of. This transformation requires the active participation of various stakeholders who play distinct but interconnected roles in creating and maintaining a circular economy. Fig. 2.17 illustrate key stakeholders in the circular economy. The key stakeholders include:

Manufacturers and producers: Manufacturers are central to the circular economy, as they determine product design, material selection, and production processes. They can integrate circular principles by using recycled materials, designing for modularity, and implementing take-back schemes. For example, Apple has introduced the Daisy Robot, which disassembles old iPhones to recover valuable components for reuse in new products.

FIGURE 2.17 Key stakeholders in the circular economy.

Consumers and end-users: Consumers drive demand for sustainable products and services. Their willingness to adopt new consumption models, such as product-as-a-service or renting instead of buying, influences the success of circular initiatives. For instance, Patagonia's Worn Wear Program encourages consumers to return used clothing for repair or resale, promoting a circular consumption model.

Retailers and distributors: Retailers play a crucial role in promoting circular products and creating platforms for refurbished or recycled goods. They can implement reverse logistics systems to collect used products and channel them back into the production cycle. IKEA has adopted a buy-back program, allowing customers to return unwanted furniture for refurbishment and resale.

Waste management and recycling companies: Waste management companies are responsible for collecting, sorting, and recycling waste materials. They facilitate resource recovery and ensure that valuable materials reenter the production cycle. Terra-Cycle collaborates with businesses to create **recycling** programs for hard-to-recycle items, turning waste into reusable materials.

Government and regulatory bodies: Governments establish regulations, incentives, and standards to encourage circular practices. They can implement policies that mandate extended producer responsibility (EPR), promote ecolabeling, and set targets for waste reduction and recycling. For example, the European Union's Circular Economy Action Plan sets binding targets for recycling and waste reduction across member states.

Investors and financial institutions: Investors can fund circular projects, incentivize sustainable business models, and encourage companies to adopt circular practices. Sustainable finance initiatives, such as green bonds and circular economy funds, support businesses that prioritize resource efficiency and waste reduction.

Technology providers and innovators: Technology providers develop solutions to enhance product traceability, optimize resource use, and facilitate recycling. Digital platforms and blockchain technology can track product lifecycles, monitor material flows, and prevent waste. For instance, Circular uses blockchain to trace recycled materials throughout the supply chain, ensuring transparency and accountability.

Educational institutions and research organizations: Academic institutions and research organizations conduct studies on circular economy models, develop innovative materials, and provide data on the economic and environmental impact of circular systems. Ellen MacArthur Foundation is a prominent research organization that promotes circular economy principles through education and policy advocacy.

NGOs and advocacy groups: Non-governmental organizations raise awareness about the environmental and social benefits of the circular economy. They also advocate for regulatory changes and collaborate with businesses to implement sustainable practices. Waste Aid, for example, partners with local communities to develop waste management solutions and promote recycling.

Community and social enterprises: Social enterprises and community groups often lead grassroots initiatives to reduce waste, repair products, and promote recycling. These stakeholders help build local circular systems and encourage sustainable consumption. Repair Cafés are community-driven initiatives where people bring broken items to be fixed, fostering a culture of repair rather than disposal.

Each of these stakeholders plays a critical role in closing the loop and ensuring that products, components, and materials circulate through the economy as long as possible. Collaboration and shared responsibility among these stakeholders are vital for transitioning from a linear to a circular economy effectively (Table 2.11).

2.7 Linear versus circular economy

The linear economy follows a traditional model of take, make, dispose, where raw materials are extracted, transformed into products, used, and then discarded as waste. This model is inherently resource-intensive, leading to environmental degradation, increased waste generation, and depletion of finite resources. In contrast, the circular economy adopts a closed-loop approach that prioritizes resource efficiency, waste minimization, and the continuous circulation of materials. Instead of discarding products at the end of their lifecycle, the circular economy focuses on reuse, recycling, remanufacturing, and regenerative design. This model not only conserves natural resources but also reduces environmental impact and fosters economic resilience by creating value from waste.

For example, in a linear economy, a plastic bottle is produced from virgin plastic, used once, and then disposed of in a landfill. However, in a circular economy, that same plastic bottle is designed to be easily recycled and transformed into new products, thus maintaining its value and reducing waste. This shift from a linear to a circular model requires systemic changes in product design, manufacturing processes, and consumer behavior, supported by policies that incentivize sustainable practices and extended product lifecycles (Table 2.12).

2.7.1 Sector-specific examples: Linear versus circular economy

2.7.1.1 Manufacturing sector

The manufacturing sector has traditionally operated on a linear economy model, which follows a straightforward path: extract raw materials, manufacture products, use them, and dispose of them. This model is resource-intensive and often results in significant environmental degradation. A notable example of this linear approach is automobile manufacturing. In the

TABLE 2.11 Comparison of key stakeholders in a circular economy.

Stakeholder	Primary role	Key contribution	Examples of initiatives	Challenges faced
Manufacturers & producers	Design and production of goods	Design for reuse, recycling, modularity; use of recycled materials	Apple's Daisy robot for iPhone disassembly	High cost of redesign, supply chain complexity, legacy systems
Consumers & end-users	Use and disposal of products	Responsible consumption, embracing reuse and repair models	Patagonia's worn wear program	Limited awareness, convenience preference, affordability of sustainable options
Retailers & distributors	Product availability and after-sales systems	Promote refurbished goods, implement reverse logistics	IKEA's buy-back and resell program	Managing take-back logistics, inventory of used products, consumer participation
Government & regulators	Policy creation and enforcement	Mandating EPR, waste targets, funding circular innovation	EU circular economy action plan	Political cycles, enforcement mechanisms, balancing economic interests
Waste management & recyclers	Collection, sorting, and recycling of waste	Recover materials, close resource loops	Terra Cycle's recycling programs	Contamination in waste streams, high processing costs, market volatility for recyclables

TABLE 2.12 Comparison between linear and circular economy.

Aspect	Linear economy	Circular economy
Model	Take → make → dispose	Make → use → reuse → recycle → regenerate
Resource use	Extracts raw materials, uses them once, then discards as waste.	Keeps resources in circulation through reuse, recycling, and remanufacturing.
Environmental impact	High environmental impact due to excessive waste and resource depletion.	Reduces environmental impact by minimizing waste and conserving resources.
Waste management	Focuses on disposal; waste is often sent to landfills or incinerated.	Focuses on resource recovery; waste is treated as a resource for new products.
Product lifecycle	Linear; products have a single, finite lifecycle.	Circular; products are designed for multiple lifecycles.
Business model	Profit-driven, dependent on continuous consumption of new resources.	Value-driven, focusing on resource efficiency, product longevity, and sustainable consumption.
Economic benefits	Short-term profits; vulnerable to resource scarcity and price volatility.	Long-term sustainability; creates economic value from waste and reduces dependency on raw materials.
Examples	Single-use plastic bottles, disposable electronics.	Modular smartphones (e.g., fairphone), refurbished furniture (e.g., IKEA).
Policy implications	Minimal regulations on resource use and waste management.	Incentivizes sustainable practices, promotes recycling, and enforces producer responsibility.

traditional process, car manufacturers extract vast quantities of raw materials such as steel, aluminum, rubber, and plastic. These materials are then processed and used to produce vehicles on a mass scale. While this system supports rapid industrial growth and meets market demand, it has serious environmental consequences. Once a vehicle reaches the end of its useful life, it is typically scrapped. Most of its components including valuable metals, plastics, and electronics are discarded. This creates a large amount of waste, much of which ends up in landfills or incinerators. Additionally, the manufacturing of new vehicles from virgin materials consumes immense energy and water resources, while releasing harmful emissions and greenhouse gases. This model contributes to resource depletion, environmental pollution, and increased carbon footprints, making it unsustainable in the long run. In contrast, the circular economy offers a more sustainable and resource-efficient alternative. This model focuses on extending the life cycle of products, minimizing waste, and reusing materials wherever possible. A prime example of the circular economy in the manufacturing sector is Renault's Re-Factory in Flins, France. This initiative represents a shift in paradigm from a consumption-based model to one based on regeneration and reuse.

At the Re-Factory, Renault disassembles old vehicles and salvages components like engines, gearboxes, batteries, and electronic systems. These components are remanufactured and upgraded to meet the standards of new vehicles, ensuring that the performance and safety of refurbished parts are on par with those of freshly manufactured components. The process not only reduces the volume of waste generated by the automotive industry but also significantly decreases the need for new raw materials, thereby conserving natural resources. Beyond component recovery, the Re-Factory also promotes vehicle refurbishment and recycling, enabling older vehicles to be revitalized and resold at affordable prices. It contributes to reducing CO_2 emissions by lowering the energy and materials required in vehicle production. Furthermore, the facility incorporates eco-design principles, sustainable logistics, and collaborative innovation with suppliers and stakeholders. The transition from linear to circular models, as demonstrated by Renault, underscores the potential of circular economy practices to transform the manufacturing sector. It proves that economic productivity and environmental responsibility can coexist. Companies adopting circular approaches benefit from cost savings, enhanced brand value, and compliance with increasingly stringent environmental regulations. While traditional automobile manufacturing epitomizes the wasteful tendencies of the linear economy, initiatives like Renault's Re-Factory illustrate how circular strategies can redefine industrial processes. By rethinking design, production, and consumption, the manufacturing sector can become a key driver of sustainable development in the 21st century.

2.7.1.2 *Electronics sector*

The electronics sector, particularly in the domain of consumer technology, has long operated under a linear economy model. This model is characterized by a "take-make-dispose" approach, where raw materials are extracted, used in the manufacturing of products, and ultimately discarded after use. A prominent example of this is smartphone production. Traditional electronics manufacturers produce smartphones using virgin raw materials such as rare earth elements, precious metals like gold and silver, and non-renewable materials such as plastics and aluminum. In the linear model, the emphasis is placed on rapid product innovation and mass consumption. Consumers are often encouraged to replace their devices frequently due to planned obsolescence, limited repair ability, or lack of software updates. As a result, once smartphones become outdated, damaged, or no longer functional, they are typically discarded. This contributes to the rapidly growing problem of electronic waste (e-waste), which is one of the fastest-growing waste streams globally.

Discarded smartphones contain hazardous materials such as lead, mercury, and cadmium, which can leach into the environment if not properly handled. At the same time, valuable resources embedded in these devices are lost forever. The extraction of raw materials for new devices continues to fuel resource depletion, environmental degradation, and social issues related to unethical mining practices. This unsustainable linear approach has raised significant concerns among environmentalists, policymakers, and consumers alike.

In contrast, the circular economy provides a more sustainable and ethical framework for managing electronic products. A leading example of circular economy innovation in the electronics sector is the company Fair phone. Fair phone has reimagined smartphone design by emphasizing modularity, durability, and ethical sourcing. Unlike traditional smartphones, Fair phone's devices are designed to be easily repairable and upgradeable. Key components such as batteries, cameras, screens, and speakers can be replaced individually using simple tools, empowering users to maintain their devices rather than discard them. This modular design significantly extends the lifespan of the product, reduces the demand for new materials, and minimizes e-waste generation.

Fair phone also prioritizes recycled and ethically sourced materials, including fair-trade gold and conflict-free tin and tungsten. By doing so, it addresses both environmental and social dimensions of sustainability. The company works closely with suppliers to improve labor conditions in the electronics supply chain, promoting transparency and responsibility. In addition to its product design, Fair phone advocates for software longevity by providing extended support and updates, allowing users to keep their phones longer without being forced into unnecessary upgrades. This further reduces the environmental impact associated with manufacturing and disposing of electronics. The contrast between the linear and circular approaches in the electronics sector demonstrates the transformative potential of rethinking product design, use, and end-of-life practices. While the linear model promotes short-term consumption at great environmental and human cost, the circular model as exemplified by Fair phone offers a regenerative approach that aligns.

2.7.1.3 Fashion and textiles sector

The fashion and textiles sector is a major contributor to global economic activity, but it is also one of the most environmentally damaging industries due to its reliance on a linear economy model. This model follows a "take-make-dispose" pattern, in which raw materials are extracted, processed into garments, consumed rapidly, and then discarded. One of the clearest illustrations of this model is fast fashion.

Fast fashion brands, such as major high-street retailers, focus on speedy production and low-cost clothing to meet rapidly changing consumer trends. Collections are updated frequently, often weekly, encouraging consumers to continuously purchase new items. These clothes are often made using cheap materials and labor, which results in garments with short lifespans. Once they fall out of fashion or wear out, they are thrown away. Millions of tons of clothing end up in landfills or incinerators each year, making the fashion industry one of the largest contributors to textile waste. Moreover, the production process consumes vast amounts of water, energy, and chemicals, leading to pollution and environmental degradation.

This linear model also has social implications. Many fast fashion garments are produced in countries where labor regulations are weak, leading to poor working conditions, low wages, and exploitative practices. The pressure to produce cheaply and quickly often results in unsafe and unsustainable manufacturing environments.

In contrast, the circular economy model presents a sustainable and responsible alternative. This model promotes the idea of designing out waste, keeping products in use for longer, and regenerating natural systems. A leading example of this model in the fashion sector is Patagonia's Worn Wear program. Patagonia, a brand known for its commitment to sustainability, has implemented circular practices that redefine the relationship between consumers and clothing. Through the Worn Wear program, Patagonia encourages customers to return their used or damaged garments. These items are then repaired, refurbished, or recycled by the company. Refurbished products are sold at a discounted price, allowing consumers to purchase high-quality clothing at lower costs while supporting environmental sustainability. This initiative helps to extend the lifespan of clothing, reduce demand for new raw materials, and divert textile waste from landfills. By treating garments as valuable resources rather than disposable items, Patagonia demonstrates how businesses can shift from a throwaway culture to one of reuse and responsibility. In addition, Patagonia incorporates recycled and responsibly sourced materials into its new product lines, including organic cotton, recycled polyester, and wool. This reduces the environmental footprint of its supply chain and supports a more circular lifecycle for textiles. The brand also offers repair guides and tools to empower customers to fix their own clothing, promoting a culture of maintenance rather than disposal.

The contrast between fast fashion and Patagonia's circular model highlights the urgent need for systemic change in the fashion industry. While the linear model fuels overconsumption and waste, the circular approach emphasizes durability, repair, and regeneration. Embracing circularity in fashion is essential for reducing environmental impact, conserving resources, and creating a more ethical and sustainable industry for the future.

2.7.1.4 Construction and real estate sector

The construction and real estate sector plays a vital role in shaping modern infrastructure and urban development. However, it is also one of the most resource-intensive and environmentally impactful industries due to its long-standing reliance on a linear economy model. In the linear system, the life-cycle of construction materials follows a simple path: extract raw materials, use them to build structures, and discard them after use—often without considering their potential for reuse or recycling.

A prime example of this linear approach is the widespread use of conventional building materials such as cement, steel, bricks, and timber. These materials are extracted from nature, processed with high energy consumption, and used in building construction without planning for their recovery or reuse at the end of the building's life. When buildings are demolished, much of the resulting construction and demolition (C&D) waste which includes concrete, wood, metals, and other valuable materials is sent to landfills or incinerated. This leads to a massive loss of resources and a significant increase in carbon emissions and environmental degradation.

This conventional process is also associated with high levels of greenhouse gas emissions, particularly from the production of cement and steel. Additionally, the extraction of virgin materials contributes to ecosystem disruption, soil erosion, and biodiversity loss, making the current linear model unsustainable in the long term. In contrast, the circular economy model in the construction sector offers a more sustainable and innovative alternative. It emphasizes reuse, recycling, and recovery of materials, designing buildings for disassembly, and reducing overall environmental impact throughout the life-cycle of a structure. One of the leading examples of circular practices in construction is found in the Netherlands, a country recognized globally for its leadership in sustainable building. Dutch construction companies such as New Horizon Urban Mining are pioneering a shift toward circular construction. Instead of demolishing buildings in a destructive manner, New Horizon focuses on careful dismantling recovering valuable materials such as steel beams, bricks, concrete panels, and wood from old structures. These materials are then repurposed and reintegrated into new building projects, significantly reducing the need for new raw material extraction. This method not only prevents construction waste from ending up in landfills but also results in substantial carbon savings by avoiding the energy-intensive processes involved in producing new materials. For example, reusing bricks or steel elements requires far less energy than manufacturing them from scratch. Furthermore, this circular approach promotes innovation in architectural design, encouraging buildings to be constructed with modular components that can be easily removed and reused in the future.

In addition to reducing waste and emissions, circular construction practices can also generate economic value by creating new markets for reclaimed materials and offering cost-effective building solutions. It fosters collaboration between architects, builders, material suppliers, and policymakers to develop resilient and resource-efficient cities. While the traditional construction model leads to excessive waste and environmental harm, the circular economy approach exemplified by the Dutch construction industry demonstrates how sustainable building practices can transform the sector. By designing for longevity, reuse, and regeneration, the construction and real estate industry can play a crucial role in building a greener and more circular future.

2.7.1.5 Food and agriculture sector

The food and agriculture sector is essential to human survival and economic stability, yet its traditional practices have contributed significantly to environmental degradation. This is largely due to the prevailing linear economy model, which follows a "take-make-waste" approach. In this model, natural resources are extracted for production, used in farming or food processing, and then discarded after consumption, often without considering the environmental consequences or the potential for resource recovery.

A clear example of the linear economy in this sector is traditional agriculture. Modern industrial farming often relies heavily on synthetic fertilizers, chemical pesticides, and monoculture crop systems. While these practices have helped boost productivity, they have also led to long-term problems such as soil degradation, loss of biodiversity, and water pollution. Over time, the excessive use of chemical inputs depletes the soil of essential nutrients, reducing its fertility and resilience. This not only affects the quality and quantity of food production but also weakens the sustainability of agricultural systems.

Moreover, the linear model generates substantial food waste. Unsold produce from farms and retailers is frequently discarded due to imperfections in appearance or overproduction. Much of this waste ends up in landfills, where it decomposes and releases methane, a potent greenhouse gas that contributes to climate change. From farm to fork, the inefficiencies in this linear chain result in the loss of valuable nutrients, energy, and labor. In contrast, the circular economy model in food and agriculture promotes sustainable and regenerative practices that aim to preserve natural resources, minimize waste, and close nutrient loops. A compelling example of this shift is Danone's Regenerative Agriculture Program, which integrates circular principles into both farming and food production processes.

Danone, a global food company, is implementing regenerative farming methods to restore and protect soil health. These practices include crop rotation, cover cropping, composting, and reduced tillage, which enhance soil fertility, improve water retention, and promote biodiversity. By focusing on the long-term health of ecosystems, regenerative agriculture not only sustains productivity but also increases resilience to climate change. In addition to sustainable farming, Danone addresses food processing waste through innovative circular strategies. Food waste generated in its plants is repurposed into biogas through anaerobic digestion or converted into animal feed, thus preventing valuable organic material from being discarded. These actions help reduce landfill waste, lower methane emissions, and recover energy or nutrients that can be reintegrated into the agricultural cycle.

This integrated approach benefits both the environment and the economy. By reducing dependence on synthetic inputs and making better use of natural processes, Danone supports carbon sequestration, improved farmer livelihoods, and ecosystem regeneration. It also exemplifies how circular economy thinking can be applied at every stage of the food system from soil to shelf. The contrast between traditional agriculture and Danone's regenerative model highlights the transformative potential of the circular economy in the food and agriculture sector. By shifting toward closed-loop systems, it is possible to reduce waste, protect the environment, and ensure sustainable food production for future generations.

2.7.1.6 Packaging sector

The packaging sector is one of the most visible and problematic contributors to global waste. In today's consumer-driven economy, packaging is essential for product safety, transportation, and marketing. However, under the linear economy model, packaging materials especially plastics are designed for single-use, leading to significant environmental challenges. This model operates on a "take-make-dispose" principle, where materials are extracted, manufactured into packaging, used briefly, and then discarded.

A prime example of this linear approach is the widespread use of single-use plastic packaging by consumer goods companies. These plastics are lightweight, cheap, and convenient, making them ideal for wrapping food, beverages, cosmetics, and other products. However, once the product is consumed, the packaging is usually thrown away. Much of this waste is not recycled due to contamination, lack of recycling infrastructure, or economic inefficiencies. Instead, it ends up in landfills, incinerators, or worse, in natural ecosystems like oceans and rivers, where it contributes to pollution and threatens marine life.

This throwaway culture not only causes long-term environmental harm but also reflects a tremendous loss of resources. Plastics are made from fossil fuels, and producing them requires significant energy. Discarding packaging after a single use wastes the embedded energy, labor, and raw materials, reinforcing the unsustainable nature of the linear model.

In contrast, the circular economy model offers a regenerative approach to packaging that focuses on reuse, durability, and material recovery. A standout example of this model is Loop, an innovative initiative launched by TerraCycle. Loop collaborates with major brands like Nestlé, Unilever, and Procter & Gamble to redesign packaging systems for consumer products.

Through the Loop platform, products are offered in durable, reusable packaging—made of stainless steel, glass, or other long-lasting materials. Consumers receive these products in specially designed containers, which they use as normal. Once finished, the empty packaging is placed in a return container and sent back to Loop. The packaging is then professionally cleaned, sanitized, and refilled, ready to be sent out to the next customer. This process creates a closed-loop system that drastically reduces the need for single-use plastics. Loop's model not only eliminates packaging waste but also shifts consumer behavior from a throwaway mindset to a reuse culture. By turning packaging into a service rather than a disposable item, Loop demonstrates how companies can align environmental responsibility with business innovation. Moreover, because the containers are more aesthetically pleasing and durable, brands can maintain their visual identity while promoting sustainability. In addition to reducing environmental impact, the Loop system contributes to resource efficiency and carbon footprint reduction. Fewer raw materials are needed, and less energy is used in producing new packaging. This circular approach also fosters collaboration among manufacturers, retailers, and consumers to design products and systems that are economically viable and environmentally sound.

The contrast between single-use plastic packaging and Loop's reuse-based model underscores the urgent need for circular solutions in the packaging sector. Moving toward systems that prioritize reuse, durability, and waste elimination is critical for addressing the global plastic crisis and building a more sustainable future.

2.7.1.7 Energy sector

The energy sector plays a crucial role in driving economic growth and supporting modern life, but it has traditionally operated under a linear economy model that is both resource-intensive and environmentally damaging. In a linear system, energy production is based on the extraction of finite natural resources, followed by energy generation and eventual waste disposal often with little consideration for environmental impact or resource recovery. A well-known example of this linear approach is the operation of fossil fuel power plants. These conventional facilities burn coal, oil, or natural gas to produce electricity. While effective in terms of energy output, they release large volumes of carbon dioxide and other greenhouse gases into the atmosphere, significantly contributing to climate change. In addition to air pollution, fossil fuel combustion also results in the generation of toxic ash and sludge, by-products that are typically buried in landfills or stored in containment ponds. These waste materials can pose severe risks to soil, air, and water quality if not managed properly.

Moreover, the infrastructure of fossil fuel plants is highly specialized and becomes obsolete once decommissioned. In many cases, these industrial sites are abandoned, leaving behind contaminated land and unused buildings that represent not only wasted space but also untapped economic potential. This linear model, driven by consumption and disposal, has proven unsustainable in light of growing environmental awareness and the global push toward decarbonization.

In contrast, the circular economy model offers a regenerative approach that focuses on resource efficiency, reuse, and long-term sustainability. One of the most innovative applications of circular economy principles in the energy sector comes from Enel, a global leader in sustainable energy solutions. Enel has embraced the circular model by repurposing decommissioned fossil fuel power plants, giving these aging infrastructures a second life in support of a greener future.

For instance, Enel is converting old coal-fired plants into solar energy parks, battery storage facilities, and sustainable urban developments. This not only avoids the environmental and financial costs of demolition and waste disposal but also maximizes the value of existing assets. By transforming these brownfield sites into renewable energy hubs, Enel reduces the need for new land use, preserves ecosystems, and supports the transition to clean energy. One notable example is the transformation of the former Eugenio Montale power plant in Italy into a renewable energy site equipped with photovoltaic panels and battery storage systems. Such initiatives represent the essence of circular thinking rethinking and redesigning systems to eliminate waste, extend asset life, and create value from what would otherwise be discarded. In addition to infrastructure reuse, Enel integrates circularity across its operations by focusing on eco-design, sustainable procurement, and life-cycle thinking. The company's circular economy projects not only reduce environmental impact but also stimulate local economies by creating new jobs and community spaces.

2.7.1.8 Healthcare sector

The healthcare sector is essential for ensuring the well-being of individuals and societies. However, it also generates a significant amount of waste due to its heavy reliance on a linear economy model. This model operates on a "take-make-dispose" approach, where resources are extracted, used to create medical products or devices, and then discarded after a single use. While this approach prioritizes hygiene and safety, it comes with considerable environmental and economic consequences.

A clear example of the linear model in healthcare is the use of single-use medical devices. Hospitals and clinics regularly use items such as scalpels, catheters, surgical instruments, syringes, and diagnostic tools made of plastic, glass, or metal. After a single use, these devices are disposed of as regulated medical waste to prevent infection or contamination. The disposal process typically involves landfilling or incineration, both of which contribute to environmental pollution. Incinerating medical waste releases harmful emissions such as dioxins, furans, and particulate matter into the atmosphere, while landfilling introduces the risk of soil and groundwater contamination.

This system not only harms the environment but also leads to high operational costs for healthcare facilities. The continuous need to purchase new devices and manage large volumes of waste adds financial strain to hospitals, especially those in resource-limited settings. Furthermore, the demand for virgin materials used in manufacturing single-use devices places additional pressure on natural resources. In contrast, the circular economy model provides a sustainable alternative by rethinking product life cycles, reducing waste, and promoting resource efficiency. An excellent example of circular practices in healthcare is the medical device reprocessing program offered by Stericycle, a global leader in healthcare waste management.

Stericycle's reprocessing program collects used single-use medical devices from healthcare facilities and transports them to specialized centers. There, the devices undergo rigorous cleaning, sterilization, testing, and refurbishment to ensure they meet strict regulatory and safety standards. Once reprocessed, these devices are returned to hospitals for reuse, allowing them to be safely used multiple times. This circular approach offers several important benefits. Firstly, it significantly reduces the volume of medical waste sent to landfills or incinerators, thereby minimizing environmental impact. Secondly, it helps healthcare facilities cut costs, as reprocessed devices are generally less expensive than brand-new ones. Thirdly, it reduces the demand for virgin materials and energy associated with manufacturing, lowering the overall carbon footprint of the healthcare sector.

Moreover, the program supports a shift in mind-set within the medical community, encouraging healthcare providers to embrace sustainable practices without compromising patient safety. Regulatory agencies such as the U.S. FDA oversee and approve reprocessed devices, ensuring that they are just as effective and safe as new ones.

In summary, the contrast between single-use medical devices and Stericycle's reprocessing program illustrates how the healthcare sector can benefit from adopting circular economy principles. By extending the life of medical equipment, reducing waste, and conserving resources, the circular model promotes not only environmental sustainability but also economic efficiency and responsible healthcare delivery. This transition is crucial for building a more resilient and sustainable future for global health systems.

2.8 Circular economy frameworks for waste management

The concept of a circular economy (CE) has emerged as a powerful alternative to the traditional linear economy, which follows a "take-make-dispose" model. In contrast, the circular economy is designed to keep resources in use for as long as possible, extract the maximum value from them while in use, and recover and regenerate materials at the end of their service life. One of the most impactful applications of the circular economy lies in waste management. Traditional waste management systems have largely focused on disposal and recycling, often overlooking the potential of prevention, reuse, and resource recovery. Circular economy frameworks aim to restructure waste management by transforming waste into a valuable resource and integrating sustainability into every phase of a product's life cycle.

At the heart of the circular economy framework is the waste management hierarchy, which prioritizes actions according to their environmental benefits. The top of the hierarchy is waste prevention, which focuses on designing products and processes that minimize waste generation in the first place. This might involve using fewer raw materials, selecting recyclable or biodegradable materials, or developing products with longer life spans. Prevention strategies not only reduce the burden on landfills and incineration plants but also decrease greenhouse gas emissions associated with production and disposal.

The second level of the hierarchy is reuse. In a circular economy, reuse extends beyond simple repurposing and involves the development of business models that support product life extension, such as repair services, refurbishment programs, and second-hand marketplaces. These strategies encourage consumers to see products not as disposable items but as assets with long-term value. For example, in the electronics industry, companies are increasingly offering refurbished devices and take-back schemes to encourage the return and reuse of used products.

Recycling, which comes next in the hierarchy, is a critical component of circular waste management, but it is not the ultimate solution. While recycling allows materials to re-enter the production cycle, it still requires energy and can involve material degradation. Thus, recycling should be seen as a complementary, not primary, approach. Effective circular economy frameworks focus on designing products for disassembly and material separation, making the recycling process more efficient and reducing contamination. Closed-loop recycling systems, where waste materials are recycled into the same product (e.g., aluminum cans into new cans), are especially encouraged in a circular economy.

Recovery, often in the form of energy recovery, is lower in the waste hierarchy and is considered only when prevention, reuse, and recycling are not feasible. This process involves converting non-recyclable waste materials into usable heat, electricity, or fuel through combustion or other technologies. While this process reduces the volume of waste, it also has environmental drawbacks, such as emissions and the loss of material resources. Hence, it is generally viewed as a last-resort strategy within circular frameworks.

Disposal, including landfilling and incineration without energy recovery, is at the bottom of the hierarchy and should be minimized as much as possible. A successful circular economy framework focuses on reducing reliance on disposal methods by promoting systemic changes in production, consumption, and waste handling.

Implementing circular economy frameworks for waste management requires coordinated efforts among policymakers, businesses, and consumers. Governments play a vital role by enacting regulations and incentives that promote circular practices, such as extended producer responsibility (EPR), landfill taxes, and eco-design requirements. EPR policies hold producers accountable for the entire lifecycle of their products, encouraging them to design with end-of-life recovery in mind. Businesses, on the other hand, need to rethink their operational models. This includes investing in sustainable product design, establishing circular supply chains, and adopting service-based models such as product-as-a-service, where customers use a product without owning it. These strategies shift the focus from volume-based growth to value-based sustainability.

Consumers also have a crucial role to play. Circular waste management relies on behavioral change, such as choosing sustainable products, separating waste properly, participating in reuse schemes, and supporting companies that prioritize circularity. Education and awareness campaigns are essential to shift mindsets and cultivate a culture of sustainability. Technology and innovation are key enablers in this transition. Digital tools such as the Internet of Things (IoT), blockchain, and artificial intelligence can enhance transparency, track materials, optimize logistics, and support decision-making in waste management systems. These technologies can facilitate real-time monitoring of waste flows, improve sorting processes, and enable predictive analytics for waste reduction.

At the end, circular economy frameworks provide a comprehensive and sustainable approach to waste management by prioritizing resource efficiency, reducing environmental impacts, and creating economic opportunities. They require systemic changes across all levels of society from product design and industrial processes to policy and consumer behavior. As the global population and consumption rates continue to rise, adopting circular economy principles in waste management is not just desirable it is imperative for achieving long-term sustainability.

2.8.1 Extended producer responsibility (EPR)

The transition from a linear to a circular economy represents a significant paradigm shift in how societies view production, consumption, and waste. One of the most powerful policy tools supporting this transition is **Extended**

producer responsibility (EPR), which places the responsibility for the end-of-life management of products squarely on the shoulders of producers. In the context of circular economy frameworks for waste management, EPR serves as a strategic mechanism to encourage sustainable product design, reduce waste generation, and improve recycling and reuse rates. Under traditional waste management systems, local authorities and taxpayers have primarily borne the cost and responsibility of waste collection and disposal. However, EPR changes this dynamic by making producers financially or physically responsible for the treatment or disposal of post-consumer products. In doing so, EPR directly aligns with the **principles of the circular economy**, where the focus is on designing out waste, keeping products and materials in use, and regenerating natural systems.

Circular economy frameworks advocate for closed-loop systems in which materials are continuously cycled through production and consumption without becoming waste. EPR supports this by creating a strong incentive for producers to **design products with longer lifespans, greater reparability, and higher recyclability**. For example, a manufacturer subject to EPR regulations is more likely to avoid complex material mixes, hazardous substances, or non-recyclable components, because doing so would increase the cost and difficulty of managing waste products under the scheme. One of the key advantages of EPR in a circular economy framework is that it promotes **innovation and eco-design**. When producers are held accountable for the post-consumer phase of their products, they are more motivated to innovate in ways that reduce environmental impact and material waste. This could involve designing modular electronics that can be easily repaired or reused, or packaging made from single materials that are easier to recycle. Over time, this systemic change can shift entire industries toward more sustainable and circular business models. EPR also plays a crucial role in enhancing **material recovery and recycling systems**. By obligating producers to set up or contribute to take-back schemes, collection centers, and recycling infrastructure, EPR ensures a steady and reliable flow of materials back into the production cycle. This reduces dependency on virgin raw materials and supports the creation of secondary markets for recycled content. In countries with well-implemented EPR policies, there has been a marked improvement in recycling rates for packaging, batteries, electronics, and vehicles.

Moreover, EPR fosters **shared responsibility and collaboration** among various stakeholders within the waste management ecosystem, including producers, consumers, local governments, recyclers, and non-profits. This collaborative approach fits neatly within circular economy frameworks, which emphasize systems thinking and interdependence. For instance, while producers are responsible for funding and managing waste systems, consumers must actively participate by returning used products, and municipalities may play a role in providing infrastructure support. EPR also enables better **tracking and transparency** in waste flows, a feature increasingly essential in modern circular systems. With the help of digital technologies like blockchain and IoT sensors, producers can monitor the lifecycle of products and materials, ensuring accountability and compliance with EPR mandates. This data-driven approach not only boosts the effectiveness of waste management but also contributes to broader sustainability goals, such as carbon footprint reduction and resource efficiency.

Extended Producer Responsibility is a vital pillar of circular economy frameworks for waste management. It shifts the burden of waste from the public to the producers, drives eco-innovation, strengthens recycling infrastructure, and fosters collaboration across the value chain. When implemented effectively, EPR not only helps reduce the environmental impact of products but also accelerates the transition toward a regenerative and sustainable economic model. As global resource constraints intensify and waste volumes increase, embracing EPR within circular frameworks is both a necessity and an opportunity for long-term environmental stewardship.

2.8.2 C.I.R.C.L.E. framework for circular waste management

The C.I.R.C.L.E. Framework is a strategic model developed to advance circular waste management, ensuring sustainability by minimizing waste generation, maximizing resource efficiency, and fostering long-term environmental stewardship. The framework stands for Conscious Design & Consumption, Integrated Waste Hierarchy Application, Regenerative Systems & Recovery, Collaborative Governance & Incentives, Loop-Driven Technologies & Traceability, and Education, Empowerment & Ecosystem Thinking. Each of these six pillars represents a fundamental approach to transforming the linear "take-make-dispose" economy into a regenerative circular system.

2.8.2.1 C—conscious design & consumption

At the heart of the circular economy lies the conscious design of products and services, combined with informed and responsible consumption patterns. This stage influences the entire lifecycle of a product, determining how it is made, used, and eventually discarded or reused.

- **Design for circularity**: Product design must incorporate principles such as durability, modularity, and disassembly. This enables easy repair, upgrading, and eventual recycling. Companies are now prioritizing cradle-to-cradle design, where end-of-life considerations are planned during the design stage.
- **Consumer awareness**: Behavioral change plays a critical role in reducing waste. Through educational campaigns, labeling schemes (such as eco-labels or environmental product declarations), and behavioral nudges (e.g., default sustainable options), consumers can be guided toward sustainable purchasing and usage behaviors.

- **Eco-material innovation**: Sustainable design involves the use of materials that are biodegradable, recyclable, or derived from renewable resources. Innovations such as bio-based plastics, mushroom packaging, or recycled textiles allow products to be produced with a minimal ecological footprint while remaining functional and appealing.

2.8.2.2 I—Integrated waste hierarchy application

This pillar emphasizes the application of the waste hierarchy, not as a fixed model but as a dynamic and context-driven framework. It moves beyond recycling to include refusing, reducing, reusing, and recovering materials.

- **The 5rs**: The principle of Refuse, Reduce, Reuse, Recycle, and Recover provides a roadmap for individuals, businesses, and governments. It prioritizes upstream solutions (like refusing single-use plastics) before downstream solutions (like energy recovery through incineration).
- **Dynamic optimization**: The waste hierarchy should be adapted based on real-time data and local conditions. For example, if recycling a specific material becomes energy-intensive or economically unviable, alternatives like reuse or material substitution can be explored.
- **Micro and macro circular loops**: Local (community-level) loops such as composting and tool-sharing libraries, and global (industrial-level) loops like closed-loop manufacturing systems, must be developed in parallel. This dual approach ensures systemic efficiency and inclusivity.

2.8.2.3 R—Regenerative systems & recovery

True circularity goes beyond reusing materials it regenerates natural and technical systems to create value from waste and reintegrate resources into economic cycles.

- **Biological cycles**: Organic waste, if managed effectively, can become a valuable resource. Composting organic materials enhances soil health, while anaerobic digestion produces biogas and nutrient-rich digestate. Agricultural and food systems benefit from these regenerative processes.
- **Technical cycles**: In industries, the recovery of materials such as rare earth elements from e-waste or plastic polymers from packaging waste enables secondary resource creation. Industrial symbiosis—where waste from one sector serves as input for another—supports sustainable value chains.
- **Zero-waste manufacturing**: Businesses are shifting to closed-loop production where every output (including waste) is repurposed as an input. This involves systems thinking, where processes are interconnected, and feedback loops are embedded to minimize residual waste. Fig. 2.18 illustrate circle framework for circular waste management.

FIGURE 2.18 Circle framework for circular waste management.

2.8.2.4 C—Collaborative governance & incentives

Circular economy implementation is impossible without cross-sector collaboration and effective governance structures. All stakeholder's governments, private sector, academia, and civil society must be involved.

- **Public-private-people partnerships (4Ps)**: These partnerships ensure inclusivity, resource pooling, and diverse perspectives in decision-making. Citizen involvement ensures that policies are grounded in real-world needs and practices.
- **Policy levers**: Regulatory tools like Extended Producer Responsibility (EPR) enforce accountability by making manufacturers responsible for the end-of-life management of their products. Tax incentives for circular businesses and green public procurement further promote circularity.
- **Local circular hubs**: Decentralized facilities for collection, repair, reuse, and recycling foster circular activity at the community level. These hubs can also serve as centers for training, awareness, and innovation, especially in rural and underserved areas.

2.8.2.5 L—Loop-driven technologies & traceability

Technology plays a central role in making circular systems efficient, scalable, and transparent. By digitizing the lifecycle of products and materials, stakeholders can track, monitor, and optimize their use and reuse.

- **Digital twins & IoT sensors**: These tools create real-time digital replicas of physical systems, enabling precise monitoring of resource use, product wear, and waste generation. IoT sensors can also facilitate smart bin systems, monitoring waste levels and optimizing collection routes.
- **Blockchain for transparency**: Blockchain ensures immutability and traceability of data, especially crucial in verifying recycled content or the origin of raw materials. This builds consumer trust and regulatory compliance.
- **AI-based sorting & analytics**: Artificial Intelligence can enhance material recovery by automating waste sorting processes. It can also analyze consumption and disposal patterns to predict and minimize waste generation, enabling proactive circular planning.

2.8.2.6 E—Education, empowerment & ecosystem thinking

The final pillar underscores the need for a cultural and intellectual shift toward sustainability through education, empowerment, and holistic ecosystem-based thinking.

- **Circular literacy**: Circular economy concepts should be embedded across educational levels schools, colleges, and professional training. Interdisciplinary curricula that integrate science, economics, and ethics are essential to build a future-ready workforce.
- **Behavioral incentives**: Practical reward systems can promote circular behaviors. Examples include reverse vending machines for recycling plastics, digital platforms for sharing unused items, and reward points for sustainable actions.
- **Community empowerment**: Local communities are often sources of innovative circular solutions. Initiatives such as zero-waste societies, repair cafés, urban mining cooperatives, and community composting groups empower people to take ownership of waste management at the grassroots level.

The C.I.R.C.L.E. Framework presents a comprehensive, actionable roadmap for embedding circular principles into waste management systems across scales and sectors. It promotes a balance between technological advancement, community participation, regulatory action, and environmental regeneration. Moving toward a circular waste economy is not merely a matter of improved recycling it requires a transformative redesign of production, consumption, and governance models. Through this framework, societies can work together to ensure sustainable growth, environmental resilience, and social well-being for future generations.

2.9 Metaverse based circular economy framework for waste management

The integration of the metaverse into sustainable waste management offers a transformative opportunity to revolutionize how societies approach environmental responsibility, particularly through the lens of the circular economy. The metaverse defined as an immersive, interconnected virtual environment powered by technologies such as extended reality (XR), blockchain, artificial intelligence (AI), and the Internet of Things (IoT) can serve as a dynamic and interactive framework to foster circularity in waste management systems. By simulating real-world scenarios, enhancing stakeholder engagement, and enabling data-driven decision-making, the metaverse can bridge the gap between theory and practice in sustainable resource management.

At the core of this framework lies the digital twin concept, where physical waste streams, infrastructure, and ecosystems are mirrored in a virtual model within the metaverse. This allows waste management authorities, urban planners, and businesses to visualize waste generation patterns in real-time, analyze inefficiencies, and forecast the environmental and economic outcomes of various interventions. For instance, through virtual simulations, stakeholders can test the impacts of material substitution, redesign packaging for recyclability, or evaluate the logistics of decentralized composting hubs without incurring physical costs or disruptions. Such proactive planning is essential to implement the 5Rs hierarchy (Refuse, Reduce, Reuse, Recycle, Recover) in an optimized, context-specific manner.

Moreover, the metaverse can enable interactive education and training platforms for citizens and professionals alike. Virtual environments can gamify sustainability practices, allowing users to experience the consequences of their consumption choices or waste disposal behaviors. These immersive learning modules can significantly improve circular literacy, especially in urban populations where waste generation is highest. Simulated environments such as smart homes, zero-waste supermarkets, or sustainable manufacturing plants can showcase best practices and encourage behavioral change through experiential learning. In rural areas or underserved communities, the metaverse can offer low-cost access to technical training for waste workers, empowering them with skills for high-value circular activities such as e-waste dismantling or plastic upcycling.

A key feature of this metaverse framework is blockchain-powered traceability, where every material's lifecycle is logged and authenticated. From the origin of raw materials to end-of-life recovery, stakeholders including producers, recyclers, and consumers can track and verify the circular compliance of products in a secure and transparent manner. This ensures trust in Extended

Producer Responsibility (EPR) systems, facilitates eco-labeling, and supports green procurement policies. Additionally, AI and predictive analytics embedded within the metaverse can continuously assess user behavior, optimize collection routes, automate material sorting, and predict supply-demand patterns for recycled materials, thereby improving overall system efficiency.

The metaverse also supports the creation of decentralized digital marketplaces for waste-to-resource exchange. Citizens can trade reusable items, access repair services, or donate surplus food virtually, connecting supply with local demand in real time. Municipalities and industries can list available recyclables or by-products, enabling industrial symbiosis across sectors and regions. Through token-based incentive systems, users can earn digital rewards such as NFTs or carbon credits for sustainable actions like returning e-waste, using refill stations, or attending sustainability workshops in the virtual world. These incentives, linked with real-world benefits, can drive wide-scale participation and behavioral transformation. From a governance perspective, the metaverse can host collaborative virtual platforms for multi-stakeholder engagement. Policymakers, academics, entrepreneurs, NGOs, and citizens can convene in immersive virtual forums to co-design policies, pilot innovations, and share data transparently. These platforms support inclusive governance, ensuring that waste management decisions reflect the diverse perspectives and needs of all stakeholders, particularly marginalized communities often excluded from planning processes. A metaverse framework aligned with circular economy principles offers an innovative pathway to make waste management more inclusive, transparent, efficient, and impactful. By blending immersive technology with environmental stewardship, it facilitates not only system-level optimization but also grassroots empowerment. As cities, industries, and governments aim to transition from linear to circular systems, integrating the metaverse can play a catalytic role in building a regenerative future—where waste is not discarded, but continuously reimagined as a valuable resource in both the virtual and physical world. Each component of the acronym M.E.T.A.V.E.R.S.E. represents a strategic pillar that integrates metaverse capabilities with circular economy principles for sustainable waste management.

M—Mirror systems (digital twins for waste streams): Mirror Systems refer to the use of digital twin's virtual representations of real-world systems to model and monitor the flow of waste across cities, industries, and institutions. These systems allow municipalities and waste management companies to replicate their physical infrastructure (e.g., landfills, recycling centers, waste bins) in the virtual metaverse. Through real-time data integration from IoT sensors, AI, and GIS tools, stakeholders can track waste generation hotspots, material contamination, and collection inefficiencies. This virtual modeling enables scenario testing and rapid decision-making for system optimization, without incurring the cost or disruption of real-world experimentation. It is the foundational layer where data meets design to simulate sustainable waste flows.

E—Extended circularity simulation: Extended Circularity Simulation allows stakeholders to virtually test and forecast the performance of circular interventions in a risk-free metaverse environment. For example, city planners can simulate the environmental and financial impacts of redesigning packaging, implementing source-separation campaigns, or launching decentralized composting programs. These simulations help evaluate multiple "what-if" scenarios such as banning single-use plastics or increasing reuse infrastructure and understand how these changes affect recycling rates, landfill diversion, and material recovery. By providing predictive analytics and virtual outcomes, this pillar supports evidence-based, strategic transitions from linear to circular systems.

T—Tokenized incentive systems: Tokenized Incentive Systems involve the creation of digital reward mechanisms, typically powered by blockchain technology, to encourage sustainable behavior in both virtual and physical spaces. Individuals can earn tokens, NFTs, or digital points for performing environmentally positive actions—such as depositing recyclables in smart bins, attending sustainability events in the metaverse, or donating reusable goods. These digital rewards can be redeemed for real-world benefits like discounts, utility credits, or public recognition. By gamifying sustainability and linking it to tangible incentives, this system fosters widespread community participation and makes waste reduction both engaging and rewarding.

A—Avatar-based community engagement: Avatar-Based Community Engagement empowers users to interact within the metaverse using digital avatars, enabling immersive and inclusive participation in the circular economy. Citizens can role-play as "waste warriors," participate in virtual clean-up missions, visit zero-waste digital marketplaces, or engage in repair challenges. They can also attend virtual town halls and workshops where they contribute ideas and co-create localized waste solutions. This interactive environment builds environmental awareness, empathy, and collaboration while giving voice to individuals who might otherwise be excluded from traditional waste governance processes—such as youth, rural populations, or marginalized groups.

V—Verified material traceability: Verified Material Traceability uses blockchain technology and AI to log the journey of materials from production to end-of-life recovery, ensuring accountability, transparency, and authenticity. Each item or material is tagged with a digital identity, recording its origin, composition, usage, recycling, and disposal details. This builds trust in systems like Extended Producer Responsibility (EPR) and enables eco-labeling, green audits, and fair circular certification. It also helps prevent greenwashing, where companies falsely claim sustainable practices. Through traceability, consumers and regulators can verify that products are part of a genuine circular loop.

E—Extended reality (XR) for training & awareness: Extended Reality (XR), including Virtual Reality (VR), Augmented Reality (AR), and Mixed

Reality (MR), plays a vital role in capacity building and behavioral change. XR tools can provide hands-on training for informal waste workers on safe sorting, dismantling e-waste, or operating machinery. In households and schools, AR-based apps can guide users on how to segregate waste or compost food scraps properly. Students can tour smart recycling plants or participate in zero-waste simulations using VR. These immersive tools make education engaging and contextually relevant, thereby improving circular literacy across demographics.

R—Resource exchange marketplaces (Virtual): This pillar envisions decentralized digital marketplaces in the metaverse for exchanging, donating, or selling surplus materials and products. Users can list second-hand clothes, electronics, or furniture for swap, while manufacturers can trade by-products or excess materials with industries that need them creating virtual industrial symbiosis. These exchanges can happen locally or globally, depending on the user's virtual environment. Repair artisans and refurbishes can offer virtual consultations, linking their services to physical repairs. Such marketplaces increase product lifespans, reduce material waste, and provide economic value, especially in urban areas with fast consumption cycles.

S—Scenario-based decision support: Scenario-Based Decision Support involves the use of data-driven dashboards, AI models, and simulation tools within the metaverse to assist policymakers and city managers. By inputting real-time data and KPIs (e.g., material recovery rate, carbon savings, cost per ton diverted), decision-makers can evaluate the consequences of different waste policies or interventions. This might include comparing centralized versus decentralized collection models, or choosing between composting and anaerobic digestion. These insights allow for smarter investments, strategic urban planning, and adaptive policy frameworks tailored to evolving needs.

E—Ecosystem integration & governance: The final component ensures the integration of all stakeholders government bodies, citizens, industry players, NGOs, and academia into a common virtual governance platform. Through collaborative virtual spaces, these actors can co-design waste management strategies, monitor progress through open data dashboards, and resolve conflicts in real-time. The metaverse enables equitable access for marginalized communities and supports participatory planning. It also supports knowledge sharing across regions, enabling cities to replicate successful circular models. This holistic, ecosystem-oriented governance approach ensures that circular economy solutions are inclusive, context-sensitive, and democratically developed.

The M.E.T.A.V.E.R.S.E. Framework offers a bold imagination of sustainable waste management in the age of digital transformation. By uniting metaverse technologies with circular economy thinking, this model promotes efficiency, transparency, equity, and engagement. It not only makes waste visible, interactive, and traceable but also turns every citizen into a conscious contributor toward a regenerative and circular future.

2.10 Waste management and circular economy toward carbon neutrality

In the global effort to combat climate change, waste management and the circular economy have emerged as critical tools in driving the transition toward carbon neutrality. Carbon neutrality refers to achieving a net-zero balance between the greenhouse gases (GHGs) emitted into the atmosphere and those removed or offset. Traditional, linear waste systems characterized by "take, make, use, and dispose" models have contributed significantly to global emissions through landfilling, incineration, and inefficient resource use. By contrast, a circular economy model aims to eliminate waste and continuously use resources through reuse, recycling, and regeneration, thereby playing a pivotal role in reducing carbon emissions.

The link between waste and carbon emissions: Waste contributes to greenhouse gas emissions in multiple ways. Landfills release methane, a greenhouse gas over 25 times more potent than CO_2, as organic waste decomposes anaerobically. Incineration, especially of plastics and mixed waste, produces significant amounts of carbon dioxide and toxic gases. Additionally, the energy used in extracting raw materials, manufacturing goods, and transporting them contributes heavily to carbon footprints. When these products are disposed of after single use, the embodied energy and carbon are lost, resulting in a continuous demand for virgin resources. By redesigning waste systems around circular principles, we not only reduce the volume of waste but also conserve energy, extend product lifespans, and drastically cut emissions associated with raw material extraction, processing, and transportation. The circular economy offers several strategies that directly support the goal of carbon neutrality:

- **Waste prevention and eco-design:** Preventing waste is the most effective way to reduce emissions. Designing products for durability, repair ability, and modularity reduces the need for replacement and conserves resources.
- **Recycling and material recovery:** Recycling materials like aluminum, plastics, paper, and glass saves significant energy compared to producing them from virgin sources. For example, recycling aluminum uses 95% less energy than creating new aluminum, directly cutting CO_2 emissions.
- **Composting and anaerobic digestion:** Organic waste can be composted or processed in anaerobic digesters to produce biogas and nutrient-rich soil amendments. This diverts methane-producing waste from landfills while creating renewable energy.
- **Industrial symbiosis:** In industrial ecosystems, the waste or by-product of one process becomes the raw material for another. This reduces material demand, minimizes transportation emissions, and lowers production energy use.

- **Reuse and refurbishment:** Extending the life of products through repair, refurbishment, and resale prevents the emissions associated with producing new items and reduces the burden on waste systems.
- **Energy recovery:** While not the most circular option, waste-to-energy technologies can recover energy from non-recyclable waste, displacing fossil fuels and reducing methane emissions from landfills.

Governments around the world are increasingly recognizing the intersection between waste, circularity, and climate. Extended Producer Responsibility (EPR) policies compel manufacturers to manage the entire lifecycle of their products, including take-back schemes and recycling mandates. Carbon pricing mechanisms and incentives for green procurement and eco-labeling also encourage businesses to adopt circular strategies. Cities like Amsterdam, Copenhagen, and San Francisco are integrating zero-waste goals with climate action plans, establishing targets for landfill diversion, material reuse, and low-carbon recycling infrastructure.

Emerging technologies further accelerate the potential of circular waste systems in reducing emissions. Artificial Intelligence (AI) and IoT sensors optimize waste sorting, collection routes, and material recovery processes, enhancing efficiency and reducing carbon-intensive operations. Blockchain technology can trace the lifecycle of materials, ensuring transparency and accountability in recycled content and carbon impact. Digital platforms enable waste exchanges, connecting supply and demand for secondary materials in real time, thereby promoting local circular economies and reducing emissions from transportation. Adopting circular waste practices not only aids carbon neutrality but also stimulates green jobs, local entrepreneurship, and resource independence. Repair services, recycling enterprises, and upcycling businesses create employment while encouraging community participation in sustainability. Moreover, reducing reliance on imported raw materials enhances national resilience and helps stabilize economies against volatile global commodity markets.

Despite the clear benefits, several challenges hinder the full alignment of waste management and circular economy with carbon neutrality. These include fragmented infrastructure, lack of awareness, policy inertia, and financial constraints, especially in developing regions. Addressing these requires a systems-level approach: aligning stakeholders across the value chain, harmonizing policies, investing in innovation, and building citizen awareness. Public-private partnerships, community-led zero-waste initiatives, and cross-sector collaborations are essential to scale circular solutions. Moreover, integrating circular economy metrics into national carbon accounting frameworks will ensure that the emission reduction potential of waste transformation is recognized and maximized.

Waste management and the circular economy are indispensable pillars in the journey toward carbon neutrality. By reimagining waste as a resource and shifting from linear to regenerative systems, societies can dramatically reduce emissions, conserve natural resources, and build resilient, low-carbon economies. The circular economy is not just a waste strategy it is a climate strategy, a resource strategy, and a social equity strategy, all rolled into one. As the global climate crisis accelerates, embedding circularity into waste systems is no longer optional it is essential for a sustainable and carbon-neutral future.

2.11 Data analysis of circular economy

The Circular Economy Dataset provides a comprehensive overview of waste management performance across multiple geographic zones, focusing on diverse material types. It is designed to evaluate the effectiveness of circular economy practices specifically reuse, recycling, composting, energy recovery, and landfill reduction and the corresponding economic value recovered through these processes.

Collected from 10 distinct zones (labeled Zone A through Zone J), the dataset spans seven key material categories: Textiles, E-waste, Plastic, Organic, Metal, Glass, and Paper. For each zone-material combination, it captures six critical performance indicators:

1. Reused (%): The proportion of material reused, reducing the need for virgin inputs.
2. Recycled (%): The percentage of material successfully reprocessed into new products.
3. Composted (%): The share of biodegradable waste converted into compost, relevant mainly to organic materials.
4. Landfilled (%): The portion of waste discarded in landfills, reflecting inefficiency in circular recovery.
5. Energy Recovered (%): The amount of waste diverted for energy generation through methods like incineration or anaerobic digestion.
6. Economic Value Recovered ($): The estimated financial value realized from circular processing efforts.

This dataset serves as a diagnostic tool for analyzing the circularity performance of different materials across varied regional contexts. It allows stakeholders researchers, policymakers, sustainability practitioners, and industry leaders to:

- Benchmark circular performance by material and location.
- Identify zones or materials that exhibit best practices.
- Detect inefficiencies or gaps in circular waste handling.
- Understand how material recovery efforts translate into economic returns. The Circular Economy Dataset is a valuable resource for advancing data-driven strategies toward sustainable waste management and resource optimization in both urban and industrial ecosystems (Table 2.13).

TABLE 2.13 Circular economy dataset.

Location	Material type	Reused (%)	Recycled (%)	Composted (%)	Landfilled (%)	Energy recovered (%)	Economic value recovered ($)
Zone B	Textiles	88	91	29	33	11	368.49
Zone H	Textiles	56	95	10	31	41	829.31
Zone F	E-waste	76	55	10	7	38	843.83
Zone J	Plastic	31	87	47	4	8	392.74
Zone F	Plastic	39	61	35	12	45	510.64
Zone J	Organic	30	64	3	35	11	741.15
Zone B	Plastic	21	91	31	18	53	979.19
Zone E	Metal	26	84	31	20	48	461.89
Zone F	Organic	78	48	45	4	29	626.49
Zone F	Metal	22	76	27	9	36	507.53
Zone E	Textiles	71	42	18	4	40	295.84
Zone I	Plastic	80	35	43	8	21	254.73
Zone H	Organic	75	82	32	11	23	833.88
Zone B	Plastic	21	46	48	35	59	210.69
Zone D	Plastic	80	34	11	19	10	798.51

Zone E	Glass	20	91	34	1	28	111.51
Zone J	Metal	45	21	29	23	57	992.99
Zone B	Plastic	86	70	20	18	37	225.64
Zone C	Metal	54	82	0	40	15	160.58
Zone C	Glass	79	29	10	8	43	607.43
Zone I	Metal	38	89	17	15	43	381.02
Zone E	Plastic	19	20	44	7	14	379.8
Zone I	Textiles	46	49	39	12	15	721
Zone C	E-waste	34	85	20	32	4	316.16
Zone D	E-waste	39	65	20	33	59	673.94
Zone E	Metal	25	29	35	10	49	242.71
Zone A	Organic	42	23	4	14	54	331.54
Zone B	Plastic	24	26	4	28	32	503.9
Zone A	Metal	88	95	20	25	53	538.89
Zone C	Paper	17	93	13	14	4	761.49

- **Key insights by material type**

Textiles

- Highest reuse and recycle rates are in Zone B (88% reused, 91% recycled).
- Zone H shows higher energy recovery (41%) and the highest economic value of $829.31.
- Variability indicates differing regional textile circularity strategies.

Plastic

- Zone B (entry 1) shows highest recycling (91%) and energy recovery (53%), also yielding the highest value of $979.19.
- Zone I has very high reuse (80%) but low recycling (35%), suggesting reuse-led circularity.
- Significant differences in performance across zones indicate potential policy gaps.

E-waste

- Zone F stands out with high reuse (76%) and economic value of $843.83.
- Zone C leads in recycling (85%), while Zone D has the highest energy recovery (59%).

Organic waste

- Zone H exhibits the highest recycling (82%), composting (32%), and economic return ($833.88).
- Zone F also shows high reuse (78%) and composting (45%), ideal for sustainable organics management.

Metal

- Zone A displays top-tier reuse (88%) and recycling (95%), achieving a value of $538.89.
- Zone J has highest energy recovery (57%) and economic value ($992.99) among metals.

Glass

- Glass in Zone E is 91% recycled, but Zone C shows highest reuse (79%) and value ($607.43).
- Both zones exhibit minimal landfilling, indicating success in glass circularity.

Paper

- Only one entry in Zone C, but impressive recycling (93%) and high value ($761.49), though reuse is minimal (17%) (Table 2.14).

TABLE 2.14 Cross-material observations.

Metric	Top performing zone (material)	Key value
Highest reuse (%)	Zone A (metal)	88%
Highest recycle (%)	Zone A (metal) and Zone B (textiles)	95%
Highest composting (%)	Zone F (organic)	45%
Lowest landfilling (%)	Zone E (glass)	1%
Highest energy recovery (%)	Zone J (metal)	57%
Highest economic value	Zone J (metal)	$992.99

Patterns & trends

- High Circular Performance → High Economic Return Zones with strong reuse and recycling (e.g., Zones A, B, J) generally show higher recovered economic value. This affirms the financial viability of circular systems.
- Organic Waste shows wide variability. While some zones excel in composting and energy recovery (Zone F, H), others underperform, indicating the need for better organic waste infrastructure.
- Plastic shows potential but also inconsistency. While certain zones have high recycling and reuse, others still rely heavily on landfilling (e.g., Zone B's second entry: 35% landfill).
- Glass and Metal materials demonstrate high recyclability and low landfilling, making them ideal targets for robust circular initiatives.

Areas for improvement

- Landfilling remains a challenge in several zones (e.g., Zone C - Metal: 40%; Zone B Plastic: 35%). These areas may lack recycling capacity or suffer from poor waste segregation.
- Energy recovery is underutilized in high-value waste streams like E-waste or Textiles in some regions. More investment in energy recovery systems may enhance efficiency.
- Zone-Level Disparity: The same material type (e.g., Plastic or Organic) shows vastly different circular outcomes across zones, highlighting unequal infrastructure or policy implementation.

Recommendations for policymakers & practitioners

1. Replicate Best Practices: Zones with high performance (e.g., Zone J for Metal, Zone B for Textiles) can serve as models.
2. Target Low-Performing Zones: Especially for materials with high landfill or low reuse rates.

3. Invest in Organic Waste Management: Composting and biogas offer untapped value.
4. Leverage Economic Potential: Demonstrate how circular strategies improve not just sustainability, but also local economies.
5. Technology & Education: Use AI for sorting and data analytics; train citizens and stakeholders on circular practices.

The two visualizations correlation heatmap (Fig. 2.19) and the pair-plot (Fig. 2.20) provide valuable insights into the relationships among key circular economy metrics within the dataset. The correlation heatmap illustrates how strongly each pair of variables is related, with values ranging from −1 to +1, where positive values indicate direct relationships and negative values indicate inverse ones. Notably, the strongest negative correlation (−0.41) is observed between composted and landfilled percentages, suggesting that as composting efforts increase, landfilling decreases an encouraging sign of effective organic waste diversion. However, a surprising slight positive correlation (+0.22) exists between recycled and landfilled waste, implying that higher recycling does not necessarily correspond to lower landfill use, possibly due to overlapping or inefficient practices. Economic value recovered shows only weak positive correlations with reuse (+0.084) and energy recovery (+0.074), while showing a mild negative correlation with composting (−0.17), indicating that economic gains are not strongly tied to any single waste management strategy but rather result from a complex interplay of factors.

The pair-plot further supports these observations by displaying scatter plots of every variable against each other, along with individual histograms. It reveals that reused and composted percentages are fairly evenly distributed, whereas recycled and landfilled values are more skewed, highlighting that

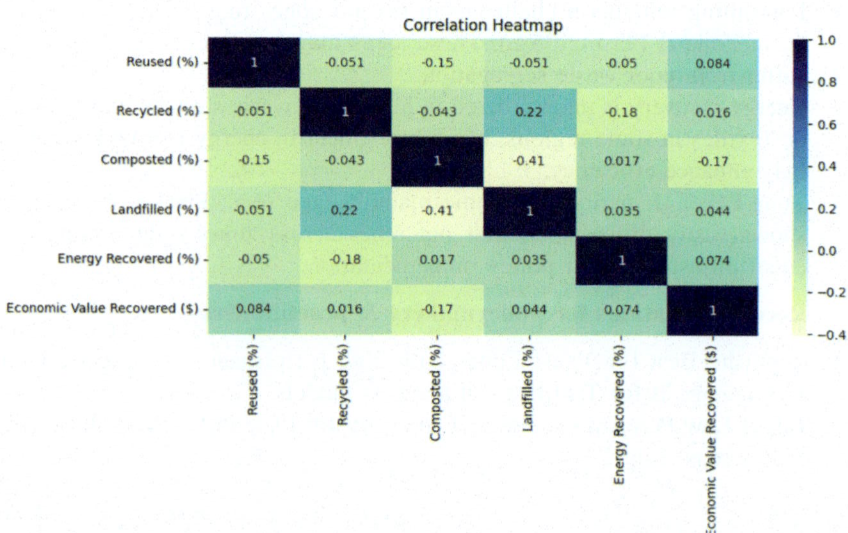

FIGURE 2.19 Heatmap of the circular economy dataset.

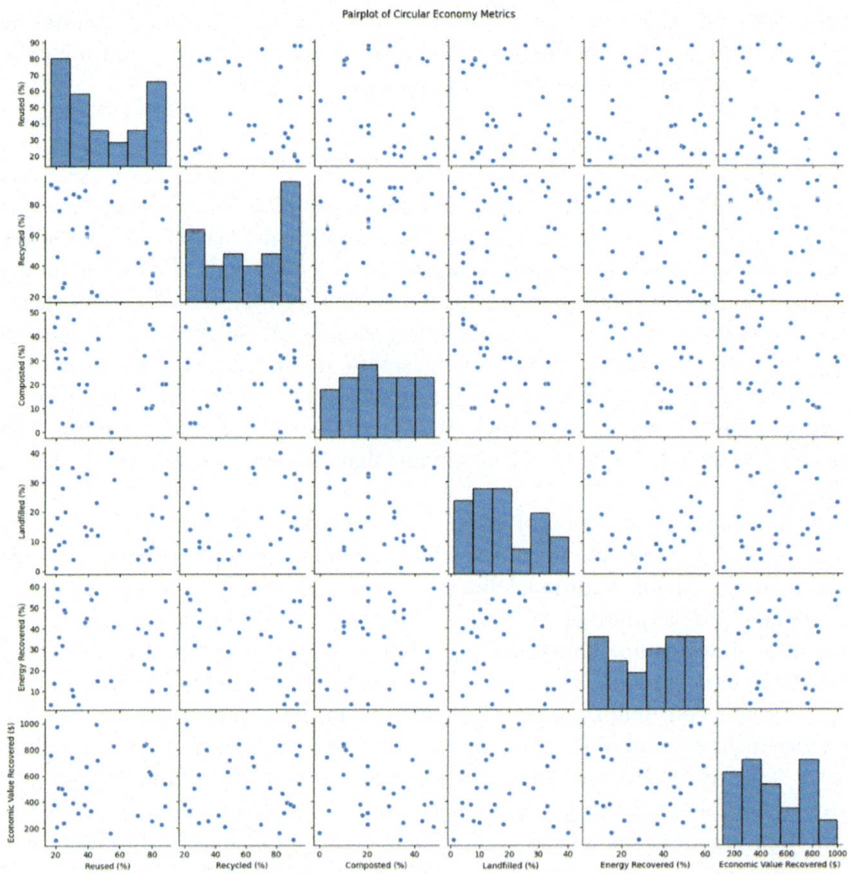

FIGURE 2.20 Pair plot of circular economy metrics.

certain zones may be heavily reliant on specific practices. The economic value histogram is right-skewed, indicating that while many zones recover moderate economic value, only a few achieve very high returns. Scatter plots show a weak inverse trend between composting and landfilling, which visually confirms the heatmap's correlation. Most other relationships appear random and lack strong patterns, suggesting that the circular economy performance across zones is highly variable and likely influenced by local infrastructure, policies, and material-specific practices. Overall, these visualizations indicate that no single metric guarantees high circularity or economic returns. Instead, a multifaceted strategy combining reuse, recycling, composting, and energy recovery—tailored to each zone and material type—is essential. The scattered nature of the data points underscores the need for targeted, data-driven interventions to improve performance. While composting is clearly effective in reducing landfill dependency, optimizing combinations of circular actions and addressing regional disparities will be key to unlocking greater environmental and economic value in circular waste management systems.

2.12 Conclusion

In conclusion, the chapter "*Waste Management Foundations with Circular Economy Practices*" offers a detailed and multidimensional exploration of the interrelationship between modern waste management and circular economy (CE) strategies. It establishes a firm foundation by first defining and categorizing various types of waste, including municipal solid waste, industrial waste, hazardous waste, and the increasingly significant waste derived from renewable energy technologies such as photovoltaic panels and wind turbine blades. This categorization lays the groundwork for understanding the magnitude and diversity of waste streams that challenge sustainability across the globe.

The historical evolution of waste management from rudimentary disposal methods to complex systems of recycling and energy recovery underscores the inadequacies of the traditional linear economy model. The waste hierarchy, introduced as a guiding principle, reflects a maturing understanding of resource stewardship. This hierarchy advocates for prevention and reuse over mere disposal, aligning closely with the tenets of the circular economy.

A major contribution of this chapter lies in the comparative analysis between linear and circular economic models. While the linear economy thrives on a "take-make-dispose" model, the circular economy promotes a regenerative system aimed at minimizing waste and maximizing resource efficiency. Through practical frameworks and case studies, the chapter shows how transitioning to CE models can significantly improve waste handling processes, reduce environmental impacts, and create sustainable economic opportunities.

Innovatively, the chapter introduces the concept of a Metaverse-Based Circular Economy Framework, demonstrating how virtual simulations, digital twins, and immersive visualization can support efficient urban waste management planning. This cutting-edge perspective aligns technological innovation with sustainability, offering novel pathways for system design, stakeholder education, and participatory governance.

Furthermore, the chapter makes a strong case for the alignment of waste management and circular economy with broader sustainability goals, especially carbon neutrality. By reducing reliance on virgin materials and minimizing emissions from waste disposal, CE practices serve as essential tools in climate action strategies. The data analysis of both traditional urban waste management and circular economy applications validates the ecological and economic gains of this paradigm shift. Ultimately, this chapter envisions a future where waste is no longer a burden but a valuable resource. It calls upon governments, industries, academia, and communities to embrace circular economy principles and invest in data-driven, technology-enhanced frameworks. The integration of CE practices into policy and urban planning is not only feasible but imperative for a sustainable and resilient future. With mounting environmental challenges and resource constraints, the transformation of waste management through circular economy is not just a desirable option it is a necessary evolution.

2.13 Case study

The case of Amsterdam, Netherlands

Title: Transforming Urban Waste: Amsterdam's Journey Toward Circular Waste Management.

Amsterdam, the capital city of the Netherlands, has been recognized globally for its progressive environmental policies. In the early 2000s, Amsterdam faced increasing urban waste volumes due to rapid population growth, increased consumption, and limitations in landfill space. Traditional linear waste management approaches "take, make, dispose" were proving unsustainable both economically and environmentally.

In response, the city set an ambitious goal: To become a fully circular city by 2050, with waste reduction and resource reuse as key pillars.

Foundational framework of waste management

1. **Waste classification and data analysis:**

Amsterdam's waste is categorized into organic waste, paper, plastic, metal, textiles, construction debris, and hazardous waste. A centralized digital platform was developed to collect and analyze real-time data on waste generation patterns, collection efficiency, and recovery rates.

2. **Hierarchical waste management approach:**

The city applied the waste management hierarchy:

- Prevention
- Reuse
- Recycling
- Energy Recovery
- Landfill Disposal

Through this structure, Amsterdam embedded the foundations of modern waste management: sorting at the source, regulated collection, community awareness, public-private partnerships, and advanced treatment facilities.

Circular economy integration

1. **Reuse and repair centers:**

The city established "Repair Cafés" and Reuse Hubs where citizens can repair or donate used items such as furniture, electronics, and clothing. These centers became community-driven initiatives promoting sustainability and job creation.

2. **Circular construction practices:**

   Amsterdam introduced circular procurement rules for public buildings, encouraging contractors to use recycled concrete, modular design principles, and biodegradable insulation. One notable project is Buiksloterham, a neighborhood developed entirely with circular principles using reclaimed materials, on-site composting, and rainwater harvesting.

3. **Organic waste and bio-energy:**

   Organic waste collected from homes, restaurants, and green spaces is converted into biogas and compost via anaerobic digestion. The AEB Amsterdam Energy Plant plays a key role in energy recovery from non-recyclable waste, converting residuals into district heating and electricity.

4. **Digitalization and circular platforms:**

   Amsterdam leverages digital twin technologies and IoT-based waste bins that signal when they are full, optimizing collection routes and reducing fuel emissions. Furthermore, digital material passports track resources in buildings, enabling future reuse and recycling.

Outcomes and impact

- **Waste reduction:** Household waste generation reduced by over 25% between 2010 and 2020.
- **Increased recycling rates:** Recycling rate rose from 35% in 2010 to 65% in 2023.
- **Economic value creation:** Creation of over 3,000 green jobs in repair, recycling, and circular design sectors.
- **Citizen engagement:** Over 200 community projects initiated under the "Amsterdam Circular 2025" initiative.

Challenges faced

- **Behavioral resistance:** Some citizens were initially resistant to separating waste.
- **Cost barriers:** High upfront investments in circular infrastructure and technologies.
- **Policy harmonization:** Aligning local, national, and EU-level waste policies required extensive coordination.

Amsterdam's waste management transformation exemplifies how a city can move beyond traditional waste handling to adopt a circular economy model, fostering resource efficiency, community involvement, and environmental stewardship. By institutionalizing waste management foundations classification, hierarchy, and data and coupling them with circular strategies, the city sets a benchmark for sustainable urban development.

2.14 Exercise

1. Discuss the major types of waste generated in urban and rural areas. How do they differ in terms of composition and management?
2. Explain the concept of Circular Economy (CE). How does it contrast with the Linear Economy model?
3. Analyze the types and characteristics of waste produced by renewable energy technologies (solar panels, wind turbines, etc.) and suggest suitable management solutions.
4. Trace the historical development of waste management practices. Highlight key global milestones.
5. Describe the waste management hierarchy. Why is it considered a guiding principle in sustainable waste governance?
6. How does data analysis help in urban waste management decision-making? Present a case study or hypothetical model.
7. Evaluate the role of circular economy in achieving carbon neutrality goals through effective waste management.
8. Design a framework for integrating circular economy principles into municipal solid waste (MSW) management.
9. How can the metaverse contribute to waste management practices? Propose a conceptual model combining virtual platforms and CE.

10. What are the challenges faced in implementing circular economy practices in developing countries' waste management systems?

11. Compare and contrast the environmental and economic implications of linear and circular waste systems.

12. Critically examine the limitations of current waste management policies in addressing the waste generated from renewable energy infrastructure.

13. Discuss how stakeholder involvement and public participation enhance the effectiveness of circular waste management systems.

14. Propose a data analysis model that can be used to measure the effectiveness of circular economy initiatives in urban waste management.

15. Explain how circular economy frameworks align with the Sustainable Development Goals (SDGs), especially in the context of waste management.

2.15 Objective type question

1. Which of the following is NOT a principle of the Circular Economy?
   - **a.** Reduce
   - **b.** Reuse
   - **c.** Incinerate
   - **d.** Recycle
2. Fill in the blank: The topmost priority in the waste management hierarchy is__________.
3. True or False: In a circular economy, waste is considered a resource.
4. Which renewable energy technology is known for generating significant e-waste at the end of its lifecycle?
   - **a.** Wind
   - **b.** Hydropower
   - **c.** Solar PV
   - **d.** Geothermal
5. Which of the following is a characteristic of linear economy?
   - **a.** Resource efficiency
   - **b.** Take-Make-Dispose model
   - **c.** Closed-loop system
   - **d.** Sustainable sourcing
6. Fill in the blank: __________ is a visualization platform proposed for simulating circular economy frameworks in the digital realm.
7. True or False: Waste-to-Energy is considered more sustainable than recycling in the waste management hierarchy.
8. Data analysis in urban waste management can help in:
   - **a.** Predicting waste volumes
   - **b.** Optimizing collection routes
   - **c.** Identifying illegal dumping
   - **d.** All of the above
9. Which of the following is NOT a type of waste?
   - **a.** Organic
   - **b.** Inorganic
   - **c.** Semantic
   - **d.** Hazardous
10. Circular economy frameworks promote__________.
    - **a.** Short-term economic gains
    - **b.** Single-use products
    - **c.** Resource regeneration and lifecycle extension
    - **d.** High consumption patterns
11. Fill in the blank: The concept of a circular economy was popularized by the__________ Foundation.
12. True or False: The linear economy encourages maximum utilization and reuse of products.

13. What is the final stage in the traditional waste management hierarchy?
  a. Recycling
  b. Reuse
  c. Landfilling
  d. Recovery

14. Which of the following waste management methods contributes least to carbon emissions?
  a. Landfilling
  b. Incineration
  c. Recycling
  d. Open dumping

15. Fill in the blank: One of the major goals of Circular Economy is to move toward__________ neutrality.

Chapter 3

Recycling technologies and innovations

3.1 Introduction

The rapid industrialization and urbanization of the modern world have significantly increased the demand for raw materials, leading to the depletion of natural resources and the accumulation of waste at an unprecedented scale. Traditionally, most economies have followed a linear model of consumption extract, produce, consume, and dispose resulting in a "take-make-waste" system that is inherently unsustainable. In contrast, the circular economy emerges as a transformative model that aims to decouple economic growth from resource consumption by designing out waste and keeping products and materials in use for as long as possible. Recycling is a cornerstone of this model, playing a critical role in closing the material loop and enabling the transition from linear to circular systems.

At its core, recycling refers to the process of collecting, processing, and reusing materials that would otherwise be discarded as waste. However, within the context of the circular economy, recycling is more than just a waste management strategy with value retention process. By recovering valuable materials from end-of-life products and re-introducing them into production cycles, recycling helps conserve natural resources, reduce greenhouse gas emissions, and lower the environmental footprint of industrial activities. The objective is not only to minimize the quantity of waste that ends up in landfills or incinerators but also to design products and systems that facilitate effective material recovery and re-integration.

The circular economy classifies recycling into various forms based on the level of processing and the integrity of the material retained. These include mechanical recycling, where materials are physically reprocessed without altering their chemical structure (e.g., shredding plastic bottles into flakes); chemical recycling, where materials are broken down into their chemical constituents to create new, virgin-equivalent products; and biological recycling, such as composting, where organic waste is returned to the environment through natural degradation processes. Each type has distinct applications,

The Circular Economy and their Management. https://doi.org/10.1016/B978-0-443-44617-7.00011-5

limitations, and environmental implications, and choosing the right method often depends on the material type, contamination level, and end-use requirements.

The integration of recycling into the circular economy also requires systemic changes across the product lifecycle, from design and manufacturing to consumption and end-of-life treatment. For recycling to be effective, products must be designed for recyclability using mono-materials, avoiding toxic additives, and enabling easy disassembly. Consumers and businesses must actively participate in source segregation, while governments and municipalities must invest in infrastructure, technologies, and policies that support collection, sorting, and processing. The success of recycling efforts thus depends on a multi-stakeholder collaboration involving manufacturers, policymakers, waste management companies, researchers, and the public.

Innovations in recycling technologies are reshaping how materials are recovered and reused. Advances in artificial intelligence, robotics, biotechnology, and digital platforms are enabling more efficient sorting, cleaner recovery, and smarter material flows. Moreover, policy instruments such as Extended Producer Responsibility (EPR), material labeling standards, and recycling incentives are being introduced worldwide to create accountability and drive systemic change. As such, recycling is no longer a reactive measure to deal with waste but a strategic pillar of resource management and environmental stewardship.

However, recycling alone is not a panacea. While it mitigates the environmental burden of waste, it must be integrated with other circular strategies such as refuse, reduce, reuse, refurbish, and remanufacture. Together, these strategies create a hierarchy of circular actions that prioritize material value retention and resource efficiency. In this broader framework, recycling represents the final loop before materials are reprocessed and reintroduced, making it a critical, though not solitary, component of the circular economy.

In conclusion, recycling in the circular economy is about creating closed-loop systems where materials continually flow through value chains with minimal loss. It supports sustainability by conserving resources, reducing environmental impacts, and enabling circular business models. As economies around the world strive for green transitions and climate resilience, strengthening and innovating recycling systems will be indispensable for building a truly circular and sustainable future.

Recycling plays a vital role in enhancing resource efficiency by maximizing the utility derived from raw materials while minimizing waste, energy use, and environmental impact. In essence, resource efficiency aims to achieve more with less delivering greater value with fewer inputs. Recycling directly supports this goal by recovering valuable materials from end-of-life products and reintroducing them into the production cycle, thereby reducing the need for virgin resource extraction. For example, recycling metals like aluminum or copper consumes significantly less energy compared to mining and refining them from ores. Similarly, recycled plastics and paper reduce dependence on fossil fuels and forestry resources, respectively. By diverting materials from landfills and incinerators, recycling also helps reduce the environmental burden associated with waste disposal. Moreover, efficient recycling systems decrease the pressure on ecosystems, conserve biodiversity, and contribute to lower carbon emissions. In the broader context of the circular economy, recycling ensures that resources remain in use for as long as possible, creating closed-loop systems where the material lifecycle is extended, and resource input is minimized. Thus, recycling is not just a method of waste management, it is a strategic approach to achieving sustainability, reducing costs, and improving material productivity across industries.

3.2 Historical evolution of recycling technologies

The evolution of recycling technologies is deeply intertwined with the historical development of human civilization, industrial growth, environmental awareness, and technological advancement. From ancient practices of reusing materials out of necessity to today's high-tech, AI-powered recycling systems, the journey of recycling technologies reflects changing societal values, economic pressures, and ecological concerns. A critical examination of this evolution reveals both progress and persistent challenges.

1. **Ancient and pre-industrial era–recycling by necessity:** In ancient civilizations, recycling was largely a survival strategy rather than an environmental concern. Metals such as bronze and iron were melted and reused due to their high value and limited availability. Archeological findings suggest that Roman and Greek societies practiced metal and glass recycling. Similarly, in ancient China and India, paper and fabric scraps were repurposed. However, these practices were localized and informal, driven by scarcity and manual labor rather than systemic environmental or economic policies.
   Critical Insight: While early recycling was resource-efficient, it lacked scale and formalization. It was reactive and need-based, without technological innovation or long-term sustainability planning.
2. **Industrial Revolution - rise in waste, lag in recycling:** The Industrial Revolution in the 18th and 19th centuries introduced mass production, resulting in increased resource extraction and waste generation. Although scrap metal and textile reuse continued, the focus shifted toward linear production models. Recycling during this era was mostly driven by economic motives during periods of resource shortages, such as during wars rather than environmental concerns.
   Critical Insight: Despite technological progress in manufacturing, recycling technologies stagnated. The growth of industries outpaced waste management systems, leading to the proliferation of disposable culture

and unsustainable consumption patterns. Fig. 3.1 illustrate historical evolution of recycling technologies.

3. **Early 20th century – institutionalized recycling during wars:** World War I and II marked a temporary resurgence in recycling due to material

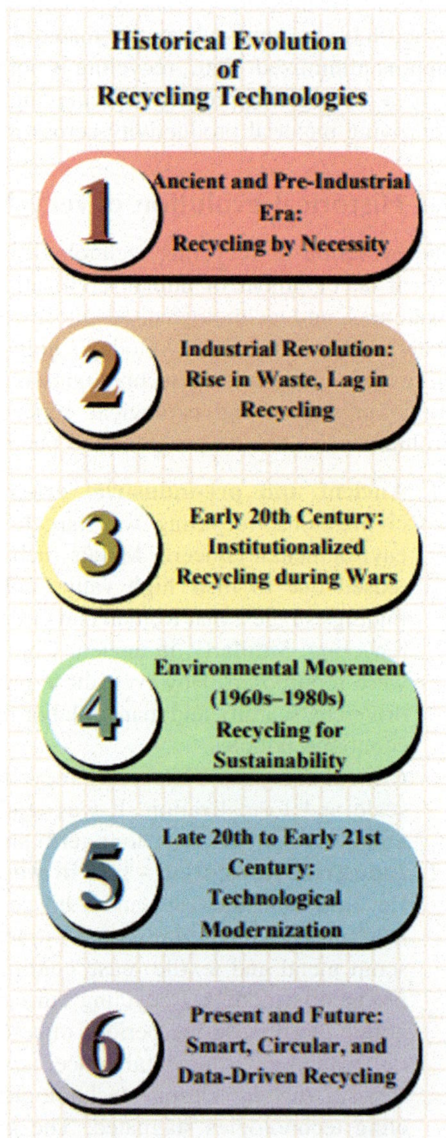

FIGURE 3.1 Historical evolution of recycling technologies.

shortages. Governments encouraged citizens to collect rubber, metal, and paper. Systems were established for centralized collection, sorting, and reprocessing. However, these efforts faded in post-war years as economic recovery and consumerism took precedence.

Critical Insight: Wartime recycling showed that coordinated, large-scale recycling was possible, but also highlighted its fragility dependent on external pressures rather than embedded systems or values.

4. **Environmental Movement (1960s–1980s) – recycling for sustainability:** The rise of the environmental movement in the 1960s and 70s marked a paradigm shift. Landmark publications like Silent Spring and events like Earth Day (1970) raised awareness of pollution, resource depletion, and the need for sustainable practices. This period saw the introduction of curbside recycling programs, material recovery facilities (MRFs), and the widespread adoption of the chasing arrows symbol for recyclable materials. Technologies began to emerge for sorting, compacting, and processing paper, glass, and plastics.

   Critical Insight: Recycling transitioned from being economically driven to environmentally motivated. Yet, it remained limited by contamination, inefficiencies in collection systems, and lack of incentives for manufacturers to design recyclable products.

5. **Late 20th to Early 21st century – technological modernization:** With advancements in automation, robotics, and sensor-based sorting, recycling technologies became more efficient. Materials like e-waste, composite plastics, and multilayer packaging began to be targeted through mechanical and thermal processes. The emergence of chemical recycling (e.g., depolymerization of plastics) aimed to address limitations of mechanical methods. Biological recycling such as anaerobic digestion also gained traction for organic waste.

   Critical Insight: Despite innovations, recycling systems often lag behind the complexity of modern products. Technologies remain costly, and recycling rates vary widely across regions due to inconsistent regulations, lack of infrastructure, and economic barriers.

6. **Present and future – smart, circular, and data-driven recycling:** Today, recycling technologies are being integrated into smart waste management systems powered by AI, IoT, and blockchain. Robotic sorters use computer vision to separate materials with high precision. Digital platforms track materials through their lifecycle, supporting circularity and extended producer responsibility (EPR). Innovations in green chemistry and biodegradable materials are also reshaping the recycling landscape.

   Critical Insight: While we are closer than ever to realizing efficient, circular recycling systems, critical issues remain: limited scalability, high capital costs, consumer behavior, and the continued use of nonrecyclable materials in product design. Moreover, global inequalities in recycling infrastructure persist, especially between developed and developing nations.

The historical evolution of recycling technologies reflects a transition from manual, necessity-based reuse to automated, data-driven material recovery. Although significant strides have been made, especially in recent decades, the recycling ecosystem remains constrained by economic, technological, and behavioral challenges. The future of recycling depends not just on innovation but also on policy alignment, producer accountability, and consumer participation within a fully realized circular economy.

3.3 Classification of recycling technologies

3.3.1 Mechanical recycling

Mechanical recycling is one of the most widely used methods for recovering materials particularly plastics, metals, and paper from postconsumer or industrial waste. Unlike chemical recycling, which alters the molecular structure of materials, mechanical recycling retains the original polymer or material composition, relying instead on physical processes to recover and reprocess waste into usable raw materials. This process is a crucial component of the circular economy, aiming to reduce the need for virgin materials, lower energy consumption, and divert waste from landfills or incinerators. However, despite its widespread application and cost-effectiveness, mechanical recycling has several technical, environmental, and economic limitations, which merit critical examination. Mechanical recycling involves physical processes such as collection, sorting, cleaning, shredding, and re-melting of waste materials to form new products. Unlike chemical recycling, it does not break down polymers into monomers or modify the chemical structure. Instead, it preserves the polymer structure and reprocesses it into raw forms like pellets or flakes, which can be used in manufacturing.

This method is most commonly used for thermoplastics plastics that can be melted and reshaped multiple times without significant degradation. Common mechanically recycled materials include polyethylene terephthalate (PET), high-density polyethylene (HDPE), polypropylene (PP), and low-density polyethylene (LDPE).

3.3.1.1 Stages of mechanical recycling

Collection and transportation: The first step in the mechanical recycling process is the collection of recyclable materials. These materials originate from a wide array of sources, including households, businesses, industries, and institutions. The efficiency and success of this stage depend heavily on well-structured collection systems such as:

- Curbside pickups provided by municipal waste services.
- Drop-off centers where individuals voluntarily bring their recyclables.
- Deposit-return schemes that incentivize consumers to return used packaging like bottles.

Once collected, the recyclables are transported to Material Recovery Facilities (MRFs). Effective transportation logistics ensure that materials reach the facility in a timely manner without contamination or excessive degradation. Fig. 3.2 illustrate different stages of mechanical recycling.

Sorting: After arrival at the recycling facility, the mixed waste stream undergoes sorting to separate different materials based on type, weight, and composition. Sorting is critical as it ensures that only pure and uncontaminated materials proceed to the next stages. Both manual and automated techniques are used in combination for maximum precision.

Key technologies involved in automated sorting include:

- Air classifiers, which separate items based on weight lighter materials like paper or plastic are lifted by air currents, while heavier items fall.
- Magnetic separators, used to remove ferrous (iron-containing) metals from the waste stream.
- Eddy current separators, which create magnetic fields to repel non-ferrous metals like aluminum and copper.
- Near-infrared (NIR) spectroscopy, which identifies different types of plastics based on their unique light absorption patterns.

Proper sorting prevents cross-contamination and ensures the production of high-quality recycled materials that meet industry standards.

Cleaning and drying: After sorting, materials especially plastics undergo cleaning and drying to eliminate impurities. This step is essential for improving the esthetic, chemical, and physical qualities of the recycled output. Contaminants such as food residue, dirt, adhesives, inks, and labels are removed using a combination of:

- Mechanical washing with water and detergents,
- High-speed friction washers that scrub materials,
- Flotation tanks to separate materials based on density.

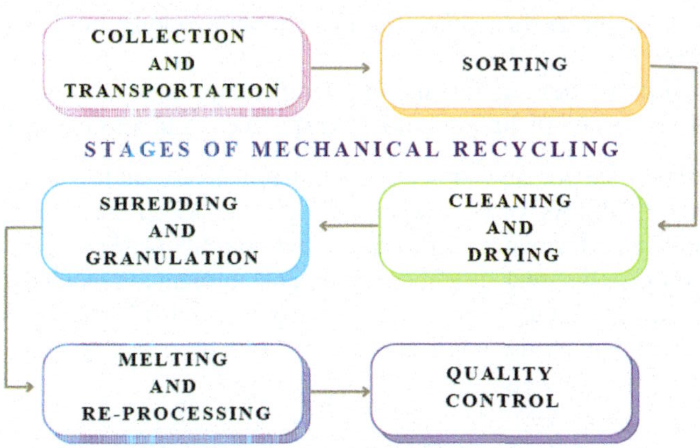

FIGURE 3.2 Stages of mechanical recycling.

After washing, the materials are passed through centrifuges or drying tunnels to remove moisture, which is crucial to avoid degradation or malfunction during reprocessing. Clean and dry input ensures better compatibility and performance in the subsequent processing steps.

Shredding and granulation: Once cleaned, materials are fed into shredders or granulators to be mechanically reduced into smaller, uniform particles such as flakes or granules. This step serves multiple purposes:

- It increases the surface area of the material, making it easier to melt uniformly.
- It improves the handling and storage of materials.
- It facilitates consistent feeding into extruders or molding machines.

Industrial shredders, rotary cutters, and granulators are designed to handle various materials and adjust output sizes to suit specific downstream applications.

Melting and reprocessing: Shredded materials are then subjected to melting and reprocessing. This typically involves:

- Feeding the granules into an extruder, where they are heated to their melting point.
- The molten material is then forced through a die to form long strands, which are cooled and cut into uniform pellets.
- These pellets are the final raw form used in manufacturing processes such as injection molding, blow molding, or film extrusion to create new products like containers, textiles, or packaging films.

Melting must be carefully controlled to prevent thermal degradation, particularly in sensitive polymers like PET or LDPE.

Quality control: The final stage in mechanical recycling is quality control. This step ensures that the recycled material meets technical and safety specifications. Quality checks are conducted to assess:

- Mechanical properties like tensile strength, elongation, and impact resistance.
- Thermal properties such as melting temperature and stability.
- Visual properties including color, transparency, and surface texture.

To compensate for any loss in material properties, additives such as stabilizers, UV inhibitors, or compatibilizers may be added during reprocessing. These enhance performance, appearance, and usability of the recycled product. In advanced systems, real-time monitoring and automated feedback loops help maintain consistent quality, allowing recycled materials to compete with virgin alternatives in demanding applications.

3.3.1.2 *Types of mechanical recycling technologies*

Mechanical recycling technology is a widely used process for managing plastic waste by physically transforming discarded materials into reusable raw forms without altering their chemical structure. The process typically involves collection, sorting, cleaning, shredding, melting, and re-molding plastics into pellets or new products. Mechanical recycling is cost-effective, energy-efficient, and environmentally friendly compared to chemical methods, making it suitable for high-volume recycling of materials such as PET, HDPE, and PP. However, the quality of recycled output may degrade over repeated cycles, highlighting the need for proper waste segregation and advanced sorting technologies. Fig. 3.3 illustrate types of mechanical recycling. Following are the description of different types of mechanical recycling technology.

3.3.1.2.1 Plastics recycling technologies

Plastic recycling technologies refer to the various methods and processes used to convert plastic waste into reusable materials or products, reducing the need for virgin plastic production and minimizing environmental harm. These technologies can be broadly categorized into mechanical, chemical, and other emerging methods, depending on how they process and transform plastic waste. Plastics are the most diverse category in mechanical recycling,

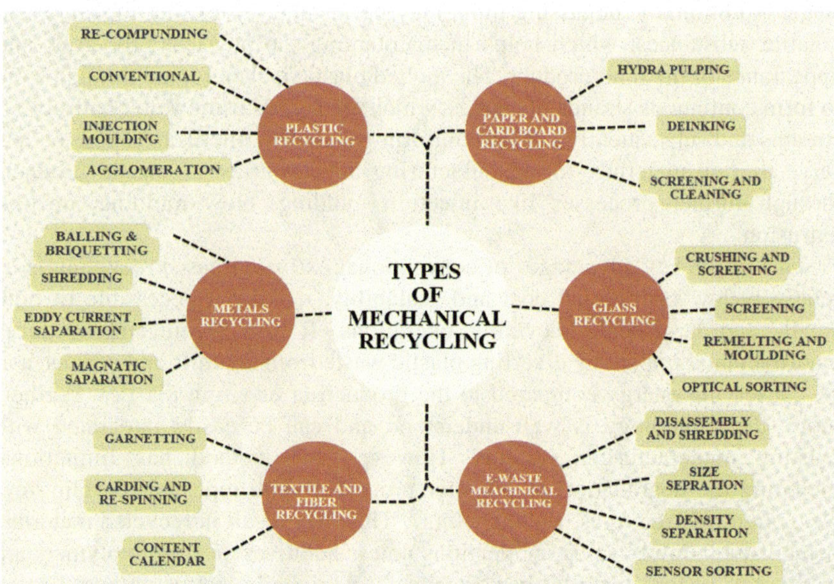

FIGURE 3.3 Types of mechanical recycling.

requiring different technologies depending on the polymer type, contamination level, and end-use application.

a. Conventional extrusion-based recycling

Conventional extrusion-based recycling is one of the most established and widely used mechanical recycling methods for processing thermoplastic waste materials, especially polyethylene terephthalate (PET), high-density polyethylene (HDPE), polypropylene (PP), and low-density polyethylene (LDPE). This method is based on the principle of melting plastic waste and reshaping it into new products without altering the fundamental chemical structure of the polymers. The process begins with the collection and segregation of plastic waste, followed by thorough cleaning to remove contaminants such as food residues, labels, adhesives, and other non-plastic materials. This is crucial because the presence of contaminants can degrade the quality of the recycled material. After cleaning, the plastic is dried and then fed into shredders or granulators, where it is reduced into small, uniform flakes or granules to facilitate even melting during the extrusion process.

These plastic granules are then introduced into an extruder typically a single-screw or twin-screw machine where they are subjected to controlled heat and mechanical shear. As the plastic moves through the heated barrel of the extruder, it melts and is homogenized into a continuous molten stream. During this phase, filtration systems like screen changers and melt filters are used to remove any residual impurities that may have survived the washing stage. Additionally, degassing units may be installed to extract moisture and volatile substances, which helps in maintaining the physical properties and appearance of the final product. The melted plastic is then forced through a die to form continuous strands or shapes, which are cooled using water baths or air streams and subsequently cut into uniform pellets by pelletizers. These pellets serve as raw materials for manufacturing a wide array of plastic products through further processes like injection molding, blow molding, or film extrusion.

One of the key advantages of conventional extrusion-based recycling is its relatively low operational cost and scalability, making it accessible to both small- and large-scale recycling operations. It also significantly reduces environmental impact by diverting plastic waste from landfills and oceans, and by conserving energy compared to the production of virgin plastics. Furthermore, the technology is well-understood and can be easily integrated with existing manufacturing systems. However, the method has limitations, including the degradation of polymer properties with repeated recycling cycles, often referred to as "downcycling." This may result in recycled materials being used in lower-value applications unless additives or virgin polymers are blended to restore material performance. Additionally, extrusion-based recycling is typically limited to clean, single-polymer streams, as mixed or multilayer plastics pose challenges in processing and quality control.

Despite these limitations, advancements in extruder design, filtration technologies, and the integration of additives have improved the efficiency and quality of recycled output. Closed-loop systems, where plastic products are recycled back into the same applications (like bottle-to-bottle recycling), are increasingly being adopted to ensure material purity and performance. In conclusion, conventional extrusion-based recycling continues to play a vital role in the global effort to manage plastic waste sustainably and support the transition to a circular economy.

b. Injection molding and blow molding

Injection molding and blow molding are two essential manufacturing processes that also play a significant role in the mechanical recycling of plastics. Once plastic waste is collected, sorted, cleaned, shredded, and converted into pellets through processes like extrusion, these pellets can be reused to manufacture new products using injection or blow molding techniques. Both methods are widely adopted in industries due to their ability to produce high-quality, uniform, and durable plastic items from recycled materials. Injection molding is a high-precision process used to manufacture complex and rigid plastic parts. In this method, recycled plastic pellets are fed into a hopper, where they are conveyed into a heated barrel by a rotating screw. The pellets melt as they move through the barrel and are then injected under high pressure into a mold cavity that defines the shape of the desired product. Once the plastic cools and solidifies, the mold opens, and the finished product is ejected. This process is widely used for making items such as electronic housings, automotive components, containers, and furniture parts from recycled plastics like PP, HDPE, and ABS. Injection molding offers excellent dimensional accuracy and allows for high-volume production, making it highly suitable for reusing recycled plastic pellets in consistent and scalable applications.

Blow molding, on the other hand, is primarily used for producing hollow plastic items like bottles, drums, and containers. The process begins similarly, with recycled plastic being melted and formed into a parison (a tube-like piece of molten plastic) or preform. In extrusion blow molding, the parison is clamped into a mold, and air is blown into it, inflating it against the inner walls of the mold to form the final shape. In injection stretch blow molding, a preform is first made through injection molding, then reheated and inflated into a mold cavity. Blow molding is especially useful for recycling PET and HDPE, which are commonly used in beverage bottles, cleaning product containers, and personal care packaging. It allows for lightweight, durable, and cost-effective production from recycled feedstock. Both injection and blow molding are compatible with recycled plastic resins, although the quality of the end product depends significantly on the purity and consistency of the recycled material. Challenges such as variation in melt flow index, presence of contaminants, and degradation of polymer chains can affect the mechanical properties of the molded products. To address this, manufacturers often blend recycled plastics with virgin polymers or add stabilizers and compatibilizers to enhance strength, flexibility, and durability. Despite these challenges, the integration of injection and blow molding in plastic recycling streams contributes greatly to a circular economy by enabling the transformation of post-consumer waste into functional and commercially valuable products. These technologies support material recovery, reduce dependency on virgin plastic production, and help lower the environmental footprint of plastic-based goods. As sustainability gains prominence, the refinement and expansion of these molding processes using recycled materials will be critical for achieving large-scale plastic reuse and waste minimization.

c. Agglomeration

Agglomeration is a mechanical recycling process used to convert soft, lightweight, and often hard-to-handle plastic waste such as films, foils, and fibers into dense, compact, and manageable forms called agglomerates. These agglomerates serve as an intermediate material, which can be further processed through extrusion, injection molding, or other manufacturing techniques. Agglomeration is particularly valuable for handling low bulk density plastics such as low-density polyethylene (LDPE), polypropylene (PP) films, polyethylene terephthalate (PET) fibers, and foamed polystyrene. The main purpose of agglomeration is to increase the bulk density of plastic materials that are otherwise too fluffy, voluminous, or slippery for efficient handling, transport, or feeding into conventional recycling machinery like extruders. Without agglomeration, such materials may clog equipment, cause inconsistent melting, or be economically unviable to process. This method simplifies material management and acts as a pre-treatment step before further recycling or reuse. Fig. 3.4 illustrate process of agglomeration. The agglomeration process involves several steps:

Feeding and pre-heating: Lightweight plastic waste is fed into an agglomerate, also known as a densifier. Inside the agglomerate, high-speed rotating blades or knives generate frictional heat by agitating the plastic.

Softening and fusing: The plastic begins to soften or partially melt due to the heat generated by mechanical friction. However, it is not fully melted as in extrusion. The particles fuse together at their surfaces and form soft lumps or clusters.

Cooling and cutting: Once the material reaches the desired consistency, cold water or air is introduced into the chamber to rapidly cool the plastic, stopping the fusing process. The cooled mass is then broken into smaller,

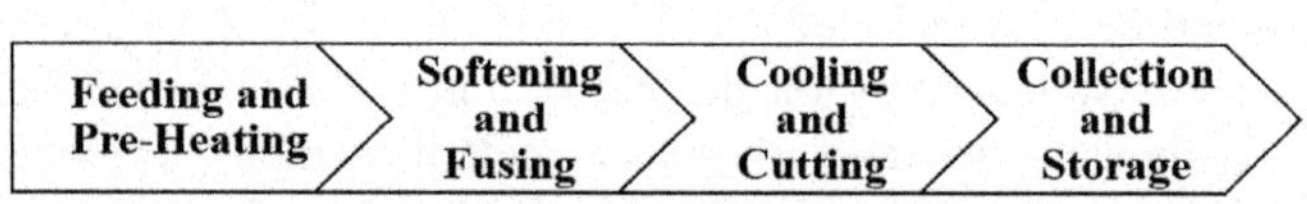

FIGURE 3.4 Process of agglomeration.

granular-sized agglomerates using internal cutters or discharged and shredded externally.

Collection and storage: The final agglomerates are collected, dried if needed, and stored or conveyed for further processing. These agglomerates have higher density, improved flow properties, and are much easier to manage in downstream recycling operations.

Agglomeration is commonly used in:

- Plastic film recycling, including packaging waste, shopping bags, and agricultural films.
- Fiber and textile recycling, particularly polyester and polypropylene fibers.
- Foam recycling, such as expanded polystyrene (EPS) waste.
- In-house industrial scrap processing, especially where film trimming or extrusion waste is generated.

The agglomerated material can be:

- Directly fed into extruders to make recycled pellets.
- Used in injection or blow molding after being compounded.
- Mixed with virgin materials to maintain product quality.

Advantages of agglomeration:

- Increases bulk density, making storage and transport more efficient.
- Improves process ability of soft plastic materials.
- Reduces the risk of material loss or contamination during handling.
- Energy-efficient compared to full-scale extrusion processes.
- Cost-effective and simple in design, making it accessible for small recycling setups.

Limitations:

- Not suitable for rigid plastics or thermosets.
- Material properties may degrade slightly due to heat and partial melting.
- Agglomerates often require further processing for final product manufacturing.
- Presence of contaminants or multilayer films can reduce effectiveness.

Agglomeration is a crucial support technology in plastic recycling, especially for handling difficult-to-process soft and lightweight plastics. It enables efficient material handling, improves feeding consistency into extruders, and prepares waste for higher-value applications. While it does not produce final consumer products, agglomeration plays an essential intermediate role in the circular plastics economy, ensuring more types of plastic waste can be recycled effectively and sustainably.

d. Re-compounding

Re-compounding is an advanced mechanical recycling process that goes beyond basic extrusion by enhancing the quality and performance of recycled plastics through the addition of additives, fillers, stabilizers, or virgin polymers. It transforms post-consumer or post-industrial recycled plastic into high-grade compounds with properties tailored to specific industrial or commercial applications. This method plays a vital role in increasing the usability of recycled plastics in demanding sectors such as automotive, electronics, packaging, construction, and consumer goods.

The core idea of re-compounding is to restore, improve, or modify the physical, mechanical, or thermal properties of recycled polymers, which often degrade due to repeated processing, contamination, or exposure to heat, UV light, and oxygen. Without re-compounding, many recycled plastics suffer from reduced impact resistance, color consistency, flexibility, or durability making them unsuitable for high-performance applications. Re-compounding offers a solution by re-engineering recycled material into a competitive alternative to virgin plastics. Fig. 3.5 illustrate process of re-compounding. The re-compounding process involves several integrated steps, usually carried out in twin-screw extruders due to their superior mixing capabilities:

Feeding and dosing: Recycled plastic flakes or pellets are fed into the extruder along with precisely dosed additives. These may include:

- Impact modifiers to improve toughness,
- Color master batches for uniform coloration,
- UV stabilizers and antioxidants to enhance environmental resistance,
- Fillers like calcium carbonate or talc to improve stiffness or reduce cost,
- Compatibilizers to enhance blending of mixed or incompatible plastics.

Melting and mixing: Inside the extruder, the materials are heated and mechanically mixed. The twin-screw mechanism ensures homogeneous dispersion of all ingredients and thorough blending of recycled and virgin polymers if used.

Degassing: Volatile substances such as moisture, oils, or residual contaminants are removed through vacuum venting, improving the thermal and mechanical performance of the compound.

Filtration and extrusion: Melt filters screen out impurities. The molten, well-compounded plastic is then extruded through a die and cut into uniform pellets using a strand or underwater pelletizer.

Cooling and packaging: The pellets are cooled and dried, then packaged for shipment and use in various manufacturing applications.

FIGURE 3.5 Process of re-compounding.

Applications of re-compounded plastics

Re-compounded plastic pellets are used in:

- Automotive parts (dashboards, bumpers, trims),
- Consumer electronics (TV casings, printer parts),
- Rigid packaging (bottles, containers),
- Construction materials (pipes, panels),
- Household items (furniture, bins, garden tools).

Industries favor re-compounded materials because they offer consistent performance, meet regulatory standards, and reduce environmental impact without compromising product quality.

Advantages

- Improved properties: Strength, flexibility, UV resistance, thermal stability.
- Wider usability: Enables recycled plastics in high-performance applications.
- Cost-efficiency: Reduces reliance on virgin raw materials.
- Customization: Tailored to customer-specific requirements.
- Circular economy support: Keeps materials in use longer.

Challenges

- Requires technical expertise in formulation and processing.
- Sensitive to contamination and input variability.
- More capital-intensive than basic mechanical recycling.
- Quality depends heavily on input consistency and equipment capability.

Re-compounding is a sophisticated and essential technology in the modern plastic recycling ecosystem. By turning low-quality recycles into engineered compounds with desired characteristics, it bridges the gap between sustainability and performance. As industries strive to reduce their environmental footprint without compromising functionality, re-compounded plastics are emerging as a sustainable and competitive alternative to virgin materials, supporting the broader goals of a circular and resource-efficient economy.

3.3.1.2.2 Paper and cardboard recycling technologies

Paper recycling relies on mechanical pulping and cleaning technologies to produce new paper from wastepaper.

a. Hydra pulping

Hydrapulping is a fundamental mechanical process used in the recycling of paper and cardboard. It involves breaking down waste paper materials into individual fibers using water and mechanical agitation. The resulting pulp, known as recycled fiber slurry, is then cleaned, refined, and reprocessed into new paper products. Hydrapulping is the first and most critical stage in paper recycling, enabling the transformation of printed, used, or discarded paper into reusable raw material while preserving the cellulose fiber structure. Fig. 3.6 illustrate process of hydra pulping.

The process takes place in a machine called a hydrapulper, which resembles a large industrial blender or agitator. It consists of a large tank (pulper vat) filled with water and rotating blades or rotors at the base. Waste paper is fed into this tank, where it is subjected to continuous agitation and turbulence. The mechanical action of the rotor, in combination with water, disintegrates the paper into small fibrous components, separating ink, adhesives, coatings, and non-fibrous contaminants in the process. Depending on the type and grade of the waste paper—such as newspapers, office paper, corrugated boxes, or coated papers—different hydrapulping settings, rotor speeds, and water-to-paper ratios are used.

Hydrapulping can be conducted as a batch process or a continuous process, depending on the scale and type of recycling facility. In batch pulping, a set amount of waste paper is pulped, cleaned, and discharged before starting a new batch. In continuous systems, waste paper is continuously fed into the pulper, and pulp is steadily drawn off for downstream processing. Often, chemicals like pulping aids, defoamers, or dispersants are added to assist in breaking down tough or coated papers and to improve pulp consistency.

After hydrapulping, the slurry undergoes screening and cleaning to remove contaminants such as staples, plastic films, inks, and adhesives. Coarse screens, pressure screens, and centrifugal cleaners help eliminate unwanted particles, ensuring that only clean, separated fibers proceed to the next stages such as deinking, refining, or thickening. Hydrapulping plays a particularly critical role when recycling mixed office waste, corrugated boxes (OCC), or

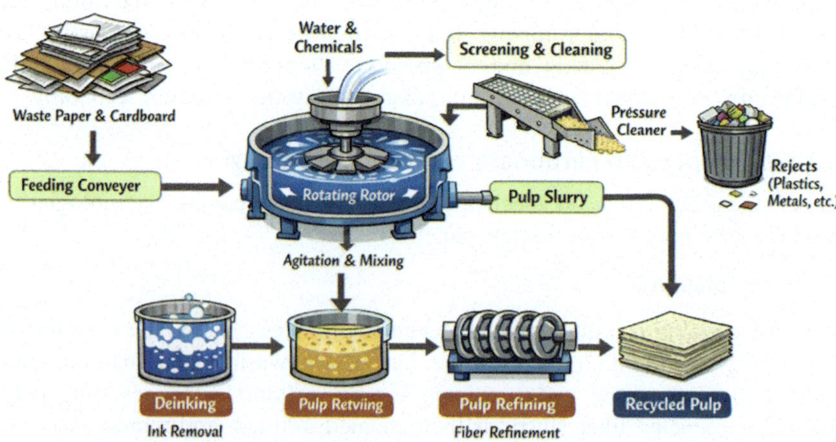

FIGURE 3.6 Process of hydra pulping.

printed materials, as it initiates the fiber recovery process while also liberating contaminants that can be filtered out. Hydrapulping is used in the recycling of:

- Old newspapers (ONP) and magazines
- Old corrugated containers (OCC)
- Mixed office waste (MOW)
- Printed or coated paperboard
- Paper mill broke (internal recycling)

The recycled pulp from hydrapulping is used to produce:

- Corrugated boards
- Tissue papers
- Newsprint
- Packaging papers
- Recycled writing and printing papers

Advantages

- Preserves fiber integrity, allowing multiple recycling cycles.
- Versatile: Works with a wide range of waste paper types.
- Enables bulk processing of large paper volumes efficiently.
- Supports sustainable paper production and reduces deforestation.
- Lowers energy consumption and carbon footprint compared to virgin pulp production.

Challenges

- Cannot effectively remove all inks or coatings—requires follow-up processes like deinking.
- Fiber quality may decline with repeated recycling cycles.
- Generates solid waste from contaminants and non-paper materials.
- Water usage and wastewater treatment are critical environmental concerns.

Hydrapulping is the cornerstone of paper recycling, enabling the disintegration of waste paper into reusable pulp while separating contaminants for further processing. It offers a practical and efficient way to recover valuable cellulose fibers, reduce landfill waste, and promote a circular economy in the paper industry. With ongoing innovations in pulper design and process optimization, hydrapulping continues to support sustainable paper manufacturing worldwide.

b. Deinking technology

Deinking technology is a critical step in the paper recycling process that focuses on removing printing inks, adhesives, and other contaminants from recycled paper pulp. This process is essential for producing high-quality recycled paper, particularly for applications that require brightness, cleanliness, and esthetic appeal such as office paper, tissue paper, newsprint, and packaging. Deinking enhances the visual and mechanical quality of the recycled pulp, allowing it to compete with virgin fiber in many end-use applications. Fig. 3.7 illustrate process of deinking technology.

The deinking process typically follows hydrapulping, where waste paper is broken down into a fiber slurry. The hydrapulped material often contains residual ink particles, glue from labels, coating materials, fillers, and other contaminants. Deinking technologies are designed to liberate, detach, and remove these unwanted substances from the pulp to ensure a clean and usable final product. Two of the most widely used methods in deinking are flotation deinking and washing deinking, and often, a combination of both is used in modern recycling systems. In flotation deinking, the pulp slurry is treated with surfactants or deinking chemicals (such as soaps, fatty acids, or alkalis) that help loosen ink particles from the fibers. Air is then injected into the slurry in a flotation cell, creating bubbles that attract and trap the detached ink and other hydrophobic particles. These ink-laden bubbles rise to the surface and form a froth, which is skimmed off and discarded. This method is especially effective for removing laser and offset printing inks, as well as toners from photocopied or digitally printed papers. Washing deinking, on the other hand, relies on diluting the pulp slurry and passing it through a series of screens, filters, and washers to physically wash away fine ink particles and small contaminants. Washing is more effective for removing water-soluble inks and small particles but can also lead to fiber loss if not carefully controlled. Modern systems often integrate both flotation and washing to maximize ink removal while preserving fiber yield and strength. Deinking also involves additional treatments, such as:

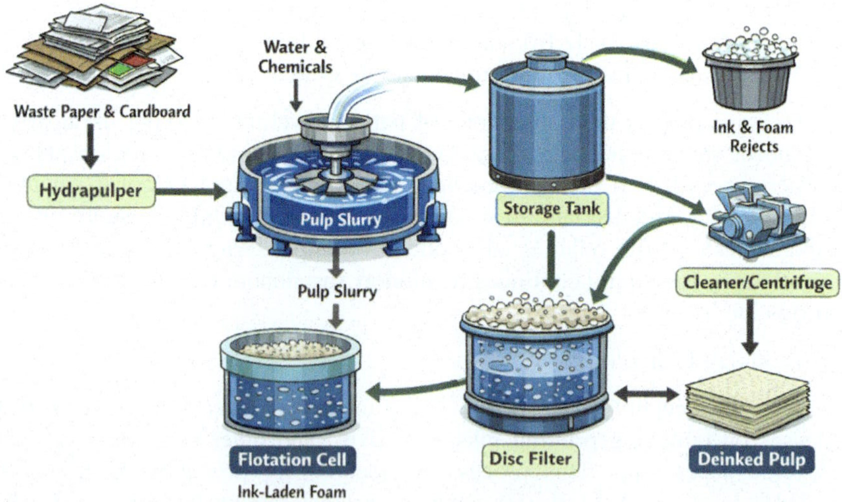

FIGURE 3.7 Process of deinking technology.

- Dispersion: A high-shear mixing process that breaks down remaining ink specks into smaller particles, improving their removal in flotation or washing.
- Bleaching: Chemicals like hydrogen peroxide may be added to improve the brightness and whiteness of the recycled pulp.
- Thickening: Excess water is removed after deinking to concentrate the cleaned pulp before it is sent to the paper-making process.

Deinked pulp is used to produce:

- Office and printing papers
- Tissue and toilet papers
- Newsprint
- Recycled cartons and packaging papers

Advantages

- Produces brighter, cleaner recycled paper.
- Removes a wide variety of inks, adhesives, and fine contaminants.
- Enhances the market value and usability of recycled fiber.
- Enables multiple cycles of paper fiber reuse, reducing deforestation.
- Compatible with modern printing technologies and ink formulations.

Challenges

- Chemical usage and water demand are relatively high.
- Not all ink types (especially UV-curable and water-based inks) are easily removed.
- Residual sludge from flotation froth and washing requires proper disposal.
- Requires precise control of pH, temperature, and chemical dosing for optimal results.

Deinking technology is a pivotal component of high-quality paper recycling, enabling the transformation of printed waste paper into clean, bright, and reusable pulp. By removing ink and other contaminants, it significantly improves the aesthetics and functionality of recycled paper products. As demand for recycled paper grows and environmental standards tighten, continued advancements in deinking chemicals, process efficiency, and closed-loop water systems will further strengthen its role in sustainable paper production and circular resource management.

c. Screening and cleaning systems

Screening and cleaning systems are essential processes in the paper recycling industry, serving the critical role of removing physical contaminants and impurities from recycled pulp. After waste paper is pulped and optionally deinked, it contains not only reusable fibers but also a variety of unwanted materials such as plastic films, metal staples, sand, glass, stickies (adhesives), ink particles, and undissolved paper fragments. Screening and cleaning systems ensure that these contaminants are removed before the pulp is used to make new paper, thereby improving the quality, strength, appearance, and machinability of the final product.

These systems operate on physical separation principles, such as size, shape, density, and centrifugal force. The process is usually divided into two main stages: screening, which removes larger and coarser impurities, and cleaning, which removes finer and denser contaminants. The efficiency of these systems directly influences the operating life of paper-making machinery, the consistency of the final paper product, and the overall economic viability of the recycling operation.

Screening systems

Screening involves passing the pulp slurry through various mechanical barriers like perforated plates, slotted screens, or rotating drums that allow only fibers and small particles to pass through, while retaining larger unwanted materials.

- Coarse screens are used initially to remove large debris like plastic wrappers, cardboard pieces, staples, and string.
- Fine screens are employed later to remove smaller contaminants such as ink particles, coating residues, and micro stickies.
- Pressure screens are commonly used, where pulp is pumped under pressure through slotted or holed cylinders. Good fiber passes through the slots while contaminants are retained and discharged as rejects.

Modern screening systems are highly efficient and include self-cleaning mechanisms, rotors to maintain flow, and automated controls to monitor performance and prevent clogging.

Cleaning systems

Cleaning typically refers to the removal of small, heavy contaminants that cannot be effectively removed by screening. These systems work on the principle of centrifugal separation.

- Centric leaners (hydro cyclones) are the most widely used cleaners. They spin the pulp slurry in conical chambers at high speeds, creating a vortex. Due to differences in density, heavier particles like sand, glass, metal, and some inks are thrown to the outer wall and collected at the bottom, while cleaner pulp flows out from the center.
- Reverse cleaners operate similarly but are designed to remove lighter contaminants such as air bubbles, wax, or plastics.

Cleaning systems are often arranged in multi-stage cascades, allowing maximum contaminant removal while minimizing fiber loss. Rejects from one stage can be reprocessed or refined further to recover usable fibers.

Applications and benefits

- Used in the recycling of various grades of waste paper: newspaper, office waste, corrugated boxes, and coated papers.
- Enhances product quality by eliminating contaminants.
- Protects machinery and equipment from damage due to hard impurities.
- Reduces maintenance downtime and improves production efficiency.

Challenges

- Needs careful calibration to balance contaminant removal and fiber retention.
- Generates waste rejects that must be treated or disposed of properly.
- High energy and water requirements in some configurations.

Screening and cleaning systems are indispensable in the paper recycling process, ensuring the removal of both coarse and fine impurities that can compromise paper quality and production efficiency. By separating contaminants based on physical properties like size and density, these systems help prepare clean, fiber-rich pulp suitable for high-quality paper manufacturing. Continuous advancements in screening technology, energy-efficient cleaning equipment, and process automation are further enhancing their role in supporting a sustainable and circular paper economy.

3.3.1.3 Metals recycling technologies

Metals recycling is a critical component of the global circular economy, conserving natural resources, reducing greenhouse gas emissions, and lowering energy consumption compared to primary metal production. Modern metal recycling relies on a combination of mechanical, magnetic, and separation technologies that allow for the efficient recovery and reuse of ferrous and non-ferrous metals from various waste streams, including municipal solid waste, industrial scrap, and end-of-life products. The key technologies involved in the metals recycling process include shredding, magnetic separation, eddy current separation, baling, and briquetting. Each of these methods plays a unique role in improving the efficiency, purity, and economic value of recycled metals. Fig. 3.8 illustrate process of metals recycling technologies.

a. Shredding

Shredding is the initial step in the metal recycling process and involves breaking down large, bulky metal items into smaller, manageable pieces. The

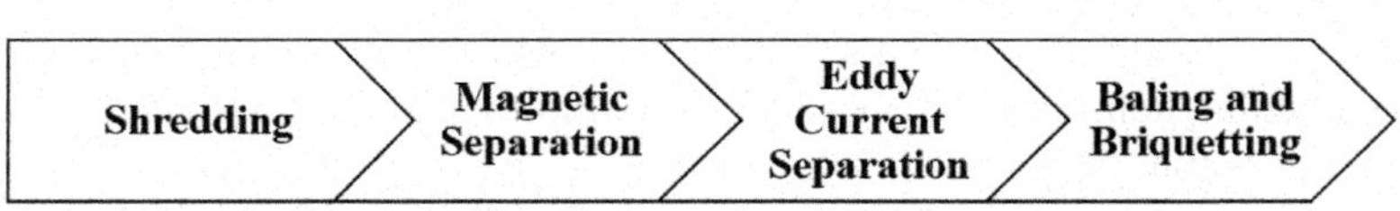

FIGURE 3.8 Process of metals recycling technologies.

purpose of shredding is to reduce the volume of metal waste, facilitate easier handling, and prepare materials for downstream separation processes. Shredders are powerful industrial machines equipped with rotating hammers, knives, or blades that crush, tear, or shear metal items such as automobiles, appliances, construction scrap, and aluminum cans. Modern shredders can process several tons of metal per hour and are capable of handling mixed material inputs.

There are two main types of shredders used in metal recycling:

- Hammer mill shredders: Utilize a rotor with swinging hammers that crush metal against anvils. They are effective for large, dense materials like car bodies and heavy scrap.
- Shear shredders: Use counter-rotating knives to cut materials. These are more suitable for lighter metals, wires, and smaller scrap.

Shredding increases the surface area of the metal, making it easier to separate different metal types and contaminants. It also allows for the recovery of valuable non-metallic materials such as plastics and rubber in further separation stages. Moreover, by creating uniform fragment sizes, shredding enhances the consistency of the recycling stream, which is essential for smelting and refining operations.

b. Magnetic separation

Magnetic separation is a technique used to separate ferrous metals (iron and steel) from mixed metal waste streams. Ferrous metals are attracted to magnets, enabling their extraction from non-magnetic materials with high precision and speed.

Magnetic separators come in various configurations:

- Overhead magnets: Suspended over conveyor belts, they lift ferrous metals from the material flow.
- Drum magnets: Cylindrical magnets that rotate and pick up ferrous materials as the waste passes by.
- Crossbelt magnets: Mounted across conveyor belts, they remove magnetic materials as waste moves beneath.

This method is widely used in automotive recycling, electronic waste processing, and municipal solid waste facilities. Magnetic separation is efficient, cost-effective, and non-invasive. It reduces the labor needed for manual sorting and ensures that steel, a highly recyclable material, is effectively reclaimed.

Recovered ferrous metals can be melted and reformed into new steel products, with recycled steel requiring up to 74% less energy than producing virgin steel. The use of magnetic separation also minimizes contamination in the remaining non-ferrous material stream, improving the overall quality of the recycling output.

c. Eddy current separation

Eddy current separation is a critical technology used to separate non-ferrous metals (such as aluminum, copper, and brass) from mixed waste streams. Unlike ferrous metals, non-ferrous metals are not attracted to magnets. Eddy current separators use powerful rotating magnets to induce electric currents in conductive metals, creating a magnetic field that repels these metals away from the main stream.

The core component of an eddy current separator is a high-speed magnetic rotor positioned within a non-metallic drum. As the drum rotates, the magnetic field induces circulating electrical currents (eddy currents) in nearby conductive metals. These currents generate opposing magnetic fields that cause the non-ferrous metals to be physically repelled and ejected from the waste stream.

Eddy current separation is especially effective for recovering aluminum from beverage cans, copper from electronic components, and other valuable metals from shredded materials. The advantages of this method include:

- High throughput capacity
- Minimal labor requirement
- Enhanced metal recovery and purity
- Non-contact, low-maintenance operation

This technology is often used in conjunction with shredding and magnetic separation to achieve a complete, multi-stage metal recovery system. It ensures that even small, lightweight non-ferrous particles are efficiently separated, contributing to the economic and environmental benefits of recycling.

d. Baling and briquetting

Baling and briquetting are densification processes used to compact metal scraps into manageable, uniform shapes for easier storage, handling, transport, and smelting. These techniques are employed after metals have been sorted and cleaned to optimize logistics and improve the value of the recycled material.

Baling compresses loose scrap metal into rectangular blocks or "bales" using hydraulic presses. This is commonly done for light gauge metals like aluminum cans, tin, and sheet metal. Bales are easier to stack and store and reduce transport costs due to their high density.

Benefits of baling include:

- Volume reduction by up to 90%
- Reduced handling costs
- Prevention of material loss during storage and transport
- Improved furnace charging efficiency in remelting processes

Briquetting involves compressing fine metal particles, chips, or shavings such as turnings from machining operations into dense, cylindrical or square briquettes. Briquetting helps recover valuable metal that would otherwise be difficult to reuse due to its loose and fragmented form.

Advantages of briquetting:

- High-density output with uniform shape
- Reduced oxidation and burning loss during melting
- Recovery of coolant oils and cutting fluids from metal chips
- Enhanced storage and cleanliness of scrap

Both baling and briquetting not only streamline material handling but also contribute to higher furnace efficiency and better yield during remelting. Briquettes and bales feed more consistently into furnaces, and their compact structure reduces air pockets, ensuring even melting and lower energy consumption. Metals recycling technologies have evolved to support a highly efficient, closed-loop system that significantly reduces the environmental impact of metal production. Shredding breaks down bulky items into smaller fragments, preparing materials for downstream processing. Magnetic separation effectively recovers ferrous metals, while eddy current separation enables the efficient extraction of non-ferrous metals. Finally, baling and briquetting ensure that the recovered metals are compacted for cost-effective handling and high-quality reprocessing. These technologies, when used in an integrated system, maximize the recovery of valuable metals, reduce waste, and support the sustainable use of natural resources. As demand for recycled materials increases and regulatory pressures rise, continuous innovation in metal recycling technologies will be vital for advancing global sustainability goals and promoting a circular economy.

3.3.1.4 Glass recycling technologies

Glass recycling plays a crucial role in waste management and sustainability efforts around the world. Unlike many other materials, glass can be recycled infinitely without losing its quality or purity, making it an ideal candidate for closed-loop recycling systems. The recycled glass, commonly known as cullet, can be transformed into new glass products with significantly lower energy consumption and environmental impact than using virgin raw materials. Modern glass recycling involves a series of highly specialized and efficient technologies, including crushing and screening, optical sorting, and remelting and molding. These technologies work together to ensure high-quality recycled glass that can be reused in packaging, construction, and industrial applications. Fig. 3.9 illustrate process of glass recycling technologies.

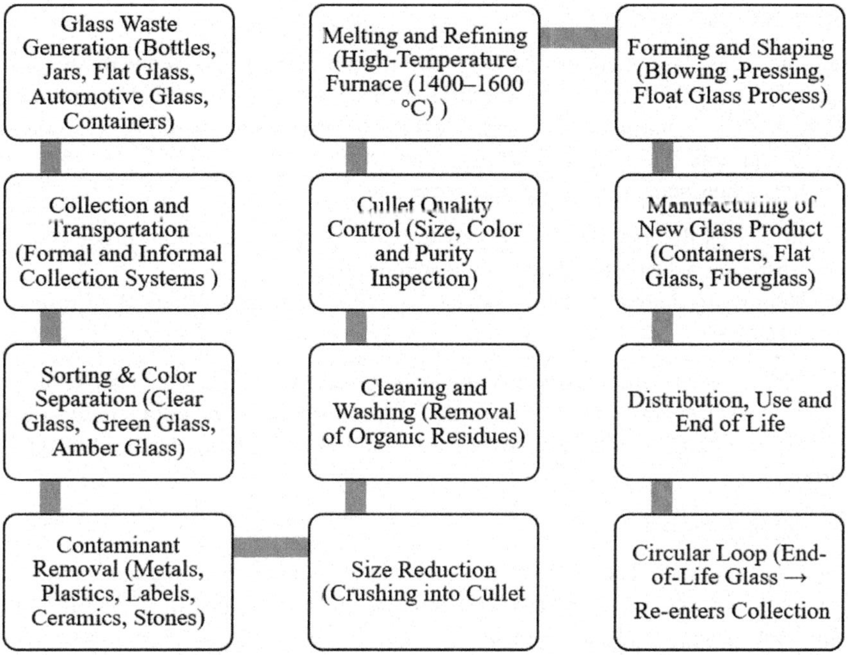

FIGURE 3.9 Process of glass recycling technologies.

a. Crushing

The first major step in glass recycling is crushing and screening, which involves reducing large pieces of waste glass into smaller fragments and separating them by size for further processing. This step is essential for preparing the glass for sorting, cleaning, and re-melting. Crushing is performed using mechanical crushers, such as hammer mills, jaw crushers, and roll crushers, that break down glass waste into smaller particles. The goal is to create cullet of uniform size, typically ranging from 3 to 50 mm, depending on the final use. The crushing process also helps remove labels, adhesives, and other surface contaminants, although some residual materials may still require further removal.

- Hammer mills use rapidly rotating hammers to pulverize the glass.
- Jaw crushers compress larger pieces of glass between two metal plates.
- Roll crushers crush glass between two rotating cylinders, producing more uniform particles.

Crushing not only reduces the volume of glass but also increases its surface area, making it easier to clean and sort in subsequent steps.

b. Screening

Once the glass is crushed, it is passed through vibrating screens or sieves to classify it into different size fractions. These screens separate fine particles (which may contain dust or non-glass materials) from medium and coarse cullet that can be used in glass manufacturing.

Screening serves multiple purposes:

- Improves the efficiency of downstream sorting and cleaning processes.
- Ensures consistent particle size for remelting.
- Removes undersized or oversized fragments that may affect glass product quality.

Effective crushing and screening set the foundation for the success of the entire recycling process. Uniform and appropriately sized cullet is easier to process, leading to higher recovery rates and better quality in the final glass products.

c. Optical sorting

Optical sorting is one of the most advanced technologies in glass recycling. Its main purpose is to identify and separate different types of glass and contaminants using cameras, sensors, and air jets. This process significantly improves the purity of recycled glass and ensures that only suitable materials are used for remelting.

Different types and colors of glass such as clear (flint), green, amber (brown), and blue glass must be separated because mixing colors can negatively affect the appearance and properties of new glass products. Optical sorting systems use a combination of near-infrared (NIR) spectroscopy, laser sensors, and high-resolution cameras to detect materials based on color, reflectivity, and chemical composition.

How Optical Sorting Works:

1. Feeding: The screened glass cullet is evenly spread onto a fast-moving conveyor belt.
2. Detection: As the glass passes under the sensor array, cameras and NIR sensors analyze the spectral and optical properties of each fragment.
3. Ejection: Precisely timed air jets are activated to blow undesirable items such as ceramics, metals, plastics, or glass of the wrong color into a separate reject bin.

Modern optical sorters can perform up to 1000 scans per second, allowing for extremely accurate and high-speed sorting. In many facilities, multiple sorting stages are used to further purify the cullet.

Benefits of optical sorting:

- Increases purity of recycled glass to above 95%
- Reduces contamination from non-glass materials

- Enables color separation, improving product quality
- Enhances efficiency and reduces labor costs

Optical sorting has revolutionized glass recycling, allowing facilities to process large volumes of mixed glass waste with high precision and minimal manual intervention.

d. Remelting and molding

The final stage in the glass recycling process is remelting and molding, where clean and sorted glass cullet is melted in a furnace and reformed into new glass products. This stage closes the recycling loop and allows for the continuous reuse of glass materials.

Remelting is carried out in high-temperature glass furnaces that reach temperatures of around 1400 to 1600°C. The clean cullet is mixed with a small proportion of raw materials—such as sand, soda ash, and limestone—if necessary. However, using cullet significantly reduces the need for virgin materials.

Advantages of using cullet in remelting:

- Lower energy consumption (up to 30% less than melting raw materials)
- Reduced CO_2 emissions and environmental impact
- Faster melting time due to cullet's lower melting point
- Improved furnace life because cullet produces less aggressive slag

During remelting, any remaining impurities float to the surface or dissolve, ensuring that the final molten glass is pure and homogeneous. Continuous monitoring and quality control are essential to maintain the integrity of the molten glass before it is molded.

After remelting, the molten glass is fed into various molding systems to form new products. Common methods include:

- Blow and blow process: Used for making bottles and jars. A gob of molten glass is dropped into a mold, blown into shape, and then cooled.
- Press and blow process: Used for wide-mouthed containers like food jars.
- Float glass process: Used to produce flat glass for windows, screens, and solar panels.

Each molding technique is designed to produce specific shapes, thicknesses, and product properties. The newly molded glass items are then gradually cooled in an annealing lehr, a type of oven that prevents cracking and strengthens the glass. Remelted and molded glass products can be virtually identical in performance and appearance to those made from virgin materials, highlighting the efficiency and sustainability of glass recycling.

Glass recycling technologies represent a mature and efficient set of systems that make it possible to reuse glass infinitely without loss of quality. Starting with crushing and screening, the waste glass is reduced to manageable sizes and separated by particle size. Optical sorting then purifies the cullet stream by removing contaminants and separating by color using sophisticated sensors and air jets. Finally, in the remelting and molding phase, the clean cullet is transformed into new glass products with significant savings in energy and raw materials.

The benefits of glass recycling are numerous. It reduces the need for raw material extraction, minimizes landfill use, saves energy, and lowers carbon emissions. Each ton of recycled glass saves over a ton of natural resources and reduces CO_2 emissions by approximately 300 kg. Moreover, the versatility of glass makes it ideal for use in a wide range of industries, including packaging, construction, and electronics.

With ongoing advancements in sorting and processing technologies, the future of glass recycling is even more promising. Innovations such as AI-integrated optical sorters, closed-loop production systems, and digital tracking of recycled content are poised to further improve the efficiency and scalability of glass recycling operations. As governments, industries, and consumers increasingly recognize the value of sustainable resource management, glass recycling will continue to play a vital role in achieving environmental and economic sustainability.

3.3.1.5 Textile and fiber recycling

The global textile and fashion industries are significant contributors to environmental pollution, resource depletion, and greenhouse gas emissions. In response, textile and fiber recycling has emerged as a critical component of sustainable production and circular economy strategies. Recycling helps recover valuable fibers from used garments, production scraps, and post-consumer textile waste, reducing the reliance on virgin materials like cotton and synthetic polymers. Among the various recycling methods, two traditional and widely adopted mechanical processes are garnetting and carding with re-spinning. These processes reclaim fibers without the use of harsh chemicals or excessive water, offering an eco-friendly alternative to conventional disposal.

a. Garnetting

Garnetting is a mechanical recycling technique that transforms post-industrial and post-consumer textile waste into fibrous material suitable for reuse. The process has been used for centuries, especially in the recycling of wool, cotton, and synthetic textiles. Garnetting is particularly effective for recycling materials that are no longer wearable or cannot be resold or donated due to damage or contamination. Garnetting involves breaking down fabric waste into loose fibers using a machine known as a garnett machine. The process follows several steps:

1. Pre-sorting and preparation: Textiles are sorted by material type (e.g., cotton, wool, polyester) and color to minimize the need for dyeing. Zippers, buttons, and non-textile parts are removed.
2. Cutting and shredding: Large pieces of fabric are cut into smaller scraps. These are then fed into shredders that begin the mechanical tearing of the textiles.
3. Garnetting: The shredded material enters the garnett machine, where it is subjected to a series of rotating cylinders with toothed rollers that open and pull apart the fabric into individual fibers. The teeth comb and fluff the material, creating a soft, fibrous mass.

The outcome of garnetting is a fiber blend that can be used as:

- Stuffing or filling for mattresses, cushions, and insulation.
- Reinforcement in nonwoven applications.
- Raw material for further spinning if fiber quality allows.

Advantages of garnetting:

- Recycles textiles without using water or chemicals.
- Conserves energy and reduces carbon emissions.
- Converts unwearable textiles into usable fiber.
- Cost-effective for low-grade textile waste.

Limitations:

- Results in shorter fibers, which may reduce the strength and durability of spun yarns.
- Fiber quality depends heavily on the input material.
- Blended textiles (e.g., poly-cotton) can be difficult to process and may not yield high-quality output.

Garnetting remains a vital method for bulk recycling of damaged textiles, particularly in industries focused on insulation, automotive interiors, and other non-apparel applications.

b. Carding and re-spinning

Carding and re-spinning is a more refined recycling process used to produce yarn from recovered textile fibers. While garnetting focuses on breaking down textiles into loose fiber, carding and spinning aim to align those fibers and convert them back into spendable yarn for creating new textiles.

Carding is a technique used in both virgin and recycled fiber processing. In recycling, it prepares fibers reclaimed via garnetting or other methods for spinning by cleaning, separating, and aligning them.

1. Feeding: Recycled fibers, sometimes mixed with virgin fibers for strength, are fed into a carding machine.

2. Opening and cleaning: Rollers open up fiber clumps, remove debris, and disentangle the fibers.
3. Aligning: The fibers are combed and straightened, forming a continuous web or sliver a soft rope-like strand of aligned fibers.

Carding is critical because it improves the uniformity and cohesion of recycled fibers, which is essential for yarn spinning. Depending on the quality of the recycled fibers, multiple carding stages may be used.

Once the fibers are carded into slivers, they undergo spinning a twisting process that converts sliver into yarn. This is done using either ring spinning, open-end spinning, or air-jet spinning:

- Ring spinning: Offers high-quality, fine yarn but may require stronger fibers.
- Open-end spinning: Suitable for shorter recycled fibers and bulk yarns.
- Air-jet spinning: Used for smooth, clean yarns, often blending recycled and virgin fibers.

The spun yarn can then be used in weaving or knitting to produce new garments, home textiles, and accessories.

Benefits of carding and re-spinning:

- Allows the creation of new high-value textile products.
- Enables closed-loop recycling in the fashion and apparel industries.
- Reduces dependency on virgin fibers and associated environmental impacts.
- Blending recycled and virgin fibers improves quality and durability.

Challenges:

- Recycled fibers are often shorter and weaker, impacting yarn strength.
- Color consistency is harder to maintain, often requiring sorting by color.
- Processing blended materials with synthetic components can be problematic.
- Cost can be higher than virgin production depending on quality demands.

Despite these challenges, many brands now incorporate recycled yarns into their products as part of sustainability commitments. Fashion leaders, including Patagonia, H&M, and Stella McCartney, use carded and spun recycled fibers to manufacture apparel with lower environmental impact.

Textile and fiber recycling through garnetting and carding with re-spinning plays an essential role in reducing textile waste and promoting a circular economy. Garnetting is ideal for transforming damaged and unwearable textiles into usable fibers for nonwoven applications, while carding and re-spinning allow for the creation of new yarns and high-quality textile products.

As the fashion industry faces increasing scrutiny over its environmental footprint, these mechanical recycling technologies offer practical, scalable solutions for reducing waste, conserving resources, and lowering carbon emissions. With growing innovation in fiber identification, sorting technologies, and machinery, the quality and efficiency of recycled fibers continue to improve, making textile recycling not just an ecological necessity but an economically viable alternative to linear production models. In the future, integrating these methods with digital tracking systems, automated sorting, and fiber-specific recycling pathways will enhance their effectiveness and help build a closed-loop textile system, where waste is continuously repurposed into new materials.

3.3.1.6 *E-waste mechanical recycling*

The exponential growth of electronic products and their rapid obsolescence has led to a global surge in electronic waste (e-waste). E-waste includes discarded computers, mobile phones, televisions, and other electronic devices. These products contain valuable metals, plastics, and rare earth elements but are also a source of hazardous substances like lead, mercury, and brominated flame retardants. Mechanical recycling has become a pivotal strategy for recovering materials and mitigating the environmental harm posed by improperly handled e-waste.

Mechanical recycling of e-waste involves a series of processes designed to dismantle, crush, and separate components and materials based on their size, density, and composition. The three major stages in this process are disassembly and shredding, size and density separation, and advanced sensor sorting. These steps work together to maximize material recovery, minimize hazardous exposure, and support a circular economy in the electronics sector.

3.3.1.6.1 Disassembly and shredding

The first step in mechanical e-waste recycling is manual or automated disassembly of electronic devices. This process involves the careful dismantling of complex products into their individual components such as printed circuit boards (PCBs), hard drives, wires, batteries, capacitors, and plastic casings.

Manual disassembly is often used when dealing with high-value or delicate items, as it allows for:

- Targeted removal of hazardous parts (e.g., mercury bulbs, batteries, CRTs).
- Preservation of intact components for reuse or resale.
- Efficient separation of high-value parts such as gold-plated connectors or RAM chips.

Automated disassembly, using robotic arms or programmable machinery, is increasingly being adopted for high-volume processing. It enhances speed and safety but may not match the precision of manual techniques for intricate items.

After disassembly, the remaining electronic components—particularly mixed material parts—undergo shredding. Shredding breaks down large items into smaller, manageable fragments for subsequent separation processes.

Shredders used in e-waste recycling are typically of two types.

- Primary shredders: Coarse shredders that break down entire devices into smaller chunks.
- Secondary shredders: Finer shredders that produce particles of uniform size (typically 10–50 mm).

Shredding is essential for:

- Reducing volume and facilitating handling.
- Liberating different materials (e.g., metal from plastic, copper from circuit boards).
- Increasing surface area, which enhances separation efficiency in later stages.

While shredding improves throughput, it can also result in the loss of component-level information and sometimes damage recoverable parts. Therefore, a balance between disassembly and shredding is crucial.

3.3.1.6.2 Size separation

Once e-waste is shredded into smaller particles, it is subjected to size and density-based separation techniques. These mechanical processes exploit the physical differences between various materials, such as weight, size, and aerodynamic properties, to sort them effectively.

Size separation uses sieving and screening machines to classify shredded e-waste particles based on dimensions. This step is vital because different separation technologies perform best within specific particle size ranges.

Key equipment includes:

- Vibratory screens: Use oscillating motion to pass smaller particles through mesh while retaining larger ones.
- Trommel screens: Rotating cylindrical drums that sort particles by size as they tumble inside.

Size separation helps in:
Optimizing feed material for downstream separation equipment.
Reducing equipment wear by filtering out oversized fragments.
Improving purity and consistency in sorted material streams.

3.3.1.6.3 Density separation

Density separation is used to differentiate materials that may be similar in size but differ in mass. Techniques employed include.

1. Air classification: Uses upward or lateral airflow to separate light and heavy particles. Plastics, papers, and foams are carried away by the air, while metals and heavier materials fall downward.
2. Gravity tables: Employ vibratory motion and inclined surfaces to sort particles based on density. Heavier items sink and move in one direction, while lighter materials rise and move in another.
3. Sink-float separation: Involves placing materials into liquid tanks where less dense materials (plastics) float and denser ones (metals like copper or aluminum) sink. Different liquid densities (e.g., water, salt solution, or heavy liquids) can be used depending on the application.

Benefits of size and density separation include:

- Non-invasive, mechanical sorting that avoids chemical treatments.
- High efficiency in separating plastics, glass, and metals.
- Preparation of homogenous material streams for refining or resale.

However, limitations exist. For example, multi-material composites and bonded parts can complicate separation, and environmental controls are needed to manage dust and airborne particles.

3.3.1.6.4 Advanced sensor sorting

As e-waste becomes more complex with miniaturized and mixed-material components, traditional mechanical methods are often insufficient for achieving high-purity recovery. This is where advanced sensor-based sorting technologies come into play. These technologies use sophisticated sensors to identify materials based on color, conductivity, chemical composition, and atomic structure, enabling high-speed, automated separation.

Types of sensor sorting technologies

1. Electrostatic separation: Uses differences in electrical conductivity to sort particles. Conductive materials like metals retain electrical charge and are attracted or repelled by an electrostatic field, while non-conductive materials like plastic are unaffected. This method is ideal for separating metal-plastic composites.
2. X-ray fluorescence (XRF): Detects elements based on their unique fluorescence signature when exposed to X-rays. It is particularly effective for sorting:
   - Different metal alloys (e.g., aluminum vs. magnesium).
   - Precious metals in circuit boards.
   - Lead-containing components.
3. Near-infrared (NIR) spectroscopy: Identifies polymers by their chemical bonds, making it useful for separating different plastic types (e.g., ABS, PVC, PET).

4. Laser-induced breakdown spectroscopy (LIBS): Provides real-time elemental analysis by firing a laser pulse that generates plasma on the material surface. The emitted light is analyzed to determine material composition.
5. Optical Cameras and Color Sorters: Used to distinguish materials based on color, gloss, or transparency. Combined with air jets, they remove undesired pieces from the conveyor stream.

Advanced sensor sorting is often integrated into automated conveyor systems, allowing for high-speed, continuous material flow. The system identifies materials on the conveyor and triggers actuators to divert them into designated bins.

For example:

- A stream of mixed shredded e-waste moves through a sensor line.
- XRF identifies copper-bearing particles and diverts them.
- NIR sorts different plastic types for polymer recycling.
- Electrostatic separation removes fine metallic dust.

Advantages of sensor-based sorting:

- Extremely high accuracy and material purity.
- Enables recovery of rare and valuable materials.
- Reduces human intervention and processing time.
- Increases economic value of recycled output.

Challenges:

- High capital and maintenance costs.
- Sensitive to surface contamination and moisture.
- Requires precise calibration and skilled operation.

Despite these challenges, sensor-based sorting is rapidly becoming a standard in state-of-the-art e-waste recycling facilities, helping to meet regulatory demands for material recovery rates and environmental compliance.

Mechanical recycling of e-waste is a vital process in the global effort to manage electronic waste sustainably and responsibly. By combining manual and automated disassembly, shredding, size and density separation, and advanced sensor-based sorting, modern facilities can efficiently recover valuable metals, plastics, and electronic components while minimizing environmental risks.

- Disassembly and shredding set the stage for recycling by reducing waste volume and isolating components.
- Size and density separation utilize physical characteristics to sort and purify material streams.

- Advanced sensor technologies allow for unprecedented precision in recovering valuable and rare materials, critical for closing the loop in electronics manufacturing.

As electronic devices continue to evolve and become more complex, the technologies used to recycle them must also advance. Future improvements in robotics, machine learning, AI-assisted sensor systems, and modular design of electronics will further enhance the efficiency and scope of mechanical e-waste recycling. This integrated approach not only conserves resources but also supports green innovation, circular economy goals, and environmental protection on a global scale. Table 3.1 illustrate comparative analysis of mechanical recycling technologies.

- **Emerging innovations in mechanical recycling:**

The global waste management landscape is undergoing a radical transformation, driven by the growing urgency of environmental sustainability and the circular economy. Traditional mechanical recycling, which involves processes such as shredding, separation, and remanufacturing, is evolving with the integration of advanced technologies and system-level design changes. These innovations are reshaping how materials are recovered, processed, and reintegrated into the economy. This explores four key emerging trends in mechanical recycling: AI-based Sorting Systems, Closed-loop Systems, Design for Recycling (DfR), and Eco-Labels and Certifications. These innovations are at the forefront of sustainable waste management and are critical to increasing recycling efficiency, reducing material loss, and promoting responsible consumption and production.

3.3.1.7 AI-based sorting systems

Artificial Intelligence (AI) is revolutionizing the recycling industry by making material sorting faster, more accurate, and more adaptable. Traditional sorting systems, which relied on mechanical and manual processes, were limited by labor costs, inefficiencies, and contamination risks. AI-based sorting technologies address these limitations by incorporating machine learning, computer vision, and robotic automation.

AI-enabled sorting systems typically include high-resolution cameras, near-infrared (NIR) spectroscopy, and machine learning algorithms. These systems.

- Analyze material characteristics (color, texture, chemical composition).
- Identify and classify waste types such as plastics (PET, HDPE, PVC), metals, glass, and paper.
- Trigger robotic arms or air jets to separate materials with high precision.

Systems like AMP Robotics and TOMRA use neural networks trained on millions of waste images to recognize and adapt to new materials in real-time.

TABLE 3.1 Comparative analysis of mechanical recycling technologies.

Recycling type	Specific technologies	Material type	Key processing steps	Contamination tolerance	Energy requirement	Output quality	Environmental impact	Major advantages	Key limitations	Typical end products
Plastic recycling	Re-compounding, conventional recycling, injection molding, agglomeration	Thermoplastics (PET, HDPE, PP, LDPE)	Sorting → shredding → washing → melting → reprocessing	Low	Moderate	Medium–High	Low emissions, resource-efficient	Cost-effective, scalable	Polymer degradation, requires clean feed	Bottles, containers, automotive parts
Paper & cardboard recycling	Hydra pulping, deinking, screening & cleaning	Paper, cardboard	Pulping → ink removal → fiber separation	Moderate	Low	Medium	Very low carbon footprint	Mature technology, high recovery rates	Fiber quality degrades over cycles	Packaging, tissue paper
Glass recycling	Crushing & screening, screening, remelting & molding, optical sorting	Container & flat glass	Crushing → sorting → remelting	Very Low	High (melting)	Very High	Significant energy savings versus virgin glass	Infinite recyclability	Sensitive to contamination	Bottles, jars, glass sheets
Metal recycling	Baling & briquetting, shredding, eddy current separation, magnetic separation	Ferrous & non-ferrous metals	Size reduction → magnetic & eddy separation	High	Moderate–High	Very High	Major energy & emission savings	Retains material properties	Capital-intensive	Construction steel, aluminum sheets
Textile & fiber recycling	Garnetting, carding & Re-spinning, Content Calendar	Cotton, wool, blended textiles	Fiber opening → cleaning → Re-spinning	Low–Moderate	Low–Moderate	Low–Medium	Reduces landfill textile waste	Simple mechanical process	Fiber shortening & quality loss	Insulation, yarns, non-wovens
E-waste mechanical recycling	Disassembly & shredding, size separation, density separation, sensor sorting	Electronic scrap	Manual disassembly → crushing → physical separation	High	Moderate	High (metal recovery)	Prevents hazardous leakage	Efficient material recovery	Hazardous residues	Copper, aluminum, precious metals

This improves sorting purity and efficiency and reduces contamination in recycling streams.

Benefits

- Increased throughput and accuracy in materials recovery.
- Reduced labor dependency and operational costs.
- Enhanced adaptability to evolving packaging trends and materials.
- Higher purity levels, making recycled materials more marketable.

AI-based sorting is particularly valuable in handling complex packaging, multi-material waste, and contaminated recyclables areas where human sorters struggle. It allows mechanical recycling plants to operate with greater consistency and reduced downtime.

3.3.1.8 Closed-loop systems

A closed-loop system is a circular model where a product, at the end of its life, is recycled and used as a raw material for the same or similar product. Unlike traditional open-loop systems where materials degrade over cycles and are eventually down cycled closed-loop recycling ensures that material quality and usability are retained across multiple lifecycles.

Closed-loop systems require an integrated approach involving.

- Product design that facilitates disassembly and material recovery.
- Consistent material streams that are easily sorted and processed.
- Clean and uncontaminated inputs, often through extended producer responsibility (EPR) programs.

Examples include:

- Aluminum can recycling, where used cans are melted and reformed into new cans with minimal loss.
- PET bottle-to-bottle recycling, where food-grade PET is recovered, cleaned, and remanufactured into new bottles.
- Electronics closed-loop, where parts like metals, glass, and plastics from smartphones or laptops are reused to make new devices.

Benefits

- Reduced reliance on virgin resources.
- Energy and cost savings in production.
- Consistency in product quality and performance.
- Supports circular economy goals and corporate sustainability targets.

Challenges

- Requires standardized materials and formats.
- Needs robust collection and reverse logistics systems.

- Faces regulatory and contamination issues (especially in food and pharma packaging).

Nonetheless, companies like Dell, Coca-Cola, and Interface are increasingly investing in closed-loop innovations, proving their feasibility in both consumer and industrial sectors.

3.3.1.9 Design for recycling (DfR)

Design for recycling (DfR) is an innovation-driven philosophy where products are created with their end-of-life recyclability in mind. DfR is a proactive approach that integrates recyclability criteria into the product development phase, addressing material selection, assembly, labeling, and disassembly.

Key principles of DfR

- Material simplification: Use of fewer types of materials or mono-material designs to ease sorting and processing.
- Detachable components: Designing products that can be easily disassembled for component reuse or material separation.
- Avoidance of hazardous additives: Using non-toxic dyes, inks, and adhesives.
- Clear material labeling: Marking plastics (PET, HDPE) and other materials for easy identification.

Application Examples

- Packaging: Replacing multi-layer flexible plastics (which are hard to recycle) with recyclable mono-polymer alternatives.
- Electronics: Designing phones and laptops with modular components and minimal adhesive use.
- Automotive: Creating bumpers, dashboards, and seats from recyclable polymers and allowing for rapid dismantling.

Benefits

- Increases recycling efficiency and material recovery rates.
- Reduces sorting errors and contamination.
- Minimizes waste going to landfills.
- Enhances product compliance with environmental standards.

Design for Recycling is increasingly mandated or encouraged by policies like the European Green Deal, Extended Producer Responsibility (EPR) schemes, and corporate ESG guidelines. It also enables companies to future-proof their products against evolving waste legislation.

3.3.1.10 Eco-labels and certifications

Informed consumers are central to the success of recycling systems. Eco-labels and certifications are emerging tools that provide transparency about a product's recyclability, environmental impact, and compliance with sustainability standards. These labels act as both consumer guides and manufacturer incentives, promoting responsible production and consumption.

Types of eco-labels

1. Recyclability certifications: Such as the How2Recycle label, which indicates the ease with which packaging can be recycled in common municipal systems.
2. Environmental product declarations (EPDs): Detailed documents that quantify a product's environmental impact over its lifecycle.
3. Third-party certifications:
   - Cradle to Cradle Certified: Assesses product safety, circularity, and climate impact.
   - Blue angel (Germany): Ensures products meet high ecological and safety standards.
   - EU ecolabel: Covers packaging, textiles, electronics, and more.

Impact on mechanical recycling

- Drives material choice and product design in favor of recyclability.
- Creates demand for products made from recycled content.
- Builds consumer trust and brand loyalty.
- Encourages innovation among manufacturers to meet labeling criteria.

Eco-labels also help harmonize recycling efforts across regions by providing a standardized language for recyclability and sustainability. For instance, a globally recognized recyclability logo ensures that consumers in any country can properly sort and dispose of their waste.

Mechanical recycling is no longer a static process of crushing and sorting; it has become a dynamic, data-driven, and innovation-led domain. Emerging trends like AI-based sorting systems, closed-loop models, design for recycling, and eco-certifications are revolutionizing how waste is managed, materials are recovered, and products are manufactured. These innovations address the core challenges of modern recycling complex materials, contamination, consumer behavior, and design inefficiencies. They provide scalable, economically viable solutions that contribute to the larger goals of the circular economy, net-zero emissions, and sustainable development. To fully leverage these innovations, collaboration across the value chain is essential. Policymakers must incentivize recyclability, manufacturers must embrace eco-design, technology developers must push the limits of AI and robotics, and consumers must be informed and engaged. In this evolving landscape, mechanical recycling stands as a testament to human ingenuity continuously adapting and innovating to turn waste into worth, and consumption into conservation.

3.3.2 Chemical recycling

Chemical recycling refers to a set of advanced processes that break down plastic waste or other complex materials into their basic chemical building blocks, allowing them to be reused to produce new materials of virgin-like quality. Unlike mechanical recycling, chemical recycling can process contaminated, mixed, or hard-to-recycle waste streams. It supports the circular economy by enabling true material-to-material (closed-loop) recycling. Chemical recycling is an advanced and increasingly vital technology in modern waste management and the circular economy. Unlike mechanical recycling, which retains the physical structure of materials, chemical recycling breaks down polymers into their chemical building blocks, allowing for the production of virgin-quality materials. This process is especially useful for handling contaminated, multi-layered, or degraded plastics that are not suitable for traditional recycling methods. The general process flow of chemical recycling includes four major stages: Feedstock Preparation, Chemical Conversion, Purification, and Re-polymerization or Refining. Each step is crucial in transforming plastic waste into reusable materials and fuels. Fig. 3.10 illustrate process of chemical recycling process.

3.3.2.1 Feedstock preparation

The process begins with the collection, sorting, and pre-treatment of plastic waste. Feedstock preparation is a foundational step that ensures the efficiency and quality of the chemical recycling process. It involves:

Sorting: Plastic waste is collected from various sources household waste, industrial scrap, packaging, etc. and sorted by polymer type (e.g., PET, PE, PP, PS) and level of contamination. Automated sorting technologies such as near-infrared (NIR) spectroscopy, AI-driven robotic arms, and manual methods are used to segregate appropriate materials.

Cleaning: Sorted plastic is washed to remove impurities such as labels, adhesives, food residues, and oils. Contaminants can interfere with chemical reactions and affect the quality of final products.

Shredding or Pelletizing: The cleaned plastics are shredded into flakes or melted and extruded into pellets to ensure uniform feedstock size. This increases surface area and improves reactivity during chemical conversion.

Importance: Proper feedstock preparation reduces energy consumption in subsequent steps, minimizes process downtime, and enhances the efficiency of chemical conversion.

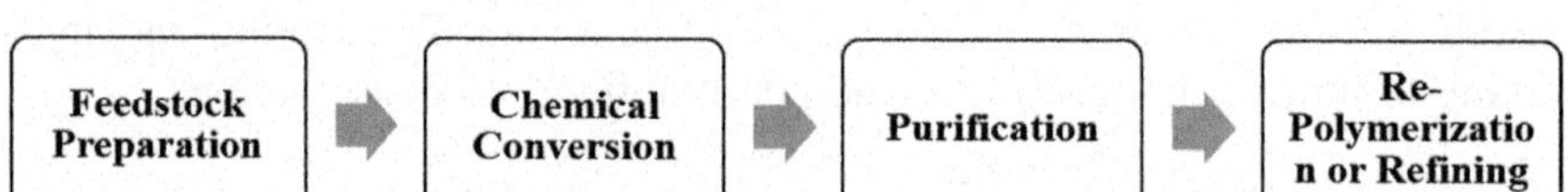

FIGURE 3.10 Process of chemical recycling process.

3.3.2.2 *Chemical conversion*

In this stage, the prepared plastic feedstock undergoes chemical or thermo-chemical transformation to break down polymer chains into monomers or other useful intermediates. The method used depends on the type of plastic and desired end products. Common conversion technologies include:

Pyrolysis: Pyrolysis involves heating plastics in an oxygen-free environment to break long polymer chains into shorter hydrocarbon molecules. It converts waste into pyrolysis oil, syngas, and char.

- Applicable to polyolefins like PE, PP.
- Output can be used for fuel or further refining into virgin plastic precursors.

Gasification: Gasification involves partial oxidation of plastic waste at high temperatures (800–1200°C) to produce syngas (a mix of CO and H_2).

- Syngas can be used for electricity generation or as a building block in chemical synthesis.

De-polymerization: This process uses solvents, catalysts, or heat to break polymers into monomers.

- Common for PET and PS.
- For example, PET can be depolymerized into terephthalic acid (TPA) and ethylene glycol (EG).

Solvolysis: Solvolysis is the decomposition of plastics using solvents like water, alcohol, or glycol.

- PET, for instance, can be broken down using glycol (glycolysis) to produce BHET, a PET precursor.

Importance: Chemical conversion determines the pathway and efficiency of recycling. It is tailored to the input material and dictates the energy requirements and quality of the final output.

3.3.2.3 *Purification*

Once polymers are broken down, the resulting products monomers, oils, gases, or chemical intermediates must be purified to remove contaminants and side products. High purity is necessary for the recovered materials to meet the standards required for food-grade packaging, medical applications, or other high-value uses.

Common Purification Techniques:

- Distillation: Separates components based on boiling points; common in pyrolysis oil and monomer purification.
- Filtration and Adsorption: Removes solid particles, residual catalysts, and dyes.

- Precipitation: Used in solvent-based recovery to isolate purified polymers or monomers.
- Solvent Extraction: Selectively dissolves and separates desired compounds.

Importance: Purification ensures the safety, quality, and compliance of recycled products. This stage also allows the recovered monomers or fuels to match or exceed the performance of virgin materials.

3.3.2.4 Re-polymerization/refining

The final stage involves the reconstruction of purified monomers or intermediates into new polymers or chemical products. This step closes the recycling loop and reintroduces materials into the production cycle.

3.3.2.4.1 Re-polymerization

Recovered monomers are polymerized using conventional chemical processes:

- TPA and EG are polymerized into PET.
- Styrene monomers are re-polymerized into polystyrene.
- Caprolactam is used to form Nylon-6.

3.3.2.4.2 Refining (for fuel applications)

If the chemical conversion yields oils or syngas, these are further refined to produce fuels like diesel, gasoline, jet fuel, or waxes.

Re-polymerized plastics can be used in:

- Packaging (including food-grade bottles).
- Automotive parts.
- Textiles and fibers.
- Consumer goods.

Importance: Re-polymerization provides high-quality, virgin-equivalent materials, enabling true circularity. It reduces reliance on fossil resources and lowers greenhouse gas emissions associated with raw material extraction.

Chemical recycling represents a transformative approach to plastic waste management, particularly for plastics unsuitable for mechanical recycling. The four major stages Feedstock Preparation, Chemical Conversion, Purification, and Re-polymerization/Refining form an interconnected process that enables high-value material recovery and supports the goals of sustainability, circular economy, and zero waste. By turning problematic waste streams into useful resources, chemical recycling can help bridge the gap between environmental responsibility and industrial productivity. However, challenges such as high energy consumption, cost, and scalability must be addressed through continued innovation, regulation, and cross-sector collaboration. As technologies mature, chemical recycling will increasingly complement traditional methods, offering a robust solution to global plastic waste challenges.

3.3.2.5 *Types of chemical recycling technologies*

3.3.2.5.1 Pyrolysis

In the quest for sustainable solutions to the global plastic waste crisis, pyrolysis has emerged as a prominent chemical recycling technology. Unlike mechanical recycling, which processes plastic through physical transformation, pyrolysis breaks down plastic polymers into smaller hydrocarbon molecules using thermal decomposition in an oxygen-free environment. This process enables the recovery of fuels and chemical feedstocks, offering a valuable route for recycling mixed and contaminated plastics that are typically non-recyclable through conventional means. Pyrolysis is a thermochemical process that decomposes organic materials at high temperatures (typically 300–900°C) in the absence of oxygen. In the context of plastic recycling, it transforms long polymer chains into smaller molecules, yielding valuable outputs like pyrolysis oil, syngas, waxes, and char. This process is often referred to as thermal cracking or plastic-to-fuel (PTF) conversion. Pyrolysis is particularly suited for recycling polyolefin such as polyethylene (PE), polypropylene (PP), and polystyrene (PS), which make up the majority of plastic packaging waste. It can also handle mixed plastic waste, contaminated plastics, multilayer films, and composite materials, which are generally rejected by mechanical recycling systems. The pyrolysis process typically involves the following stages:

a. **Feedstock preparation:** Before pyrolysis, plastic waste undergoes pretreatment, including:
   - Sorting and selection of plastic types (preferably polyolefin).
   - Shredding or pelletizing to create uniform feedstock.
   - Drying and removal of contaminants such as food residue, metals, or PVC, which can release hazardous chlorine gas.

b. **Thermal decomposition:** The prepared feedstock is heated in a pyrolysis reactor, such as a rotary kiln, fluidized bed, or batch reactor. The absence of oxygen prevents combustion, allowing for thermal decomposition.
   Key variables:
   - Temperature: 300 to 900°C.
   - Heating rate: Determines product distribution (oil vs. gas).
   - Residence time: Affects the extent of cracking.

   At elevated temperatures, polymers break down into:
   - Pyrolysis oil (liquid hydrocarbons): Used as a fuel or refined into chemicals.
   - Syngas (H_2, CO, CH_4): Used for energy within the system or as a feedstock.
   - Char (carbon-rich solid): Can be used as solid fuel or filler.

c. Condensation and recovery:
Volatile gases are cooled in condensers to extract liquid oil, while non-condensable gases are collected as syngas. Remaining solids settle as char at the reactor base.

- **Types of pyrolysis:**
Pyrolysis, as a transformative chemical recycling technology, can be categorized into several types based on operational conditions, heating methods, and intended outputs. These include Slow Pyrolysis, Fast Pyrolysis, Catalytic Pyrolysis, and Microwave-Assisted Pyrolysis. Each variant offers distinct advantages and is suited for specific applications depending on the feedstock and desired end products. Understanding the different types of pyrolysis is crucial for selecting the optimal approach in plastic waste management and resource recovery.

a. Slow pyrolysis: Slow pyrolysis is the most traditional and straightforward form of pyrolysis. It operates at relatively low temperatures, typically around 300 to 500°C, with long residence times ranging from several minutes to hours. The heating rate is gradual, allowing the thermal degradation of plastic polymers over an extended period.

Characteristics:
- Low heating rate.
- Long vapor residence time.
- Yields more solid residue (char) than liquids and gases.

Advantages:
- Simple reactor design and low energy requirement.
- Produces a stable, carbon-rich char that can be used in asphalt, bricks, or as a soil amendment.
- Useful in small-scale and decentralized waste treatment systems.

Limitations:
- Lower oil and gas yield compared to other pyrolysis types.
- Longer processing time reduces throughput.

Applications:
- Treatment of mixed plastic waste and biomass.
- Recovery of char for industrial and construction uses.

b. Fast pyrolysis: Fast pyrolysis is optimized for the production of liquid fuels and chemical feedstock's. It involves rapid heating of feedstock to temperatures around 450 to 600°C with very short vapor residence times (typically less than 2 s).

Characteristics:
- High heating rate.
- Short residence time.
- Maximum production of pyrolysis oil.

Advantages:
- Higher liquid yields compared to slow pyrolysis.

- More efficient for recovering fuel-grade hydrocarbons.
- Continuous operation is possible with fluidized bed reactors.

  Limitations:
- Requires precise temperature control and rapid quenching.
- More complex and expensive reactor systems.

  Applications:
- Conversion of polyolefins (PE, PP) into pyrolysis oil.
- Used by commercial plants focusing on plastic-to-fuel technologies.

c. **Catalytic pyrolysis:** Catalytic pyrolysis introduces catalysts into the pyrolysis process to lower the activation energy, enhance selectivity, and improve the quality of pyrolysis products. Catalysts like zeolites, alumina, silica, or metal oxides are used to promote targeted chemical reactions.

  Characteristics:
- Operates at lower temperatures than non-catalytic pyrolysis (350–550°C).
- Increases production of desired molecules (e.g., olefins, aromatics).

  Advantages:
- Produces high-quality, narrow-range products such as synthetic diesel, naphtha, or monomers.
- Reduces coke formation and unwanted by-products.
- Tailorable product composition based on catalyst type.

  Limitations:
- Catalyst deactivation due to fouling or poisoning.
- Higher operational cost due to catalyst regeneration or replacement.

  Applications:
- Closed-loop plastic recycling into virgin-quality monomers.
- Specialty chemical and fuel production.

d. **Microwave-assisted pyrolysis:** Microwave-assisted pyrolysis (MAP) uses microwave energy to heat the feedstock directly through electromagnetic radiation, rather than relying on external heat sources. This leads to rapid, uniform heating and increased energy efficiency.

  Characteristics:
- Volumetric heating mechanism.
- Operates at varying temperatures (300–700°C).
- Short reaction times.

  Advantages:
- Faster and more uniform heating.
- Higher efficiency and lower energy consumption.
- Modular and scalable, suitable for decentralized applications.

  Limitations:
- Requires microwave-absorbing materials or additives.
- Higher initial capital cost due to specialized equipment.

  Applications:

- Small to medium-scale plastic waste treatment.
- Integrated systems with real-time energy recovery.

Each type of pyrolysis presents unique benefits and trade-offs. Slow pyrolysis is ideal for char production and small-scale systems, while fast pyrolysis is suited for high-yield liquid fuel recovery. Catalytic pyrolysis offers refined outputs with lower energy needs, and microwave-assisted pyrolysis introduces innovative, energy-efficient processing methods. The selection of a pyrolysis type depends on feedstock characteristics, desired end products, operational scale, and environmental considerations. As plastic waste continues to grow, adopting the appropriate pyrolysis technology will be essential for advancing circular economy goals and achieving sustainable material management.

- **Products and applications:**
  Pyrolysis is a thermochemical recycling technology that decomposes plastic waste into valuable products in an oxygen-free environment. This process not only diverts waste from landfills and incinerators but also recovers energy and raw materials for industrial use. The main products of pyrolysis are pyrolysis oil, syngas, and char. Each has specific properties and applications in the fields of energy, chemical manufacturing, agriculture, and material science. Understanding these products and their uses is essential for evaluating the economic and environmental viability of pyrolysis as a circular economy solution.
  a. Pyrolysis oil: Pyrolysis oil, also known as bio-oil or TACOIL (thermal anaerobic conversion oil), is a complex mixture of hydrocarbons resulting from the condensation of volatile gases produced during the pyrolysis process. It resembles crude oil in appearance and composition but is derived entirely from plastic or biomass feedstock.
  Composition and properties:
  - Mixture of alkanes, alkenes, aromatics, ketones, and aldehydes.
  - Variable viscosity and heating value depending on feedstock and process parameters.
  - Can contain impurities such as chlorine, nitrogen, or oxygenated compounds.

  Applications:
  1. Fuel source:
    - Pyrolysis oil can be used directly as an industrial fuel for boilers, furnaces, and kilns.
    - It has a calorific value comparable to diesel (35–40 MJ/kg).
    - Can partially replace heavy fuel oil in marine and stationary power sectors.

  2. Chemical feedstock:
    - After refining, pyrolysis oil can be cracked into monomers such as ethylene, propylene, benzene, and toluene.

- Used as a feedstock in steam crackers for producing virgin-quality plastics.
- Enables closed-loop plastic-to-plastic recycling, reducing dependence on fossil fuels.

3. Blending component:
   - Pyrolysis oil can be blended with conventional fuels to meet sustainability quotas.
   - Some companies use it to produce low-carbon gasoline or diesel blends.
4. Road construction and lubricants:
   - Oil fractions from certain plastics can be used in bitumen formulations.
   - Waxy components are usable in the manufacture of lubricants and coatings.

Challenges:

- Needs further refining to remove chlorine, sulfur, and other contaminants.
- Inconsistent quality may affect commercial acceptance.
- Infrastructure for large-scale refining is still under development.

Despite these challenges, pyrolysis oil is central to many plastic recycling initiatives and is actively being adopted by major petrochemical firms for circular product streams.

b. Syngas: Syngas, or synthesis gas, is a mixture of hydrogen (H_2), carbon monoxide (CO), methane (CH_4), and small amounts of CO_2 and other gases. It is a product of both pyrolysis and gasification and is formed from the thermal degradation of carbon-rich materials.

Composition and properties:

- High calorific value (~10–20 MJ/Nm^3).
- Non-condensable, flammable gas.
- Composition depends on the feedstock and pyrolysis conditions.

Applications:

1. Onsite energy generation:
   - Syngas is frequently used within pyrolysis facilities to power the system.
   - Drives turbines, gas engines, or is combusted directly for heat.
   - Improves overall energy efficiency and reduces external fuel dependency.
2. Electricity production:
   - Syngas can fuel combined heat and power (CHP) plants.
   - Can be converted to electricity using gas turbines or internal combustion engines.
3. Hydrogen production:
   - Hydrogen can be separated from syngas via pressure swing adsorption (PSA).

- Used in fuel cells, ammonia synthesis, or as a hydrogenation agent in chemical processes.

4. Chemical synthesis:
   - Syngas is a precursor in the Fischer-Tropsch process, producing synthetic diesel, methanol, and olefins.
   - Enables conversion of waste plastic into base chemicals for new polymers or fuels.
5. Carbon capture and utilization (CCU):
   - CO in syngas can be captured and converted into urea, methanol, or other carbon-based products.
   - Supports low-emission industrial processes.

Challenges:

- Requires gas cleaning systems to remove tar, particulates, and acid gases.
- Storage and transport require pressurized containment and safety protocols.

Syngas adds economic and environmental value to pyrolysis systems by supporting self-sustaining energy operations and enabling integration with the broader energy grid and chemical industries.

c. **Char:** Char, also called solid residue or carbon black, is the non-volatile, carbon-rich solid left behind after pyrolysis. Its characteristics depend on the type of plastic or biomass feedstock used, as well as the temperature and duration of pyrolysis.

Composition and properties:

- Primarily composed of elemental carbon, inorganic ash, and trace heavy metals.
- Appears as a black powder or granular solid.
- Has a high surface area and porosity?

Applications:

1. Solid fuel:
   - Can be used as a low-grade fuel in cement kilns, boilers, or brick manufacturing.
   - Suitable for co-firing with coal to reduce net carbon emissions.
2. Carbon black substitute:
   - In rubber and plastic industries, char can replace carbon black, a reinforcing filler in tires, hoses, and inks.
   - Requires upgrading for quality control and consistency.
3. Construction materials:
   - Used as a filler in concrete, asphalt, or road surfacing materials.
   - Improves strength and reduces thermal conductivity.
4. Soil amendment (biochar):
   - When made from biomass or food-contaminated plastics, char may be treated and applied as biochar.
   - Enhances soil fertility, moisture retention, and microbial activity.

- Sequesters carbon in the soil, contributing to carbon neutrality.

5. Adsorbents and filtration media:
   - Due to its porosity, char can be activated and used as activated carbon for water or air filtration.
   - Removes heavy metals, dyes, and organic contaminants.

   Challenges:
   - Potential presence of toxins (e.g., heavy metals, dioxins) limits agricultural applications unless treated.
   - Requires quality standardization for industrial use.

Char transforms a waste by-product into a valuable resource, promoting zero-waste pyrolysis models and contributing to circular material loops.

The three primary products of pyrolysis pyrolysis oil, syngas, and char offer diverse applications across multiple industries. Pyrolysis oil serves as both a fuel and a chemical precursor, supporting energy needs and plastic-to-plastic recycling. Syngas is a versatile energy and chemical resource, while char offers functional uses in materials, energy, and agriculture.

Together, these products enable pyrolysis to bridge the gap between waste management and resource recovery, making it a key pillar of the circular economy. With continued technological advancements, regulatory support, and investment in refining infrastructure, the full potential of pyrolysis and its products can be harnessed for sustainable development and environmental protection.

- **Advantages of pyrolysis:**

  Pyrolysis offers several compelling benefits as a recycling technology:

  a. Handles mixed and contaminated plastics: It can process low-quality, heterogeneous, and contaminated plastics that mechanical recycling cannot handle.

  b. Reduces landfill and incineration: Diverts large volumes of plastic waste from landfills and reduces the environmental impact of incineration.

  c. Recovers valuable resources: Produces usable energy and petrochemical feedstocks, reducing dependency on virgin fossil resources.

  d. Compatible with circular economy: Liquid oils can be refined into monomers, enabling plastic-to-plastic recycling and closing the loop.

  e. Energy recovery: Pyrolysis is energy-efficient when syngas is reused for internal heating.

- **Limitations and challenges of pyrolysis:**

  Despite its advantages, pyrolysis faces several technical and commercial challenges.

  a. High capital and operational costs: Building and maintaining a pyrolysis plant requires significant investment and skilled labor.

  b. Product variability: The quality of pyrolysis oil varies based on feedstock, temperature, and process conditions, making standardization difficult.

c. Emissions and residues: Though conducted in an oxygen-free environment, pyrolysis can still produce volatile organic compounds (VOCs), dioxins, and residual char, requiring air pollution control systems.
d. Limited commercial scale: While small-scale units exist, large-scale commercial adoption remains limited due to regulatory and economic uncertainties.
e. Feedstock quality dependence: PVC, PET, and polyurethanes are problematic in pyrolysis due to release of toxic gases like HCl and cyanide.

3.3.2.6 Gasification

Gasification is a cutting-edge thermochemical recycling technology that transforms plastic waste and other carbon-based materials into a valuable energy-rich gas called syngas (synthesis gas). Unlike pyrolysis, which primarily produces liquid fuels, gasification operates in a low-oxygen or controlled-oxygen environment at extremely high temperatures, generally ranging from 800 to 1200°C. This process enables the breakdown of complex hydrocarbons into simpler molecular components, providing a clean and flexible feedstock for energy generation and chemical production.

Gasification involves heating plastic waste in a gasifier, a specially designed reactor, under limited oxygen or steam conditions. The lack of sufficient oxygen prevents complete combustion and instead promotes the conversion of waste materials into syngas, which primarily consists of carbon monoxide (CO), hydrogen (H_2), methane (CH_4), and trace gases like CO_2, H_2S, and nitrogen.

The process typically consists of four key stages:

1. Drying – Moisture is removed from the feedstock.
2. Pyrolysis – Thermal decomposition begins, forming volatile gases and char.
3. Oxidation – Partial combustion of feedstock provides heat.
4. Reduction – Gasification reactions occur, producing syngas from CO_2, H_2O, and char.

Gasification is flexible with regard to feedstock and can process a wide variety of materials:

- Mixed plastic waste (PE, PP, PS, PET, PVC in controlled amounts).
- Biomass (wood, agricultural residues).
- Municipal solid waste (MSW).
- Sewage sludge and industrial residues.

Additives such as air, oxygen, or steam are injected to control the reaction atmosphere and temperature.

Advantages:

- Handles mixed waste: Unlike mechanical recycling, gasification can process contaminated, non-recyclable, and mixed plastics without pre-sorting.

- Energy recovery: Converts waste into syngas, offering a high-energy yield and supporting internal plant operations.
- Supports circular economy: Enables transformation of plastic waste into basic chemicals and fuels.
- Emission control: Advanced gasification units have lower air pollutant emissions than incinerators.
- Carbon capture integration: CO_2 in syngas can be captured and reused in industrial processes.

Challenges and limitations

- High capital cost: Gasification plants require significant investment in reactors, gas cleaning systems, and energy integration.
- Complex operation: Requires precise control of temperature, pressure, and oxygen levels.
- Tar and particulate formation: Syngas must be cleaned of tars, sulfur, and particulates before use.
- Material sensitivity: Certain plastics, especially those with halogens (e.g., PVC), may generate hazardous emissions if not properly controlled.

Gasification represents a powerful and flexible technology in the chemical recycling landscape. By converting plastic waste into syngas and valuable downstream products, it offers a path toward reducing landfill dependency, mitigating environmental impact, and achieving resource circularity. While technical and economic hurdles remain, ongoing innovation and policy support could position gasification as a cornerstone in the transition to sustainable waste management and green energy production.

3.3.2.7 Depolymerization

De-polymerization is a form of chemical recycling that breaks down plastic polymers into their original monomeric units or other valuable chemical intermediates. Unlike thermochemical methods like pyrolysis or gasification, depolymerization is a more selective and controlled process, often carried out under milder temperatures with the help of solvents, catalysts, or enzymes. This process enables the reconstruction of high-quality plastics, creating a circular lifecycle for materials that would otherwise be discarded.

Depolymerization involves cleaving the chemical bonds (such as ester, amide, or ether linkages) that hold long polymer chains together. This is typically achieved through chemical reactions such as hydrolysis, glycolysis, methanolysis, or aminolysis. It is highly effective for condensation polymers, such as PET (polyethylene terephthalate), PS (polystyrene), and polyamides (nylons).

Depending on the method and conditions used, depolymerization can produce:

- Monomers: like ethylene glycol (EG), terephthalic acid (TPA), and styrene.
- Oligomers: intermediates for re-polymerization.
- Useful additives and solvents.

Methods of depolymerization

a. Hydrolysis
  - Breaks down polymers using water under acidic or basic conditions.
  - PET is hydrolyzed into TPA and EG.
  - Requires high pressure and temperature.

b. Glycolysis
  - Uses glycol (e.g., EG) as a solvent to break down PET into BHET (bis-hydroxyethyl terephthalate).
  - Commonly used in industrial PET recycling.

c. Methanolysis
  - Methanol reacts with PET to form DMT (dimethyl terephthalate) and EG.
  - Requires catalyst and high pressure.

d. Enzymatic depolymerization
  - Uses specially engineered enzymes to break down PET and PEF at ambient or moderate temperatures.
  - Highly selective and eco-friendly.

Depolymerization is ideal for homogeneous, sorted plastics and textiles made from PET, nylon, or polyesters. Unlike pyrolysis, it is less suited for mixed or contaminated waste.

Feedstock includes:

- Beverage bottles and food containers (PET).
- Packaging films.
- Worn-out polyester garments.
- Industrial polyamide waste.

The monomers and intermediates recovered via depolymerization are of virgin-like quality, making them ideal for.

- Repolymerization into new plastic products (e.g., PET bottles, polyester fibers).
- Use as feedstock in the chemical industry.
- Blending with virgin materials to reduce carbon footprint.
- Upcycling into specialty polymers or coatings.

Advantages of depolymerization

- High product quality: Recovered monomers are indistinguishable from virgin materials.
- Enables closed-loop recycling: Plastics can be repeatedly recycled without degradation.

- Low emissions: Especially in enzymatic processes.
- Targeted recovery: Specific polymers can be broken down selectively.

Limitations

- Requires sorted feedstock: Mixed plastics can contaminate the process.
- Energy and chemical input: Some methods need high pressure or harsh chemicals.
- Economic viability: Dependent on monomer market prices.
- Scaling enzymatic methods: Still in development and costly at commercial scale.

Depolymerization offers a promising route for achieving high-quality, closed-loop recycling for condensation plastics like PET and nylon. Its ability to recover monomers for reuse in manufacturing makes it a crucial part of the circular economy. With continued investment in catalyst and enzyme development, depolymerization is poised to become a cornerstone in sustainable plastic management strategies.

3.3.2.8 Solvolysis

Solvolysis is an advanced chemical recycling method that breaks down polymers into monomers or oligomers using a solvent as a reactant. The process targets condensation polymers such as PET (polyethylene terephthalate), polyamides (nylons), polyurethanes, and polycarbonates. Solvolysis enables the recovery of high-purity chemical constituents, allowing for the production of virgin-quality plastics and specialty chemicals. Unlike thermal processes like pyrolysis or gasification, solvolysis uses specific chemical reactions, often at moderate temperatures and pressures, making it selective and energy-efficient. This technology is pivotal for the closed-loop recycling of plastics, particularly in the textile and packaging industries. Solvolysis involves the cleavage of polymer chains via chemical reactions between the polymer and a solvent, breaking the ester, amide, or urethane linkages. Depending on the solvent used, solvolysis is classified into different types:

- **Glycolysis**: Uses ethylene glycol or other glycols.
- **Methanolysis**: Uses methanol.
- **Hydrolysis**: Uses water under acidic, neutral, or alkaline conditions.
- **Aminolysis**: Uses amines.
- **Alcoholysis**: Uses alcohols such as ethanol or propanol.

The choice of solvent and reaction condition dictates the efficiency, selectivity, and type of product obtained.

The solvolysis process typically includes the following steps:

Feedstock Preparation: Sorted, clean plastic waste (mostly PET or PU) is shredded and sometimes pre-dried.

Reaction phase: The shredded plastic is mixed with the solvent and catalyst (if needed) in a reactor, operated under controlled temperature and pressure.

Product separation: Monomers and oligomers are separated using distillation, filtration, or crystallization.

Purification and reuse: Products are purified and either repolymerized or sold as chemical intermediates. Solvents are usually recovered and reused.

Solvolysis produces high-purity outputs suitable for reuse in various industries:

- PET waste → Produces BHET (bis-hydroxyethyl terephthalate), TPA (terephthalic acid), and EG (ethylene glycol).
- Polyurethanes → Yields polyols and isocyanate derivatives.
- Nylons → Yields caprolactam (for Nylon-6) or other dicarboxylic acids and diamines.
- Polycarbonates → Can be depolymerized to bisphenol A (BPA) and diols.

These outputs are reused in:

- Bottle and packaging manufacturing.
- Textile production (polyester fibers).
- Polyurethane foams and coatings.
- Automotive and electronics applications.

Advantages:

- High purity of products: Produces monomers suitable for repolymerization into virgin-quality plastics.
- Closed-loop potential: Enables infinite recyclability of certain polymers without degrading material quality.
- Selective and tunable: Can be tailored to specific polymers using appropriate solvents.
- Moderate operating conditions: Lower energy input than pyrolysis or gasification.
- Solvent recovery: Most solvents are recyclable, reducing operational cost and environmental footprint.

Limitations and challenges

- Requires sorted feedstock: Effective only with well-separated and relatively clean plastic waste.
- Chemical handling risks: Solvents and catalysts may be hazardous, requiring careful handling and waste treatment.
- Capital intensive: Initial setup costs are high due to reactors and solvent recovery systems.
- Scalability issues: Limited commercial-scale implementations; many systems are in pilot or demonstration phases.

Solvolysis represents a highly promising and technically mature pathway for recycling condensation polymers into high-value monomers and oligomers. By enabling circularity and preserving material quality, solvolysis helps address one of the most pressing environmental challenges plastic pollution. As technology matures and infrastructure scales, solvolysis will play a critical role in the sustainable management of plastic waste in industries ranging from packaging and textiles to automotive and electronics.

3.3.2.9 *Enzymatic recycling*

Enzymatic recycling is an innovative and eco-friendly method for breaking down plastics, especially polyethylene terephthalate (PET), using specific enzymes. Unlike traditional mechanical or chemical recycling processes, enzymatic recycling operates under mild conditions, offering a sustainable and energy-efficient alternative to deal with plastic waste. This method is gaining traction due to its ability to recycle plastic without degrading its quality, enabling a truly circular economy. Enzymatic recycling is based on the biological activity of enzymes specialized proteins that catalyze chemical reactions. In the context of plastic recycling, these enzymes target the polymer chains of plastics like PET and break them down into their monomer components, such as terephthalic acid (TPA) and ethylene glycol (EG). These monomers can then be purified and reused to produce new plastics of virgin quality.

This process typically involves:

- Pre-treatment of plastic waste (e.g., cleaning, shredding).
- Enzymatic hydrolysis, where enzymes degrade the polymer chains.
- Separation and purification of monomers.
- Repolymerization into new plastic products.

Several types of enzymes have been identified for plastic degradation, with the most notable being:

- PETase: Originally discovered in *Ideonella sakaiensis*, a bacterium capable of breaking down PET. PETase initiates the degradation by hydrolyzing PET into mono(2-hydroxyethyl) terephthalate (MHET).
- MHETase: Works alongside PETase to convert MHET into TPA and EG.
- Cutinases: Found in fungi and some bacteria; also used for PET hydrolysis due to their efficiency under mild conditions.

These enzymes are optimized through protein engineering to improve activity, thermal stability, and resistance to inhibitors.

Advantages of enzymatic recycling

- Mild operating conditions: Enzymatic recycling typically occurs at temperatures between 30 and 70°C and neutral pH, significantly reducing energy requirements compared to traditional recycling.

- High purity output: The process yields high-purity monomers suitable for repolymerization into food-grade plastic, supporting a closed-loop recycling system.
- Selective degradation: Enzymes can be engineered to target specific polymers, reducing the contamination issues faced by mechanical recycling.
- Lower carbon footprint: By avoiding incineration and reducing energy consumption, enzymatic recycling emits fewer greenhouse gases.
- Scalability for mixed waste: Research is ongoing to create enzyme cocktails that can degrade complex or mixed plastic waste streams.

Limitations and challenges

- Cost and efficiency: Current enzyme production and recycling processes are not yet cost-competitive with mechanical recycling on a large scale.
- Slow reaction rates: Compared to chemical methods, enzymatic reactions are slower, though ongoing research in enzyme engineering is improving their kinetics.
- Limited substrate scope: Most successful applications focus on PET; other plastics like polyethylene (PE), polypropylene (PP), and polystyrene (PS) remain difficult to degrade enzymatically.
- Pre-treatment requirements: Plastics need to be cleaned and possibly amorphized to make them more susceptible to enzymatic attack, adding to process complexity.

Enzymatic recycling holds tremendous potential to revolutionize plastic waste management. As enzyme engineering, bioreactor design, and process economics improve, this method could become a vital pillar of circular plastic economy strategies. With its low environmental footprint and capacity for high-quality recycling, enzymatic recycling can play a pivotal role in global sustainability goals and plastic pollution mitigation. Enzymatic recycling is an emerging green technology offering a clean, efficient, and potentially scalable solution to the growing plastic waste crisis. Though currently limited in scope and scalability, continuous scientific advancements are paving the way for its wider adoption and integration into global recycling infrastructures. As society moves toward more sustainable practices, enzymatic recycling represents a transformative step toward truly circular materials management.

3.3.2.10 Hydrogenolysis

Hydrogenolysis is a powerful chemical recycling method that uses hydrogen gas to break down plastic polymers into smaller hydrocarbon molecules, such as alkanes, which can be further processed into fuels, lubricants, or new chemical feedstocks. This process represents a significant advancement in converting waste plastics especially polyolefins like polyethylene (PE) and polypropylene (PP) into valuable, usable products with minimal environmental impact. Hydrogenolysis involves cleaving carbon–carbon (C–C) or carbon–heteroatom (C–X) bonds in the presence of hydrogen gas and a metal catalyst. In the case of plastic waste, the long hydrocarbon chains are broken into smaller molecules when hydrogen atoms insert themselves into the broken bonds. This process typically occurs under.

- Elevated temperatures (150–300°C)
- Moderate to high hydrogen pressures (10–50 bar)
- Presence of metal catalysts such as platinum (Pt), ruthenium (Ru), or palladium (Pd) supported on carriers like carbon, silica, or alumina.

Hydrogenolysis is especially suited for polyolefins, which are otherwise very difficult to recycle due to their chemical inertness. These include.

- Polyethylene (PE)
- Polypropylene (PP)
- Polystyrene (PS) (with more complex mechanisms) It is less commonly used for PET and PVC due to their heteroatom content, which may poison catalysts or lead to toxic by-products.

Depending on the process conditions and catalysts, hydrogenolysis can yield.

- Short-chain alkanes: For use as gases (e.g., methane, ethane), fuels, or feedstocks.
- Liquid hydrocarbons: Such as diesel, gasoline-range fuels, and lubricants.
- Waxes and oils: Useful in various industrial and consumer applications.

This ability to tune the reaction conditions allows targeted production of desired output molecules. The choice of catalyst is crucial in determining the efficiency, selectivity, and longevity of the hydrogenolysis process. Common catalysts include:

- Ruthenium (Ru): Highly active for C–C bond cleavage in polyolefins and offers high selectivity toward desired fuel-range hydrocarbons.
- Platinum (Pt) and palladium (Pd): More expensive but offer excellent activity and stability under harsh conditions.
- Bimetallic catalysts (e.g., Ru–Ni, Pt–Re): Engineered to improve performance, selectivity, and resistance to deactivation.

Researchers are also exploring earth-abundant and low-cost catalysts to make hydrogenolysis more economically feasible.

Advantages of hydrogenolysis

- High efficiency for polyolefins: Converts even highly crystalline PE/PP into useful hydrocarbons.
- Value-added products: Produces fuel, waxes, and lubricants with significant commercial value.

- Scalable technology: Adaptable to large-scale reactors, making it attractive for industrial use.
- Cleaner process: Compared to pyrolysis or incineration, hydrogenolysis offers better control over product composition and fewer unwanted by-products.
- Versatile feedstock: Tolerates a range of plastic grades, including mixed and contaminated polyolefin waste.

Challenges and limitations

- High hydrogen consumption: Sourcing green hydrogen sustainably is essential to prevent net emissions.
- Catalyst deactivation: Catalysts can be poisoned by additives or impurities in waste plastics, requiring careful feedstock preparation.
- Cost of catalysts: Noble metals are expensive, and regeneration or recycling is needed.
- Energy demand: Requires elevated temperature and pressure, demanding energy input and reactor robustness.

Hydrogenolysis is a promising frontier in the chemical recycling of plastics, especially polyolefins that are largely neglected by mechanical and enzymatic processes. By leveraging hydrogen and catalytic chemistry, this method not only diverts plastic from landfills but also transforms it into valuable resources, supporting both environmental and economic sustainability. While the challenges of cost, catalyst life, and green hydrogen supply remain, continued innovation is pushing hydrogenolysis closer to large-scale, real-world application as a key solution to the plastic pollution crisis.

3.3.2.11 Aminolysis and alcoholysis

Aminolysis is a chemical depolymerization process in which polymers typically polyesters (like PET) or polyurethanes are broken down using amines. This process involves the cleavage of ester bonds in the polymer backbone by nucleophilic attack of an amine group ($-NH_2$), yielding amide derivatives and other useful monomers. The general reaction is: ***Polyester*** + ***Amine*** → ***Amide derivatives*** + ***Glycol***

Types of amines used

- Primary amines (e.g., methylamine, ethylamine)
- Secondary amines (e.g., diethylamine)
- Polyfunctional amines (used in polyurethane recycling)

Process conditions

- Temperature: Typically 80 to 200°C
- Pressure: Ambient to slightly elevated

- Catalysts: Sometimes used (acid or base) but often reaction proceeds without one
- Solvent: Reaction can be solvent-free or in organic solvents

Applications and products

- Recovery of monomers like terephthalamide and ethylene glycol from PET.
- Useful in producing urea-based intermediates and polyurethane precursors.
- Synthesized amides can be used in pharmaceuticals, resins, and coatings.

Advantages

- High specificity and selectivity toward ester linkages.
- Produces functionalized molecules with amine groups, useful in chemical industries.
- Suitable for functional upcycling into value-added chemicals.

Limitations

- Not suitable for all plastic types (mainly for condensation polymers).
- Amines may be costlier and more toxic than other reagents.
- Requires careful handling and purification of reaction products.

Alcoholysis is a chemical recycling method in which alcohols (like methanol, ethanol, or ethylene glycol) are used to cleave the ester linkages in plastics, especially polyesters and polycarbonates. The process breaks down the polymer into alkyl esters and diols, which can be purified and reused for polymer synthesis:

PET + Alcohol $\rightarrow$ Alkyl terephthalate + Ethylene glycol

For example, Methanolysis uses methanol to yield dimethyl terephthalate (DMT) and EG.

Types of alcoholysis

- Methanolysis (most common): PET $\rightarrow$ DMT + EG
- Ethanolysis: PET $\rightarrow$ Diethyl terephthalate + EG
- Glycolysis (variant): PET $\rightarrow$ BHET (Bis-hydroxyethyl terephthalate)

Process conditions

- Temperature: 150 to 250°C
- Pressure: Often elevated (for methanolysis)
- Catalysts: Metal acetates (e.g., zinc acetate), organic bases, or ionic liquids
- Reaction time: Ranges from 30 min to a few hours

Applications and products

- Dimethyl terephthalate (DMT) is a key monomer reused in PET production.

- Alcoholysis of polycarbonates yields diaryl carbonates, useful in green chemistry.
- Glycolysis products (BHET) are used to make new PET and polyurethanes.

Advantages

- High efficiency and purity of recovered monomers.
- Effective for colored and contaminated plastics where mechanical recycling fails.
- Enables closed-loop recycling of PET and polycarbonates.

Limitations

- High pressure and temperature requirements.
- Use of toxic or volatile solvents like methanol demands strict safety protocols.
- Catalyst recovery and separation can be challenging.

Recent developments

- Microwave-assisted alcoholysis for faster depolymerization.
- Use of ionic liquids and deep eutectic solvents as green alternatives to conventional catalysts.
- Integration with renewable alcohols to reduce carbon footprint.

Both aminolysis and alcoholysis offer promising chemical recycling routes, particularly for hard-to-recycle plastics like PET and polyurethanes. Aminolysis produces functional amide derivatives suitable for specialty applications, while alcoholysis facilitates the efficient recovery of clean monomers for virgin-grade plastic production. As research continues to refine catalysts and process conditions, these methods are becoming more viable for industrial-scale adoption, contributing to a circular economy and sustainable materials management.

3.3.2.12 Catalytic cracking

Catalytic cracking is a thermochemical process traditionally used in petroleum refining to break down large hydrocarbon molecules into smaller, more valuable compounds like gasoline, diesel, and olefins. In the context of plastic waste recycling, catalytic cracking has been adapted to depolymerize plastic polymers, especially polyolefins such as polyethylene (PE) and polypropylene (PP), into fuel-grade hydrocarbons and chemical feedstocks. The presence of a catalyst lowers the temperature required for cracking and improves the selectivity of the process toward desired products, offering a more energy-efficient and controlled approach than thermal cracking or pyrolysis. Catalytic cracking involves two primary steps.

- Adsorption and activation of the polymer on the catalyst surface.
- Cleavage of C–C and C–H bonds, leading to the formation of smaller hydrocarbons like alkanes, alkenes, and aromatics.

This process typically occurs in a fluidized bed or fixed-bed reactor, under.

- Temperature: 350 to 600°C (lower than pyrolysis)
- Pressure: Atmospheric or slightly elevated
- Catalyst: Solid acid catalysts like zeolites

The choice of catalyst plays a central role in the efficiency and product distribution. Common catalysts include:

- Zeolites (e.g., ZSM-5, Y-zeolite): Provide high surface area and acidic sites that promote cracking and aromatization.
- Silica-alumina: Used as a base catalyst with moderate acidity.
- Metal-loaded catalysts: Nickel (Ni), Iron (Fe), or Palladium (Pd) are added to enhance dehydrogenation or hydrogen transfer.
- Clay-based catalysts: Inexpensive alternatives with lower performance but viable for bulk waste conversion.

These catalysts can be regenerated and reused, though they may suffer deactivation due to coke deposition.

Catalytic cracking is most suitable for:

- Polyolefins: Polyethylene (PE), polypropylene (PP)
- Polystyrene (PS): Cracks into styrene monomers and aromatics
- Mixed plastic waste: Particularly in municipal solid waste streams

It is less effective for oxygenated polymers like PET or PVC without preprocessing, as these may form corrosive or toxic by-products (e.g., HCl from PVC).

The output of catalytic cracking varies based on the feedstock and reaction conditions but typically includes:

- Gaseous hydrocarbons: Methane, ethane, propylene, butylene
- Liquid hydrocarbons:
- Gasoline-range fuels
- Diesel-range alkanes
- Aromatics (e.g., benzene, toluene, xylene)
- Solid residues: Coke (carbonaceous material formed on catalyst)

The process can be tuned to favor different product distributions—more light olefins for petrochemicals or more liquids for fuels.

Advantages of catalytic cracking

- Lower temperature requirements: Compared to pyrolysis, catalytic cracking operates at significantly lower temperatures, reducing energy consumption.

- Higher selectivity: Catalysts direct the reaction toward valuable products like aromatics or olefins.
- Suitable for mixed waste: Can handle contaminated or composite plastics to some extent.
- Well-established technology: Leverages existing refining infrastructure and knowledge.
- Scalable process: Already used in petroleum refineries, making industrial-scale application feasible.

Limitations and challenges:
Despite its advantages, catalytic cracking also presents challenges.

- Coke formation: Carbonaceous deposits on catalysts reduce activity and require regeneration.
- Catalyst deactivation: Impurities (e.g., halogens, sulfur) in waste plastics can poison the catalyst.
- Product variability: Inconsistent plastic feedstock can lead to fluctuating product quality.
- Cost of catalyst recovery: Though reusable, catalysts must be regenerated or replaced periodically, increasing operational costs.

Several companies and research groups are working to integrate catalytic cracking with waste management systems and refinery operations, turning waste into high-value fuels or chemical feedstocks in a circular economy framework. Catalytic cracking presents a highly promising and flexible route for the chemical recycling of plastic waste, particularly polyolefins. It offers advantages in terms of energy efficiency, product selectivity, and integration with existing petrochemical infrastructure. While challenges like catalyst deactivation and feedstock variability remain, continued research in catalyst development and reactor design is pushing catalytic cracking closer to commercial viability. As part of a broader recycling and circular economy strategy, it holds strong potential for addressing the global plastic waste crisis.

3.3.3 Biological recycling technologies

Biological recycling, often referred to as biodegradation or bioconversion, is a natural and sustainable method for treating and managing waste using biological organisms such as bacteria, fungi, algae, insects, and enzymes. It plays a crucial role in recycling organic waste and biodegradable plastics, helping transition from a linear economy to a circular and regenerative waste management model. Unlike mechanical or chemical recycling, biological recycling relies on nature's own decomposition mechanisms, making it eco-friendly and less energy-intensive. With increasing environmental awareness and the need for sustainable waste processing, biological recycling technologies are being researched, developed, and deployed across various sectors, including agriculture, municipal solid waste management, and even in tackling plastic pollution. Biological recycling operates on the principle of biodegradation, wherein complex organic compounds or biodegradable polymers are broken down into simpler substances such as carbon dioxide (CO_2), water (H_2O), methane (CH_4), and biomass. This is facilitated by enzymes secreted by microorganisms or by digestive systems of organisms like earthworms or insects. Fig. 3.11 illustrate core process of biological recycling technologies. The core processes involve:

- Hydrolysis: Breaking down of large molecules using water and enzymes.
- Oxidation and reduction: Microbial metabolic reactions that release energy.
- Mineralization: Conversion of organic matter into inorganic forms.
- **Types of biological recycling technologies**

Biological recycling harnesses the power of living organisms to break down and convert waste into reusable forms, playing a critical role in sustainable waste management and circular economy models. Among the various approaches, microbial degradation stands out as a foundational process where bacteria and other microorganisms decompose complex organic substances into simpler compounds. These microbes secrete enzymes that target polymers such as plastics or plant materials, enabling degradation in soil, water, or compost environments. Building upon this, enzymatic recycling uses isolated enzymes often produced by microbes to catalyze the breakdown of specific materials, such as polyethylene terephthalate (PET) in plastics. This method is highly selective and operates under mild conditions, offering a promising solution for high-purity recycling without extensive energy input.

Composting is one of the oldest and most accessible biological recycling methods. It involves aerobic decomposition of organic waste such as food scraps, garden clippings, and manure by naturally occurring microbes. The result is nutrient-rich compost that enhances soil quality and promotes plant growth. Unlike composting, anaerobic digestion occurs in the absence of

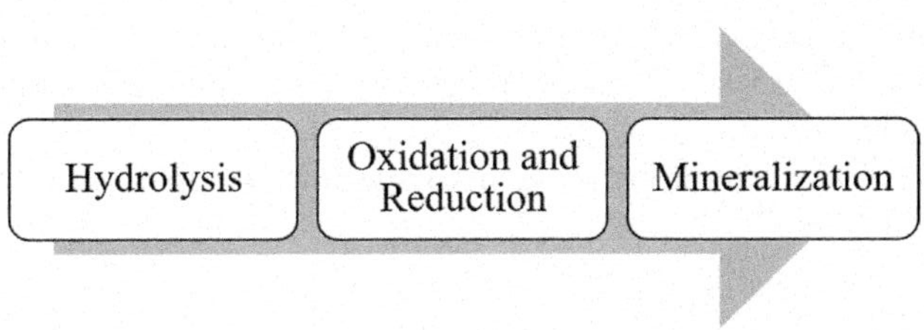

FIGURE 3.11 Core process of biological recycling technologies.

oxygen and is typically conducted in sealed digesters. Here, microorganisms decompose organic matter, producing biogas (a mix of methane and CO_2) and a nutrient-rich slurry. This dual output supports renewable energy production and organic fertilizer development.

Another emerging technique is mycoremediation, which utilizes fungi to degrade or neutralize pollutants. Fungi secrete extracellular enzymes capable of breaking down tough substances like hydrocarbons, pesticides, and even some plastics. Their mycelial networks infiltrate waste material, making them especially effective for treating contaminated soils and industrial waste. In parallel, algal biodegradation leverages algae's ability to absorb and degrade pollutants, especially in water bodies. Some algae can metabolize hydrocarbons and heavy metals, helping in wastewater treatment and the remediation of aquatic ecosystems.

Insect-based bioconversion has gained traction for converting organic waste into protein and fat through insects like black soldier fly larvae. These insects consume large quantities of food or agricultural waste and are later processed into animal feed, biofuels, or fertilizers. On a microscopic level, symbiotic gut microbiota in insects, ruminants, or even humans play a role in breaking down substances like cellulose or complex polymers. Research into replicating or modifying these microbiomes holds promise for bioreactors or future waste management systems. Lastly, bio-upcycling refers to biologically converting low-value or waste materials into high-value products. For instance, engineered microbes or enzymes can transform plastic waste into bioplastics or textile fibers. Unlike traditional recycling, which often downcycles materials, bio-upcycling offers a way to enhance the value of waste while closing the resource loop. Collectively, these biological recycling processes provide innovative, eco-friendly pathways for managing waste, reducing environmental pollution, and supporting resource circularity. Their diversity and adaptability make them integral components of next-generation waste management systems. Table 3.2 illustrate comparison of different types of biological recycling technologies.

Advantages of biological recycling

- Eco-friendly: Operates under mild, non-toxic conditions.
- Energy-efficient: Low energy input compared to pyrolysis or incineration.
- High-purity output: Especially in enzymatic depolymerization.
- Supports circular economy: Converts waste into reusable resources.
- Flexible feedstock: Handles mixed or contaminated organic waste effectively.
- Scalable for decentralized systems: Especially composting and anaerobic digestion.

TABLE 3.2 Comparison of different types of biological recycling technologies.

Technology	Source organism	Input material	Operating conditions	End products	Speed of degradation	Scalability	Current applications
Microbial degradation	Bacteria, Archaea	Organic waste, some plastics, hydrocarbons	Ambient to mesophilic (20–45°C)	CO_2, water, biomass	Moderate	High	Soil remediation, wastewater treatment
Enzymatic recycling	Enzymes from microbes	PET plastics, starch, cellulose	Mild, pH-specific, enzyme-optimized	Monomers (e.g., terephthalic acid)	Fast (for specific polymers)	Medium	Plastic recycling, biochemical synthesis
Composting	Aerobic microbes	Food waste, garden waste, paper	Aerobic, 30–60°C	Compost (fertilizer)	Moderate	High	Agricultural composting, municipal waste systems
Anaerobic digestion	Anaerobic bacteria	Organic waste, sewage sludge, manure	Anaerobic, meso/thermophilic (35–55°C)	Biogas, digestate	Moderate	High	Biogas plants, renewable energy generation
Mycoremediation	Fungi (e.g., white-rot fungi)	Toxic pollutants, petroleum, lignin	Moist, low-light, pH-dependent	Detoxified soil, fungal biomass	Slow to moderate	Medium	Environmental cleanup, soil detoxification

Continued

TABLE 3.2 Comparison of different types of biological recycling technologies.—cont'd

Technology	Source organism	Input material	Operating conditions	End products	Speed of degradation	Scalability	Current applications
Algal biodegradation	Microalgae, cyanobacteria	Wastewater nutrients, hydrocarbons	Aquatic, sunlight, CO_2	Biomass, treated water	Moderate	Medium	Water purification, biofuel production
Insect-based bioconversion	Insects (e.g., BSF larvae)	Food waste, agricultural waste	Ambient to warm (25–35°C)	Protein meal, oil, compost	Fast	High	Animal feed, biofertilizers, waste reduction
Symbiotic gut microbiota	Gut microbes in insects/ animals	Cellulose, lignin, synthetic polymers	Anaerobic, body temp (35–40°C)	Simpler organics, gases	Variable	Low (experimental)	Research in bioreactors, plastic degradation
Bio-upcycling	Engineered microbes, enzymes	Low-value plastics, agricultural residues	Tailored to process; bioreactor-based	High-value chemicals, bioplastics	Fast (engineered)	Medium to High	Sustainable materials, circular economy solutions

3.4 Recycling of materials used in renewable energy systems

As renewable energy technologies such as solar panels, wind turbines, batteries, and fuel cells grow in scale and adoption, a parallel concern emerges what happens to these systems at the end of their life? Renewable energy may be green in operation, but without proper end-of-life management, the materials used in these technologies could contribute to significant environmental problems. Therefore, recycling of materials in renewable energy systems is crucial for sustainability, resource conservation, and circular economy goals.

3.4.1 Recycling of solar energy system material

As the world transitions to cleaner energy, solar power has emerged as one of the most promising and widely adopted renewable energy sources. Solar photovoltaic (PV) systems convert sunlight into electricity and are now found on rooftops, in solar farms, and integrated into devices worldwide. However, with the average lifespan of solar panels being 25 to 30 years, the first generations of solar panels are already reaching the end of their life. This raises a critical question: What happens to solar panels after their operational phase ends? The answer lies in effective and sustainable recycling processes for solar energy system materials.

The drive for solar energy sustainability must go beyond electricity generation and address the entire life cycle, including end-of-life (EoL) management. Without proper recycling, solar panels can contribute to electronic waste, leach hazardous materials into the environment, and waste valuable resources like silver, silicon, and rare metals.

Key benefits of recycling:

- Environmental protection – Prevents landfill disposal and pollution.
- Resource conservation – Recovers valuable and scarce materials.
- Economic potential – Generates raw material supply for new solar products.
- Job creation – Development of recycling industries and technologies.

Components of solar panels and their material composition: Solar PV panels typically consist of the following main layers and components:

1. Glass (70%–80%) – Used as a front protective layer.
2. Aluminum frame – Provides structural support.
3. Silicon cells – The semiconductor layer that generates electricity.
4. Metals – Such as silver, copper, and tin used in wiring and contacts.
5. Polymers – Ethylene-vinyl acetate (EVA) used for encapsulation and a back sheet layer made of plastic or composite.

These diverse materials are tightly integrated, making disassembly and recycling a complex and multi-step process.

- **The recycling process: step-by-step**

The recycling of solar panels can be divided into three main stages: collection and transportation, dismantling and separation, and material recovery.

Stage 1: Collection and transportation

- Collection centers: EoL solar panels are collected from solar farms, residential units, or recycling drop-off points.
- Inspection: Panels are checked to determine whether they can be reused, refurbished, or fully recycled.
- Logistics: Transported safely to specialized recycling facilities. Handling is done carefully to avoid breakage or exposure to hazardous materials (e.g., cadmium in some thin-film panels).

Stage 2: Dismantling and initial separation.

At the recycling facility, panels undergo manual and automated processes to separate major components.

1. Aluminum frame removal:
    - Easily detached using mechanical tools or automated equipment.
    - Recycled separately through conventional metal recycling.
2. Glass removal:
    - The front glass layer is detached using machines that cut and lift the pane.
    - If unbroken, glass can be cleaned and reused or melted and reprocessed into new products.
    - In some processes, glass is crushed and sieved for use in tiles, sandblasting, or construction.
3. Junction box and wiring removal:
    - Cables and connectors are removed and sent for **copper and plastic recovery.**
    - Silver and tin used in soldering may also be extracted.

Stage 3: Thermal, chemical, and mechanical treatments.

After dismantling, the remaining layer consists of silicon cells, encapsulant, and backsheet. This part is complex and requires advanced processes:

Thermal process (pyrolysis or thermal delamination)

- The encapsulant (EVA) and backsheet layers are made of heat-resistant plastic that must be burned or melted off.
- Panels are heated in a furnace to temperatures around 500 to 600°C.
- Organic layers evaporate, leaving behind bare silicon cells and metals.
- Fumes are captured and treated to avoid pollution.

Advantages:

- Effective at removing tough polymer layers.
- Enables separation of high-purity silicon wafers.

Challenges:

- High energy consumption.
- Risk of emissions from burnt polymers.

Chemical processing

- Involves leaching or solvent treatments to extract materials like silver, lead, and indium (in thin-film panels).
- Acids such as nitric or sulfuric acid dissolve metal components, which are then precipitated or filtered.
- Remaining silicon wafers may be purified and reused.

Benefits:

- Recovers high-value metals.
- Can yield silicon of sufficient purity for reprocessing into new cells.

Challenges:

- Requires careful handling of hazardous chemicals.
- Expensive and slower than mechanical processes.

Mechanical Processing (Shredding and Milling)

- Panels are shredded into fine pieces using industrial mills.
- Material fragments are separated by:
- Magnetic separation – to isolate ferrous metals.
- Eddy current separation – to extract non-ferrous metals like aluminum and copper.
- Optical sorting to identify and classify glass and plastic fragments.

Pros:

- Scalable and fast.
- Minimal chemical use.

Cons:

- Reduces material purity.
- Silicon is often down cycled (not reused in electronics).

Recycling solar energy system materials is no longer optional it is a necessity to ensure the long-term sustainability of clean energy. By effectively recovering glass, metals, and semiconductors from old panels, we not only reduce waste but also create opportunities to reduce reliance on virgin materials and lower the carbon footprint of manufacturing. While technological, regulatory, and economic challenges remain, the solar industry is moving steadily toward a circular future. With continued innovation, collaboration, and investment, solar recycling can play a vital role in a truly sustainable energy ecosystem.

3.4.2 Recycling process for wind energy system materials

Wind energy has become a cornerstone of global efforts to transition to renewable and sustainable energy sources. Massive wind turbines now dot landscapes and shorelines across the world, harnessing the kinetic energy of wind to generate electricity. However, like all infrastructure, wind turbines have a limited lifespan typically 20 to 25 years. As thousands of turbines reach their end-of-life (EoL) phase, the question arises: What happens to these structures when they are decommissioned? The answer lies in systematic recycling and repurposing of wind energy system materials. Recycling wind turbine components is crucial for reducing environmental impact, conserving valuable materials, and promoting a circular economy within the renewable energy sector. Wind turbines are complex structures composed of multiple materials many of them recyclable, others more difficult to manage. Without proper recycling, these components could end up in landfills or result in significant environmental harm. Recycling supports.

- Resource recovery (e.g., steel, copper, rare earth elements)
- Waste reduction
- Cost savings and environmental benefits
- A circular approach to clean energy infrastructure

Components of a wind turbine and their material composition.
Wind turbines consist of three main parts.

1. Tower
   - Material: Steel (85%–90%), concrete in onshore turbines
   - Function: Provides height to optimize wind capture
2. Nacelle
   - Material: Copper, aluminum, plastics, steel, rare earth magnets
   - Function: Contains gearbox, generator, and electronic components
3. Blades
   - Material: Composite materials (fiberglass, carbon fiber, epoxy resins)
   - Function: Capture wind energy and transfer torque

Step-by-step recycling process for wind turbine materials.

The recycling of wind turbines is a multi-phase process, involving disassembly, material separation, and recovery using mechanical, thermal, or chemical methods.

Stage 1: Decommissioning and transportation

- Planning and safety: Before dismantling, risk assessments and safety protocols are developed.
- Disassembly: Turbines are taken apart at the site using cranes and specialized equipment.
- Transport: Components are loaded onto trucks and transported to recycling or repurposing facilities. Blade transportation is especially challenging due to their size and rigidity.

Stage 2: Material-specific recycling processes

A. Tower recycling (steel and concrete)

Process:

- Cutting and shredding: Steel tower sections are cut into manageable pieces.
- Magnetic separation: Any non-steel metals are removed.
- Melting: Steel is melted in furnaces and reused in the construction or automotive industry.

Concrete bases (if removed) can be:

- Crushed and reused as aggregate in new construction or roadwork.

Advantages:

- Steel is highly recyclable with well-established infrastructure.
- Minimal loss of material value.

B. Nacelle recycling (metals, electronics, and magnets)

Components:

- Generator: Contains copper windings, steel casing, aluminum parts
- Gearbox and Yaw System: Steel gears, oil, and lubricants
- Control systems: Electronic boards and wiring

Recycling methods.

- Mechanical dismantling: Manual or robotic systems separate components.
- Metal recovery:
- Copper and aluminum are extracted through shredding, magnetic and eddy current separation.
- Steel and other alloys are smelted and reused.

Rare earth magnets:

- Found in permanent magnet generators (especially in direct-drive turbines).

- Elements like neodymium, dysprosium are valuable and critical for high-tech applications.
- Chemical leaching and hydrometallurgical processes are used for recovery.

Challenges:

- Some electronic components contain hazardous substances (e.g., lead, mercury).
- Rare earth magnet recycling is technically complex and still under development at scale.

C. Blade recycling (composite materials)

Biggest recycling challenge in wind energy systems.
Composition.

- Fiberglass (primary)
- Carbon fiber (for advanced designs)
- Thermoset resins (epoxy, polyester)

Why are blades hard to recycle?

- Blades are made from thermoset composites which cannot be melted and reshaped like thermoplastics.
- Strong, lightweight, and durable—but these properties hinder recyclability.

Recycling methods:

1. Mechanical recycling:
   - Blades are cut and shredded into smaller pieces.
   - Ground material is used as filler in cement, asphalt, insulation materials, or construction boards.
   - Downcycling – the value and quality of recycled material is lower than the original.
2. Thermal processing:
   - Pyrolysis: Blades are heated in the absence of oxygen to decompose resins and recover fibers.
   - Fluidized bed processing: Involves burning off resins in a fluidized sand bed, recovering glass or carbon fibers.
3. Chemical recycling (solvolysis):
   - Uses solvents to dissolve the resin matrix and extract fibers.
   - Preserves fiber strength for reuse.
4. Co-processing in cement kilns:
   - Entire blade sections are burned in cement kilns.
   - Energy value of resins and silica content contribute to cement production.
   - This method is used widely in Europe.

3.5 Innovative recycling technology

In recent years, innovative recycling technologies have emerged to tackle the growing challenge of waste management and promote a circular economy. These technologies go beyond traditional mechanical recycling and aim to recover valuable materials from complex or previously unrecyclable waste streams. One of the most notable innovations is chemical recycling, where plastics are broken down into monomers or fuels using processes like pyrolysis, de-polymerization, or solvolysis, allowing for the recycling of multilayer and contaminated plastics. Companies like Carbios are also pioneering enzymatic recycling to depolymerize PET using biological enzymes. Another advancement is the use of AI-powered robotic sorting systems, developed by firms such as AMP Robotics, which use computer vision and machine learning to efficiently sort recyclables with high accuracy and speed. In the field of e-waste recycling, bioleaching using microbes to extract precious metals like gold, copper, and cobalt is gaining attention for being eco-friendly and cost-effective. Similarly, direct recycling of lithium-ion batteries preserves battery components for reuse without energy-intensive smelting. For composite waste, like wind turbine blades, innovative solutions include thermal and chemical processing, or even repurposing into construction materials and furniture.

These technologies represent a shift toward smart, sustainable, and high-efficiency recycling systems that support resource recovery, reduce environmental impact, and align with global sustainability goals.

3.5.1 AI-powered robotic sorting systems

In the era of rapidly growing waste generation, especially with the complexity of modern packaging and the rise in e-waste, traditional waste sorting methods are becoming inefficient, costly, and labor-intensive. This has led to the rise of AI-powered robotic sorting systems, a cutting-edge innovation that is revolutionizing the waste management and recycling industry. These intelligent systems combine artificial intelligence (AI), computer vision, and robotics to identify, sort, and separate waste materials with high precision, speed, and consistency—outperforming human capabilities in many cases.

At the heart of these systems lies machine learning algorithms trained on thousands of images of waste materials. These algorithms enable robots to distinguish between different types of plastics (like PET, HDPE, PVC), metals, paper, glass, and even complex multi-material products. Through deep learning, the robots continuously improve their accuracy and adaptability by learning from real-time data. The integration of conveyor belts, sensors, high-resolution cameras, and robotic arms ensures swift and efficient handling of materials in recycling facilities.

One of the key drivers for the development of robotic sorting systems is the challenge posed by contamination in recycling streams. Contaminated or improperly sorted materials reduce the quality of recycled products and often result in valuable resources being sent to landfills. AI-based sorting robots address this issue by recognizing and removing contaminants, thus increasing purity levels of sorted recyclables and improving the economic viability of recycling operations.

Below are some key features and benefits of AI-powered robotic sorting systems:

- High sorting accuracy: These systems can achieve accuracy levels of 90% or higher, distinguishing between similar-looking materials with remarkable precision.
- Speed and efficiency: Robotic sorters can process thousands of pieces per hour—far beyond human capability leading to higher throughput and reduced bottlenecks in waste facilities.
- Labor cost reduction: With the automation of sorting tasks, facilities reduce dependence on manual labor, which is often subject to fatigue, injury, and high turnover rates.
- Real-time adaptability: AI models are updated continuously, enabling systems to adapt to changing waste streams (e.g., new packaging types, seasonal variations).
- Data-driven insights: These systems collect and analyze vast amounts of data, providing waste management companies with valuable insights into waste composition, contamination sources, and material recovery rates.
- Remote monitoring and diagnostics: AI sorting systems often come with remote connectivity, allowing for predictive maintenance, diagnostics, and performance monitoring from centralized control centers.

Several technology companies have pioneered the development and deployment of these intelligent systems. For instance, AMP Robotics (USA) has developed the AMP Cortex system, which uses AI to identify and sort various recyclable materials at high speed. Similarly, ZenRobotics (Finland) has introduced heavy-duty robotic systems for sorting construction and demolition waste. These systems are capable of recognizing over 20 types of materials and can be programmed to prioritize high-value recyclables.

AI-powered robotic sorting systems are also being adopted in municipal solid waste (MSW) plants, material recovery facilities (MRFs), electronic waste recycling centers, and plastic recovery operations. The scalability and modularity of these systems make them suitable for both large-scale urban recycling centers and decentralized waste management solutions in smaller municipalities or industrial zones.

Despite their immense benefits, there are some challenges and limitations to be addressed. The high initial investment cost can be a barrier for small or underfunded facilities. In addition, frequent retraining of AI models may be necessary when introducing new materials or packaging types into the waste stream. However, these issues are gradually being resolved through innovations in cloud-based AI training, subscription-based robotic systems, and collaborations between governments and private tech companies.

From a sustainability perspective, AI-powered robotic sorting plays a crucial role in achieving circular economy goals by improving the quality and quantity of recovered materials. It reduces waste sent to landfills or incinerators and enables closed-loop recycling, where materials are reused repeatedly with minimal quality degradation. Furthermore, by increasing the efficiency of recycling operations, these systems contribute to lower greenhouse gas emissions, reduced resource extraction, and conservation of natural ecosystems.

AI-powered robotic sorting systems represent a significant leap forward in modern waste management. By merging the precision of robotics with the intelligence of AI, these systems enhance recycling efficiency, reduce environmental impact, and support a more sustainable and circular economy. As technology continues to evolve and costs decrease, we can expect wider adoption of these systems, leading to smarter, cleaner, and more efficient waste handling solutions around the world.

3.5.2 Green concrete from recycled materials: A sustainable innovation in construction

As global construction activities continue to grow, the demand for concrete a fundamental building material has surged dramatically. However, traditional concrete production is one of the largest contributors to carbon dioxide (CO_2) emissions, primarily due to the energy-intensive cement manufacturing process. In response to environmental challenges and increasing awareness about sustainable development, researchers and industry leaders have developed Green Concrete, an innovative solution that incorporates recycled materials to reduce environmental impact while maintaining structural performance.

Green Concrete is essentially a sustainable alternative to conventional concrete where components such as cement, aggregates, and water are partially or wholly replaced by recycled or waste materials. These materials can include industrial by-products, demolition waste, recycled aggregates, fly ash, silica fume, slag, recycled plastic, and even shredded rubber. The main goal of green concrete is to reduce carbon emissions, conserve natural resources, minimize landfill waste, and support a circular economy in construction.

Key features and benefits of green concrete:

- Use of recycled aggregates: Crushed concrete from demolished buildings is used as coarse or fine aggregates, significantly reducing the need for virgin materials.

- Supplementary cementitious materials (SCMs): Waste products like fly ash (from coal plants), ground granulated blast-furnace slag (GGBS), and silica fume are used to partially replace Portland cement, reducing CO_2 emissions.
- Waste plastic and rubber integration: Shredded plastic or tire rubber is used to improve flexibility and reduce brittleness in some structural and non-structural applications.
- Reduced carbon footprint: Green concrete can reduce CO_2 emissions by up to 30% to 40% depending on the materials used.
- Improved durability: Certain green concrete formulations exhibit better resistance to chloride penetration, sulfate attack, and thermal cracking, increasing the life of the structure.
- Cost efficiency: In some regions, using waste materials is more cost-effective than sourcing virgin materials, especially when waste disposal costs are considered.

A prominent example is the use of fly ash, a by-product of coal combustion, as a partial cement replacement. It not only reduces emissions but also enhances the workability and long-term strength of concrete. Similarly, recycled concrete aggregate (RCA) is increasingly used in road bases, pavements, and even structural elements, proving that recycled materials can be structurally viable.

Despite the many advantages, green concrete faces a few challenges:

- Variability in waste material quality: Inconsistent composition and impurities in recycled materials may affect strength and performance.
- Limited awareness and acceptance: Contractors and developers may hesitate to adopt green concrete due to lack of knowledge or performance concerns.
- Regulatory and standardization gaps: Building codes in some regions still do not recognize or support green concrete practices fully.

Green concrete from recycled materials is a powerful step toward sustainable construction and environmental conservation. By turning waste into a valuable resource, it not only addresses the global waste problem but also significantly lowers the environmental footprint of infrastructure development. As research continues and awareness spreads, green concrete has the potential to become the new standard in eco-friendly construction.

3.5.3 Mycoremediation and mycorecycling: Fungal solutions for sustainable waste management

In the quest for sustainable and eco-friendly waste management practices, mycoremediation and mycorecycling have emerged as promising biological technologies. Both are based on the use of fungi, particularly their mycelium (the vegetative root-like structure), to break down, detoxify, or transform waste materials. These processes are not only low-cost and low-energy but also align perfectly with the principles of the circular economy and biological recycling.

Mycoremediation is a form of bioremediation that uses fungi to decompose or neutralize hazardous substances such as pollutants, heavy metals, pesticides, petroleum products, and plastics in the environment. Certain fungi have evolved enzymes capable of breaking down complex and toxic molecules, including hydrocarbons, dyes, and even explosives.

Key characteristics:

- Utilizes white rot fungi like *Pleurotus ostreatus* and *Phanerochaete chrysosporium*, which produce powerful enzymes such as lignin peroxidase, manganese peroxidase, and laccase.
- Capable of detoxifying soil, water, and air pollutants.
- Works well in both in situ (on-site) and ex situ (off-site) applications.
- Supports ecosystem restoration and natural habitat rehabilitation.

Applications:

- Oil spill clean-up: Fungi can degrade hydrocarbons found in petroleum.
- Heavy metal removal: Mycelium binds to metals like lead and mercury.
- Pesticide degradation: Breaks down agricultural chemicals in contaminated soil.
- Landfill treatment: Reduces leachate and hazardous waste levels.

Mycorecycling extends the concept of mycoremediation by using fungi to recycle and repurpose waste materials into valuable products. It focuses not just on degradation, but on transforming waste into functional materials such as biodegradable packaging, construction materials, textiles, or even food supplements.

Key characteristics:

- Involves feeding fungi with organic waste like sawdust, agricultural residue, paper, or plastic.
- Mycelium grows into the shape of a mold or form, creating lightweight, strong, biodegradable materials.
- Integrates waste streams into biomanufacturing cycles.

Applications:

- Mycelium packaging: Eco-friendly alternative to Styrofoam; developed by companies like Ecovative.
- Building materials: Insulation panels, bricks, and furniture made from hardened mycelium.
- Textiles and leather alternatives: Fungal leather (*mycelium leather*) used in fashion.

- Plastic degradation: Certain fungi can consume polyurethane and polyester-based plastics.

Benefits of mycoremediation and mycorecycling

- Natural and eco-friendly: No toxic by-products, low energy usage.
- Sustainable waste management: Converts waste into useful products.
- Supports circular economy: Promotes reuse, reduces landfill waste.
- Cost-effective: Less expensive than industrial remediation techniques.
- Versatile and scalable: Applicable in urban, rural, and industrial settings.

Mycoremediation and mycorecycling offer nature-inspired, low-impact solutions to some of the world's most pressing environmental challenges. By harnessing the incredible digestive power of fungi, these technologies not only clean up pollutants but also transform waste into useful, sustainable materials. As society moves toward greener alternatives, fungi may very well become a key ally in achieving a regenerative and circular economy.

3.5.4 3D printing with recycled materials: A sustainable manufacturing innovation

The integration of 3D printing and recycling technologies is redefining how we manage waste and manufacture products. This innovation involves using recycled materials primarily plastics but also metals and composites—as raw input for 3D printers. It supports sustainability by reducing dependence on virgin resources, minimizing landfill waste, and enabling localized, cost-effective production. The process generally includes collecting post-consumer or industrial waste, cleaning and sorting it, then shredding it into flakes or pellets. These are melted and extruded into filament, which can then be used in Fused Deposition Modeling (FDM) 3D printers to produce new items. Advanced systems now combine all these steps into closed-loop desktop units for schools, makerspaces, and research labs.

Common materials used include recycled PET (from bottles), PLA (from failed prints), ABS (from electronics), and even metal powders in industrial setups. Some filaments also include natural waste additives like coffee grounds or wood dust for eco-friendly composite printing.

The benefits of this technology are numerous:

- Environmental sustainability: It reduces plastic pollution and lowers landfill waste.
- Cost efficiency: Recycled filament is cheaper than virgin materials.
- Circular economy: Encourages reusability and local material loops.
- Custom properties: Recycled composites can be engineered for strength, flexibility, or biodegradability.
- Educational value: Demonstrates practical recycling and sustainability to students and communities.

Real-world applications are expanding rapidly. Construction industries are creating 3D-printed structures from recycled plastic or cement, consumer goods like furniture and eyewear frames are being printed from waste, and aerospace firms are using recycled metal printing for lightweight, high-performance parts. Projects like Print Your City turn local plastic waste into urban furniture, while startups like ReFlow, Filabot, and re:3D are pioneering filament recycling systems. 3D printing with recycled materials represents a powerful synergy between technology and sustainability. It transforms waste into opportunity, enabling a future where production is not only smarter but also significantly eco-friendlier.

3.6 Conclusion

The development of recycling technologies has undergone a significant transformation, progressing from rudimentary reuse practices to sophisticated industrial systems that can handle the complexity and scale of modern waste streams. By categorizing these technologies into mechanical, chemical, and biological methods, we gain a structured understanding of their applications and integration into current sustainability efforts. The recycling of components from solar and wind energy systems underscores the urgent need to design renewable infrastructure with end-of-life management in mind. As we move forward, innovative technologies such as AI-driven sorting, bio-based degradation, and material upcycling through 3D printing are proving instrumental in overcoming the limitations of traditional recycling approaches. These advancements not only enhance resource recovery and reduce environmental footprints but also lay the foundation for a truly circular and regenerative economy. Thus, continued research, investment, and policy support are essential to scaling these innovations and ensuring they are accessible across industries and geographies.

3.7 Case study: Integrated recycling approaches by Veolia and first solar – A path toward circular economy in renewable energy and urban waste

As renewable energy installations (especially solar and wind systems) reach their end-of-life phase, the recycling of associated materials has become a major sustainability concern. Traditional recycling methods such as mechanical shredding or landfill disposal are insufficient for managing the complex composition of solar panels, wind turbine blades, and electronic waste. Recognizing this, companies like Veolia Environment (France) and First Solar (USA) have implemented advanced recycling strategies integrating mechanical, chemical, and biological technologies to reduce environmental footprints and recover valuable resources.

Veolia, a global leader in resource management, has evolved from conventional waste collection services in the 1980s to a multi-technology recycling innovator. By 2018, it had launched Europe's first dedicated solar panel recycling facility in Rousset, France. The plant demonstrates how the industry has transitioned from simple mechanical separation techniques to advanced multi-stage material recovery. Similarly, First Solar, one of the world's largest producers of thin-film solar modules, has pioneered closed-loop recycling for its Cadmium Telluride (CdTe) panels. Their process combines mechanical dismantling, chemical leaching, and material purification, recovering over 90% of semiconductor material and 90% of the glass by weight.

Application of mechanical, chemical, and biological recycling

Mechanical recycling:

- Veolia's solar panel facility utilizes shredders and separators to crush the panels and segregate materials like glass, aluminum, and plastics.
- These materials are then used in construction and packaging industries.

Chemical recycling:

- First Solar uses chemical leaching to dissolve semiconductor layers (CdTe) without damaging glass substrates.
- The recovered CdTe is purified and reintegrated into new photovoltaic modules, reducing the need for virgin raw materials.

Biological recycling (emerging practice):

- Research by Fraunhofer Institute (Germany) and startups like Bio-Cellection are investigating enzyme-based and fungal treatments to break down organic polymers in composites (used in wind blades).
- Although not yet industrialized, these technologies are promising for next-generation biological recycling of epoxy and resin-based materials.

Recycling of materials used in solar and wind energy systems

Solar energy system (Case: Veolia and First Solar):

- Solar panels contain valuable and hazardous materials like glass, aluminum, silver, silicon, cadmium, and tellurium.
- The recycling process focuses on separating glass panels, removing toxic layers, and recovering semiconductors.
- First Solar's recycling plants in the US, Malaysia, and Germany operate on a closed-loop model, where recovered materials are fed back into production.

Wind energy system (Case: Siemens Gamesa & Re-Wind Project):

- Wind turbine blades are largely made of fiberglass and thermoset resins, which are notoriously difficult to recycle.

- Siemens Gamesa has developed the RecyclableBlade, which uses a new resin system that can be chemically dissolved, enabling full recycling of blade components.
- The re-wind project (a collaboration among researchers from Ireland, US, and the UK) repurposes old wind blades into footbridges, bike shelters, and urban structures—a pioneering example of upcycling.

Innovation in recycling technologies

AI-powered sorting:

- AMP Robotics (USA) has partnered with waste management companies to install AI-guided robotic arms that identify and sort materials like plastics, metals, and glass from municipal solid waste with high precision.
- These technologies drastically reduce contamination in recycling streams and increase throughput.

3D printing with recycled materials:

- In the Netherlands, The New Raw (a design studio) runs the Print Your City initiative where municipal plastic waste is turned into urban furniture using large-scale 3D printers.
- This model supports local circular economies and citizen engagement in recycling.

Mycorecycling:

- Ecovative Design (USA) uses mycelium (fungal roots) to convert agricultural and packaging waste into biodegradable products such as packaging, insulation panels, and furniture.
- Though still niche, mycoremediation and mycorecycling hold promise for organic and composite waste treatment.

Outcomes and impact

- Veolia's facility can recycle 4000 tonnes of solar panels annually, equivalent to roughly 130,000 panels per year, and expects to scale up as global PV waste increases.
- First Solar's closed-loop process ensures 95% recyclability, greatly reducing the carbon footprint of solar module manufacturing.
- Re-Wind Project's blade repurposing has diverted tonnes of composite waste from landfills and inspired regenerative urban design.

These case studies show that integrating historical learning with cutting-edge innovation leads to more efficient, scalable, and sustainable recycling systems. Governments and industries are now investing more in eco-design, reverse logistics, and product lifecycle planning to ensure that renewable energy infrastructure remains green throughout its life—from creation to disposal.

This case study demonstrates how advanced recycling technologies—across mechanical, chemical, and biological domains are being implemented in real-world scenarios to address the growing challenges of material recovery, particularly in the renewable energy sector. The innovative approaches adopted by organizations like Veolia, First Solar, and Siemens Gamesa not only reflect the evolution of recycling technologies but also their critical role in achieving the circular economy. These examples underline the urgent need for cross-sector collaboration, policy support, and continued R&D to scale sustainable recycling solutions globally.

3.8 Exercise questions

1. Explain the historical evolution of recycling technologies from ancient times to the modern era.
2. Differentiate between mechanical, chemical, and biological recycling technologies with suitable examples.
3. Discuss the advantages and limitations of mechanical recycling in modern waste management systems.
4. Describe the process and benefits of chemical recycling. Provide real-world examples of its applications.
5. What is biological recycling? Illustrate the role of microbes and enzymes in the recycling process.
6. How is recycling managed in solar photovoltaic (PV) systems? List the main recyclable materials and their treatment.
7. Explain the challenges and opportunities involved in recycling wind turbine blades.
8. Analyze the importance of rare earth metal recovery from wind turbine components.
9. Discuss the concept of "design for recyclability" in renewable energy systems. Why is it important?
10. Evaluate the environmental impact of not recycling solar and wind energy system components.
11. What are the recent innovative recycling technologies introduced in the last decade? Discuss any two in detail.
12. Write a short note on the role of AI-powered robotic sorting in enhancing recycling efficiency.
13. Compare the recycling processes of traditional plastics with biodegradable plastics.
14. How does 3D printing with recycled materials contribute to the circular economy?
15. Propose a sustainable model integrating mechanical, chemical, and biological recycling for e-waste.

3.9 Objective types questions

1. Which of the following is an example of mechanical recycling?
   a) Pyrolysis
   b) Composting
   c) Shredding and re-molding plastics
   d) Fermentation
2. True or False: Chemical recycling alters the molecular structure of materials.
3. Which enzyme is commonly involved in the biological degradation of PET plastic?
   a) Laccase
   b) Cellulase
   c) Lipase
   d) PETase
4. The term ‘urban mining’ refers to:
   a) Mining near cities
   b) Extraction of metals from e-waste
   c) Recycling plastics
   d) Composting urban food waste
5. Which of the following is a major component in wind turbine blades?
   a) Concrete
   b) Aluminum
   c) Fiberglass composite
   d) Polypropylene
6. ________ is the process of using fungi to break down toxic waste materials.
7. Which country leads in solar panel recycling regulations?
   a) USA
   b) India
   c) Germany
   d) Australia
8. True or False: Mechanical recycling is energy-intensive compared to chemical recycling.
9. The recycling of wind turbine nacelles focuses primarily on recovering which materials?
   a) Glass and concrete
   b) Steel, copper, and rare earth elements
   c) Plastic and rubber
   d) Paper and cardboard
10. What is the main environmental concern regarding the disposal of solar panels?
    a) Excess heat
    b) Land use
    c) Toxic materials like lead and cadmium
    d) Air pollution

11. Which of the following technologies uses AI to improve sorting accuracy?
    a) Electrochemical separation
    b) Robotic sorting systems
    c) Fluidized bed processing
    d) Solvolysis
12. True or False: Mycorecycling can be used to make biodegradable furniture from waste.
13. What is the main purpose of using fly ash in green concrete?
    a) Reduce cost
    b) Improve strength
    c) Reduce CO_2 emissions
    d) All of the above
14. The process of converting recycled plastic into 3D printer filament is called ________.

Chapter 4

Circular approaches to reuse and repurposing in clean energy systems

4.1 Introduction

The global shift toward renewable energy has ushered in a new era of technological innovation and sustainability. Solar panels, wind turbines, and energy storage systems are now central to efforts to decarbonize power generation and reduce dependence on fossil fuels. However, as the deployment of renewable energy infrastructure accelerates, so too does the volume of end-of-life (EoL) components, retired equipment, and manufacturing byproducts. These emerging waste streams present both a challenge and an opportunity: while they risk undermining the sustainability of the sector if improperly managed, they also offer immense potential for recovery, reuse, and repurposing through circular economy strategies. The book chapter titled "*Circular Approaches to Reuse and Repurposing in Renewable Energy*" explores this evolving landscape by highlighting innovative practices and models that promote circularity throughout the life cycle of renewable energy systems.

At the foundation of these strategies lies the concept of circular design principles, which emphasize durability, modularity, reparability, and recyclability in renewable energy systems. Unlike the traditional linear model—–where products are manufactured, used, and discarded—circular design encourages the creation of components that can be easily dismantled, upgraded, or repurposed after their initial use. In solar and wind technologies, this includes designing photovoltaic (PV) modules with replaceable layers, using standardized parts for turbine assemblies, and employing materials that can be economically separated and recovered. By embedding circularity into the design phase, manufacturers can significantly extend product lifespans and reduce material waste, while also lowering the environmental footprint of energy production. Moving beyond design, the chapter delves into reuse strategies for renewable energy assets. Many components, particularly in solar

The Circular Economy and their Management. https://doi.org/10.1016/B978-0-443-44617-7.00007-3

and wind systems, retain significant value even after their first operational cycle. For instance, wind turbine gearboxes and nacelles can often be refurbished and redeployed in less demanding settings, such as training simulators or microgrids. Similarly, solar panels with diminished but functional efficiency can be relocated to rural areas, low-income communities, or off-grid applications. These reuse strategies not only delay the disposal of valuable components but also contribute to energy access and equity in underserved regions. They also help reduce the need for new raw material extraction, further supporting the sustainability of the renewable energy transition.

A particularly innovative domain addressed in this chapter is repurposing: innovation beyond energy. Repurposing involves finding entirely new applications for decommissioned renewable energy components. For example, wind turbine blades, typically made from fiberglass composites, are being creatively transformed into pedestrian bridges, playgrounds, and even furniture. Solar panel glass is being reused as architectural cladding, while inverters are being adapted for backup power systems. These novel uses not only prevent materials from entering landfills but also unlock new commercial value across sectors such as construction, design, and urban planning. Repurposing thus represents a bridge between circular economy principles and cross-sectoral innovation.

The realization of such circular practices is increasingly supported by technologies empowering circularity. Digital tools like artificial intelligence (AI), the Internet of Things (IoT), blockchain, and robotics are revolutionizing how renewable energy components are tracked, assessed, maintained, and disassembled. AI-driven predictive maintenance can forecast component failure and optimize reuse timelines, while IoT devices monitor system performance and feed real-time data into decision-making platforms. Blockchain enables secure tracking of materials through circular supply chains, and robotics allows for automated disassembly, reducing labor costs and improving safety. These technological enablers are not only increasing the feasibility of circular practices but also making them scalable, data-driven, and cost-effective.

Lastly, the chapter examines business models driving reuse and repurposing, which are critical for translating circular principles into real-world practice. Models such as Product-as-a-Service (PaaS) shift the focus from ownership to usage, with manufacturers retaining responsibility for the entire lifecycle of energy systems. Circular supply chains connect dismantlers, refurbishers, and manufacturers in value-generating loops, while buy-back and refurbishment programs incentivize the return and restoration of used equipment. These business innovations ensure that circularity is not only environmentally necessary but economically viable, offering new revenue streams and strengthening customer relationships.

In summary, this chapter provides a comprehensive overview of the tools, techniques, and strategies shaping the future of circularity in renewable energy. By embracing design for longevity, enabling component reuse, fostering creative repurposing, leveraging advanced technologies, and deploying circular business models, the renewable energy sector can move toward a more resilient, resource-efficient, and sustainable future.

4.2 Circular design principles in renewable energy systems

As the global transition to renewable energy accelerates, so too does the need for sustainable management of the components and materials used in clean energy systems. Wind turbines, solar panels, and energy storage systems are often built to last decades, but like all technologies, they eventually reach the end of their useful life. Traditional disposal methods, such as landfilling, incineration, or informal dismantling, pose serious environmental risks and squander valuable resources. To counteract this, the adoption of circular design principles in renewable energy systems is essential. These principles aim to extend the life of products, enable reuse and repurposing, and ensure materials remain in the economic loop for as long as possible. A cornerstone of circular design, Design for Disassembly involves engineering products in a way that allows for easy separation of components at the end of their life. This is particularly important in solar panels and wind turbines, which are typically constructed with tightly integrated parts. If these components can be easily disassembled, they can be more readily reused, repaired, or recycled. For example, modular inverters in solar systems can be replaced without dismantling the entire setup, reducing maintenance costs and material waste. Materials used in renewable energy systems must be durable yet recyclable. Circular design encourages the use of mono-materials or compatible materials to simplify recycling processes. For instance, using aluminum framing for solar panels not only ensures strength and durability but also facilitates high-value recycling. Additionally, avoiding hazardous substances like lead or cadmium in photovoltaic cells supports safer end-of-life treatment and repurposing. Standardized components and modular design allow for interchangeability and easier upgrades, leading to longer lifespans for energy systems. A modular wind turbine gearbox, for example, can be removed and replaced without dismantling the entire turbine. Similarly, standardized battery cells enable the refurbishment or substitution of failed units, promoting second-life applications and minimizing waste. Integrating circular design principles in renewable energy systems is not just an environmental necessity, it is a strategic imperative for achieving sustainability and resource efficiency. By reimagining how systems are designed, maintained, and retired, we can create a regenerative energy future that aligns with the core principles of the circular economy: reduce, reuse, repurpose, and regenerate. These innovations not only minimize waste but also unlock new economic value and social equity across the energy landscape. The following are the explanations of different design principles:

4.2.1 Design for disassembly

In a world rapidly shifting toward sustainability and resource efficiency, Design for Disassembly (DfD) has emerged as a foundational principle of the circular economy. DfD refers to designing products in a way that allows easy and nondestructive separation of components at the end of their life. This design philosophy is especially critical in renewable energy systems, such as solar panels, wind turbines, and battery storage units, which are built from complex assemblies and materials. Applying DfD principles ensures that these components can be reused, repurposed, or recycled efficiently, reducing environmental impact and enabling material circularity. The following are the core principles of design for disassembly:

Modularity: Components should be designed in self-contained modules that can be easily removed, replaced, or upgraded. For example, modular battery packs allow damaged cells to be swapped out without discarding the entire unit.

Minimized fasteners: Use of minimal, uniform, and reversible fasteners (e.g., screws instead of adhesives) allows for easy detachment. This avoids the destruction of materials during dismantling.

Standardization: Standard-sized parts and connectors ensure compatibility across different systems or generations of products, making reuse and repair more practical.

Accessible assembly: Parts should be accessible without special tools or destructive procedures. Access ports and clear disassembly pathways help technicians dismantle products safely and quickly.

Material compatibility: Using materials that can be separated mechanically or chemically without contamination supports efficient recycling or repurposing. Table 4.1 illustrate property of DfD. Fig. 4.1 illustrate correlation between different parameters of DfD (Fig. 4.2).

4.2.1.1 Application in renewable energy systems

Solar panels: Traditional solar panels use strong adhesives and encapsulants to protect photovoltaic (PV) cells, making disassembly difficult. DfD-oriented panels now focus on:

- Modular glass-glass designs that use interlocking parts instead of permanent bonding.
- Thermal release adhesives that allow safe removal of PV cells when heated.
- Panels where frames, junction boxes, and glass sheets can be separated easily and reused or recycled.

Wind turbines: Wind turbines are massive machines composed of blades, towers, and nacelles. DfD principles include:

TABLE 4.1 Property of design for disassembly (DfD).

Property of design for disassembly (DfD)	Explanation (general)	Relevance to circular design in renewable energy systems
Ease of component separation	Components should be easy to detach without damage.	Enables efficient recovery of valuable materials from solar panels, wind turbines, batteries, etc., reducing waste.
Use of reversible fasteners	Preference for screws, clips, bolts instead of adhesives or welds.	Allows maintenance, upgrading, and final disassembly of renewable energy equipment to maximize reuse and recycling.
Modular architecture	System designed in independent modules for quick removal or replacement.	Facilitates upgrading technologies (e.g., inverter modules, battery packs) and extends the overall lifecycle of renewable systems.
Material compatibility	Using materials that can be recycled together or separated easily.	Prevents contamination and enhances recovery of scarce materials (rare earths from wind turbines, silicon from PV panels).
Minimal use of permanent bonds	Avoiding glue, welding, and fused materials.	Reduces energy and cost required to dismantle systems at end-of-life and increases recyclability.
Standardization of components	Using common sizes, shapes, and interfaces.	Allows interchangeability of parts across renewable energy products, reducing the need for new materials and simplifying repair.
Clear labeling of parts	Markings to identify material type, disassembly instructions, and hazards.	Helps recyclers handle components safely and recover critical resources such as lithium, cobalt, copper, and rare minerals.
Accessible disassembly points	Disassembly points should be visible and easy to reach.	Lowers maintenance cost and time for solar farms and wind turbines, supporting long-term circularity.
Non-destructive disassembly	Disassembly should not damage components.	Enables reuse of still-functional components like PV frames, turbine gear parts, cables, and electronics.

Continued

TABLE 4.1 Property of design for disassembly (DfD).—cont'd

Property of design for disassembly (DfD)	**Explanation (general)**	**Relevance to circular design in renewable energy systems**
Optimized life extension	Products designed to enable repairs, upgrades, and refurbishing.	Extends the operational life of renewable systems, reducing the demand for raw resources and minimizing carbon footprint.
Safe disassembly	Ensures that disassembly can be done without hazards (e.g., electric shock, chemical leaks).	Supports safe handling of energy storage systems, batteries, and power electronics within renewable installations.
Material recovery efficiency	Maximizing the amount of material that can be recovered at end-of-life.	Improves economic value and sustainability of recycling solar PV, wind blades, and battery systems.

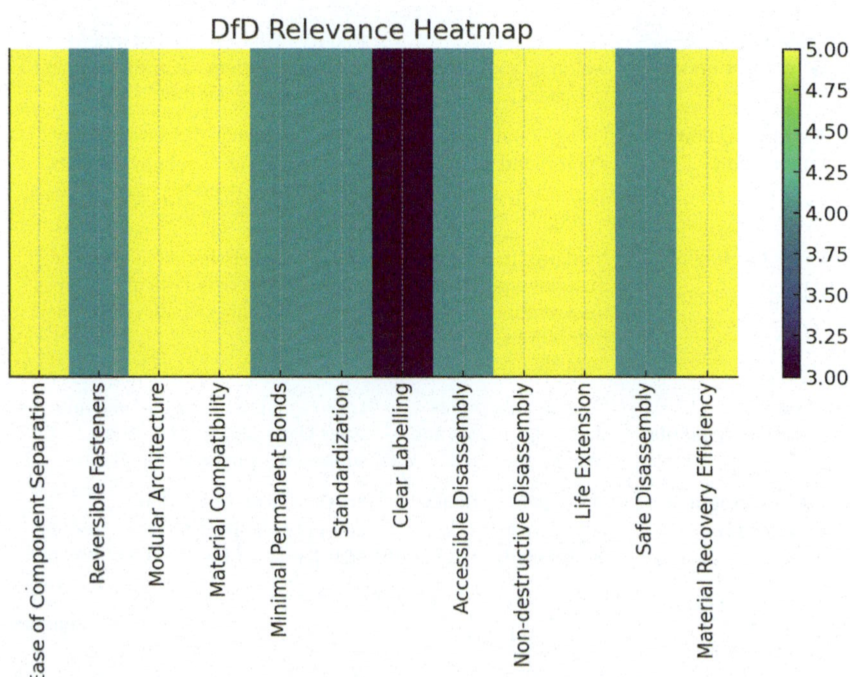

FIGURE 4.1 Correlation between different parameters of design for disassembly.

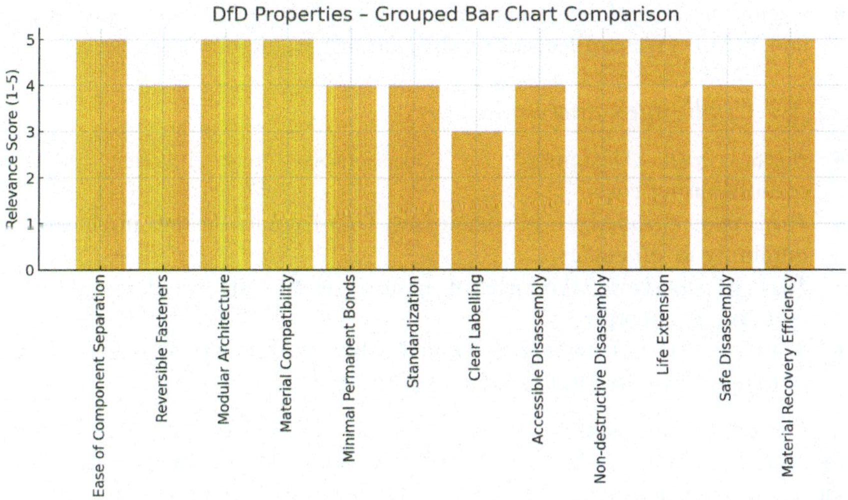

FIGURE 4.2 Design for disassembly property bar chart.

- Designing bolted instead of welded joints in turbine towers for easier dismantling.
- Using thermoplastic resins in blades that can be melted and reused, unlike thermoset composites that are hard to recycle.
- Making internal electrical systems modular for quick component replacement.

Battery energy storage systems: In energy storage, DfD ensures:

- Battery cells are arranged in removable modules, allowing faulty units to be replaced rather than scrapping the entire system.
- Use of mechanical fasteners and nontoxic adhesives to allow safe disassembly.
- Color coding and labeling for safer handling and sorting of critical materials like lithium, cobalt, and nickel.

4.2.1.2 Benefits of DfD in renewable energy

- *Resource recovery*: Enables high-value material recovery (e.g., silver from PV cells, rare earth metals from turbines).
- *Cost reduction*: Reuse of components cuts costs for manufacturers and end-users.
- *Job creation*: Disassembly and remanufacturing create green jobs in local economies.
- *Environmental sustainability*: Reduces landfill waste and carbon emissions associated with raw material extraction.

- *Regulatory compliance*: Supports Extended Producer Responsibility (EPR) and WEEE (Waste Electrical and Electronic Equipment) directives.

4.2.1.3 Challenges and limitations

- *Initial cost*: Designing for disassembly may increase upfront manufacturing costs.
- *Performance trade-offs*: In some cases, DfD-friendly designs might reduce durability or efficiency.
- *Lack of standards*: Absence of industry-wide DfD standards hampers widespread adoption.
- *Training and infrastructure*: Skilled labor and disassembly centers are required to handle DfD products effectively.

Design for disassembly is not merely a technical consideration; it is a mindset that aligns product design with the goals of sustainability and circularity. In renewable energy systems, DfD maximizes material recovery, reduces waste, and extends product life. As the sector grows, embedding DfD from the design phase will be essential to build a truly circular and environmentally resilient energy future.

4.2.2 Material recovery forecasting

As the deployment of renewable energy technologies such as solar photovoltaic (PV) panels, wind turbines, and energy storage systems continues to rise, so does the volume of materials embed within these systems. When these systems reach the end of their life, effective material recovery becomes critical to minimizing environmental impact, reducing reliance on virgin raw materials, and supporting the goals of the circular economy. One of the most strategic tools enabling this transition is Material Recovery Forecasting (MRF). Material Recovery Forecasting is the process of predicting the type, quantity, quality, and timing of materials that will become available for recovery, reuse, or recycling from renewable energy infrastructure. It integrates data from lifecycle assessments, product design, usage patterns, failure rates, and technology retirement curves to anticipate the future availability of valuable materials. The following are the significance of material recovery forecasting:

Supports circular economy goals: MRF enables the planning of reuse and recycling pathways, ensuring that materials like aluminum, copper, lithium, rare earth metals, and glass are reintegrated into production cycles.

Reduces supply chain risk: By forecasting the availability of secondary raw materials, industries can reduce dependence on imported or scarce primary materials, enhancing supply chain resilience.

Informs recycling infrastructure development: Knowing the volume and timing of material recovery helps policymakers and recyclers to develop appropriate collection, dismantling, and processing infrastructure.

Promotes sustainable product design: Insights from recovery forecasting can be fed back into the design phase, encouraging the use of materials that are easier to recover or have higher future value.

To effectively implement material recovery forecasting, it is essential to understand the key components that make accurate forecasting possible. These **key components of material recovery forecasting,** ranging from lifecycle assessments and technology retirement curves to failure analysis and market trends, serve as the building blocks for predicting when, where, and how valuable materials from renewable energy systems can be recovered. Table 4.2 illustrate property of material recovery forecasting. The following are the key components of material recovery forecasting.

4.2.3 Lifecycle assessment integration

Lifecycle Assessment (LCA) Integration is a critical component of Material Recovery Forecasting (MRF), particularly within the renewable energy sector. It provides a comprehensive methodology to evaluate the environmental impacts associated with all the stages of a product's life, from raw material extraction through processing, manufacturing, distribution, use, repair and maintenance, and disposal or recycling. In the context of renewable energy systems such as solar panels, wind turbines, and energy storage units, LCA helps in identifying and quantifying the types and quantities of materials used throughout the lifecycle. This includes metals like aluminum, copper, silver, and rare earth elements, as well as glass, polymers, and semiconductors. By understanding the material composition and flow at each stage, stakeholders can develop accurate models that predict what materials will be available for recovery and when. Fig. 4.3 illustrates properties of LCA integration. Table 4.3 illustrate property of LCA integration.

LCA integration into MRF tools enables analysts to estimate the potential for resource recovery based on actual usage patterns, system efficiencies, and degradation rates. For instance, in a solar photovoltaic system, the degradation of the panel's efficiency over time is a known factor; LCA models can incorporate this to determine when panels will likely be decommissioned and what materials they will yield. Furthermore, LCA helps identify the criticality of materials, including their economic importance and supply risk. Materials deemed critical can be prioritized in recovery strategies to enhance resource security and reduce dependency on virgin raw materials.

Moreover, LCA supports the identification of environmental hotspots—- stages in the product lifecycle that have a disproportionately high environmental impact. This information is vital for guiding improvements in product design and end-of-life management. For example, if the recycling of a certain material consumes more energy than its original production, LCA might recommend design modifications or alternative materials. By feeding these insights into the design and manufacturing process, producers can create renewable energy systems that are not only efficient during their operational

TABLE 4.2 Property of material recovery forecasting.

Property of material recovery forecasting	Explanation (general)	Relevance to circular design in renewable energy systems
Assessment of material composition	Identifying types, quantities, and quality of materials used in products.	Helps forecast recoverable materials from solar panels, wind turbines, and batteries, guiding recycling strategies.
Lifecycle material flow analysis (MFA)	Mapping material inputs, uses, and outputs across the product lifecycle.	Predicts long-term availability of recyclable metals (e.g., copper, aluminum, silicon, rare earth elements).
End-of-life (EoL) volume prediction	Forecasting when products will reach end-of-life based on lifespan data.	Supports planning for recycling capacities for PV module waste, wind blades, or battery packs.
Demand–supply alignment	Estimating future demand for recovered materials versus raw material needs.	Enables strategic planning to replace virgin materials with recycled ones in renewable energy manufacturing.
Recovery rate estimation	Predicting the percentage of materials that can be effectively recovered.	Helps evaluate the efficiency and economic viability of recycling processes for renewable technologies.
Critical material prioritization	Identifying valuable or scarce materials (e.g., lithium, cobalt, neodymium).	Ensures circular design focuses on maximizing recovery of critical and high-impact materials.
Technological forecasting	Assessing how advancements in recycling technologies affect recovery.	Aligns renewable system design with future recycling capabilities (e.g., improved PV cell separation tech).
Environmental impact forecasting	Estimating carbon savings, waste reduction, and ecological benefits of recovery.	Enhances circularity by quantifying the environmental benefits of recycling renewable energy components.
Cost–Benefit analysis	Evaluating the economic feasibility of recovering different materials.	Guides circular designs that optimize both sustainability and financial returns of renewable systems.

TABLE 4.2 Property of material recovery forecasting.—cont'd

Property of material recovery forecasting	Explanation (general)	Relevance to circular design in renewable energy systems
Geographical and market forecasting	Analyzing regional recycling capabilities and material market trends.	Helps determine where recovery systems should be set up for solar farms or wind installations.
Scenario planning	Simulating multiple recovery pathways (reuse, refurbish, recycle).	Supports strategic decision-making for circular design, considering uncertainty in future waste streams.
Regulatory forecasting	Monitoring upcoming policies that affect recycling or waste management.	Ensures renewable energy products are designed to comply with future EPR (extended producer responsibility) rules.
Stakeholder mapping	Identifying recyclers, manufacturers, and logistics partners.	Aids in building circular ecosystems around renewable energy systems.

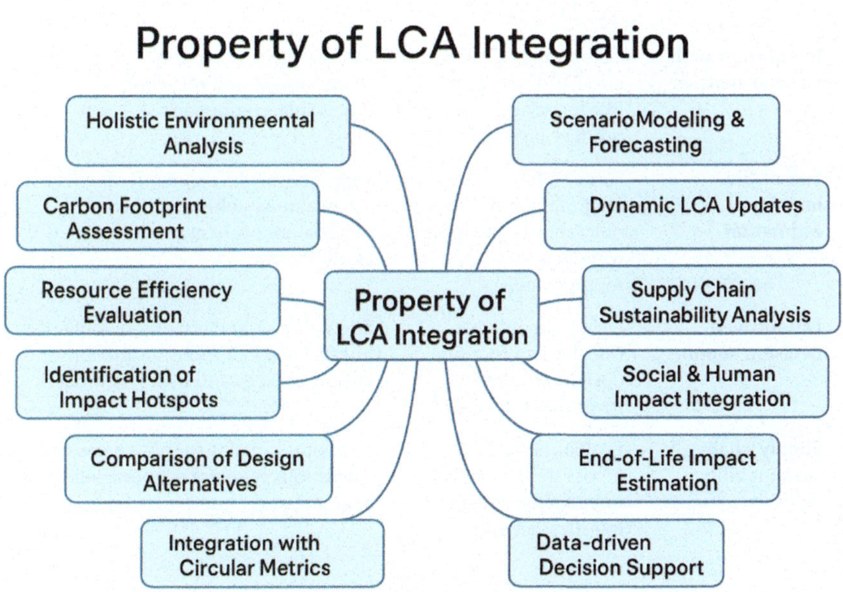

FIGURE 4.3 Properties of LCA integration.

TABLE 4.3 Property of LCA integration.

Property of LCA integration	Explanation (general)	Relevance to circular design in renewable energy systems
Holistic environmental analysis	Evaluating environmental impacts from raw material extraction to end-of-life.	Ensures renewable systems (solar, wind, batteries) are designed with minimal ecological footprint across their entire lifecycle.
Carbon footprint assessment	Measuring total greenhouse gas emissions at each lifecycle stage.	Supports low-carbon material selection and helps optimize renewable infrastructures for climate-neutral performance.
Resource efficiency evaluation	Analyzing energy, water, and material use throughout the product lifecycle.	Helps designers choose materials and processes that support resource circularity and reduce waste.
Identification of impact hotspots	Pinpointing stages with the highest environmental burden (e.g., mining, manufacturing).	Guides design modifications that lower impacts—such as using recyclable materials or modular designs.
Comparison of design alternatives	Comparing different materials, components, or system designs using quantitative data.	Helps select more circular and sustainable design pathways for renewable energy systems.
Integration with circular metrics	Combining LCA with circularity indicators like recyclability rate or product lifespan.	Ensures designs support long-term reuse, repair, and recycling of renewable components.
End-of-life impact estimation	Forecasting environmental outcomes of disposal, recycling, or repurposing.	Supports the development of renewable systems that maximize material recovery and minimize landfill waste.
Data-driven decision support	Using quantitative impact data to guide design, policy, and investment decisions.	Strengthens circularity by helping industries adopt more sustainable design choices and business models.
Supply chain sustainability analysis	Evaluating impacts across transportation, manufacturing, and logistics networks.	Promotes circular supply chains for renewable systems, encouraging local sourcing and closed-loop flows.

TABLE 4.3 Property of LCA integration.—cont'd

Property of LCA integration	Explanation (general)	Relevance to circular design in renewable energy systems
Scenario modeling and forecasting	Running simulations to compare long-term environmental outcomes under different conditions.	Helps anticipate future environmental challenges and improve the circular design resilience of renewable systems.
Dynamic LCA updates	Continuously updating assessments as new data or technologies emerge.	Ensures renewable energy products stay aligned with evolving circular design standards and regulations.
Social and human impact integration	Assessing social factors (labor, safety, community impact).	Encourages ethically responsible and socially sustainable circular designs.
Compliance with global standards	Aligning with ISO 14040/44 and sustainability reporting frameworks.	Ensures renewable systems meet legal and industry expectations for circularity and environmental responsibility.
Feedback loop for design improvement	Using LCA findings to refine product design in iterative cycles.	Enhances circularity by promoting continuous improvement in materials, processes, and recoverability.

life but also sustainable and resource-efficient at end-of-life. Fig. 4.4 illustrate comparison of different properties of LCA integration.

Another significant advantage of LCA in MRF is its ability to inform policy and regulatory frameworks. Governments and organizations can use LCA-based data to develop regulations around Extended Producer Responsibility (EPR), eco-design requirements, and material recovery targets. These policies can drive innovation in the industry and encourage manufacturers to adopt circular design principles from the outset. Lifecycle Assessment integration in Material Recovery Forecasting provides a data-driven foundation for understanding the material dynamics of renewable energy technologies. It offers a systematic way to predict the availability and value of recoverable materials, optimize recycling processes, and design for circularity. As the renewable energy sector continues to expand, incorporating LCA into forecasting tools will be essential for building sustainable and resilient energy systems that align with the principles of the circular economy.

4.2.4 Technology adoption and retirement curves

Technology Adoption and Retirement Curves are crucial for projecting the timeline and magnitude of material availability in Material Recovery

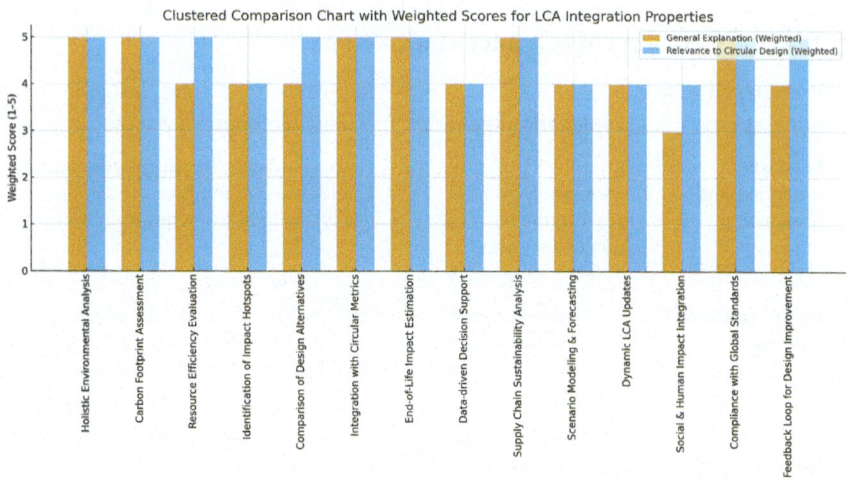

FIGURE 4.4 Comparison of different properties of LCA integration.

Forecasting (MRF). These curves model the lifecycle trajectory of renewable energy technologies from market introduction and growth, through maturity, and finally to retirement or decommissioning. Understanding this temporal aspect allows for accurate prediction of when renewable energy systems will reach their end-of-life phase, and consequently, when significant volumes of materials will become available for recovery.

Each renewable energy technology has a typical lifespan, and these lifespans vary across systems and components. For instance, solar panels generally function efficiently for 25 to 30 years before their output declines below economic thresholds. Wind turbines have operational lifespans ranging from 20 to 25 years, while lithium-ion batteries used in energy storage systems and electric vehicles often degrade significantly after 8 to 10 years. By analyzing deployment data from past decades and projecting current installation trends, MRF models can forecast retirement peaks and plan for future recovery operations.

These forecasting models rely on large datasets that include installation year, geographic location, manufacturer specifications, system capacity, and known failure or replacement rates. By aggregating this data, analysts can identify patterns and generate curves that highlight when and where the largest volumes of decommissioned equipment will emerge. For example, large-scale solar farms installed in the early 2000s are expected to begin retiring masses in the 2030s, creating opportunities and challenges for the recycling industry. Table 4.4 illustrate property of technology adoption and retirement curves.

In addition to projecting when materials will become available, these curves also provide insight into the types and conditions of materials. Early retirements due to system failure, policy changes, or technological

TABLE 4.4 Property of technology adoption and retirement curves.

Property of technology adoption and retirement curves	Explanation (general)	Relevance to circular design in renewable energy systems
Adoption rate analysis	Tracks how quickly a new technology is accepted by users or markets.	Helps forecast demand for renewable systems (solar, wind, batteries) and plan circular material flows accordingly.
S-curve/Innovation diffusion	Technologies follow an S-curve: slow start → rapid adoption → saturation.	Enables manufacturers to prepare sustainable supply chains, modular designs, and recycling systems in advance of mass uptake.
Market saturation identification	Understanding when a technology reaches maturity and slows in growth.	Guides planning for end-of-life (EoL) recycling and second-life usage as older renewable systems become obsolete.
Technology lifespan prediction	Estimates functional lifespan based on historical trends and performance data.	Supports circular design by planning for refurbishment, remanufacturing, and material recovery at the right time.
Retirement curve modeling	Predicts when users will discard or replace technology.	Helps forecast waste volumes from solar panels, wind turbines, and batteries, enabling sustainable recycling capacity planning.
Upgrade and replacement cycles	Analyses when technologies become outdated and require upgrading.	Encourages modular and upgradable designs in renewable energy systems to reduce material waste and extend product life.
Second-life potential assessment	Evaluates whether retired components can be repurposed (e.g., EV batteries → storage).	Supports circularity by enabling reuse and value recovery from partially functional renewable components.
Material flow forecasting	Predicts how much material will enter recycling/reuse streams over time.	Helps set up circular supply chains in advance for metals like lithium, copper, rare earths, and aluminum.
Economic feasibility modeling	Evaluates the cost benefits of adopting or retiring technologies at different stages.	Supports decisions on when to refurbish, remanufacture, or recycle renewable energy systems.

Continued

TABLE 4.4 Property of technology adoption and retirement curves.—cont'd

Property of technology adoption and retirement curves	Explanation (general)	Relevance to circular design in renewable energy systems
Environmental impact timing	Assesses when the greatest environmental impacts occur across adoption/retirement.	Guides interventions at key stages to reduce carbon footprint and improve circular design outcomes.
Policy and regulation sensitivity	Evaluates how policies accelerate or delay adoption/retirement.	Aligns product design with future circular regulations such as EPR (extended producer responsibility).
Risk and uncertainty analysis	Identifies uncertainties in adoption trends (economic shifts, tech breakthroughs).	Encourages flexible, adaptable circular designs that remain viable under changing market conditions.
Innovation cycle mapping	Tracks how quickly new technologies replace old ones.	Helps designers anticipate future waste streams and develop systems optimized for future disassembly and material recovery.
Stakeholder behavior understanding	Examines consumer, industry, and government adoption patterns.	Helps shape circular business models such as leasing, service-based energy solutions, and buy-back programs.

obsolescence might yield components in relatively good condition, suitable for reuse or refurbishment. On the other hand, systems retired after their full operational life might be more suitable for material recovery or recycling. Moreover, adoption and retirement curves help manufacturers and policymakers assess the long-term impacts of renewable energy deployment. They can inform infrastructure planning, ensuring that collection, transportation, and recycling facilities are built in anticipation of future waste streams. They also support strategic decisions about investments in recycling technologies and workforce training programs. Challenges in utilizing these curves include the uncertainty of real-world performance, evolving technological standards, and inconsistent reporting across regions and manufacturers. External factors such as changes in energy policy, subsidies, or market dynamics can also accelerate or delay retirement trends. Therefore, MRF tools must be dynamic and capable of incorporating new data and scenarios to remain accurate and useful. Ultimately, technology adoption and retirement curves serve as the temporal backbone of Material Recovery Forecasting. They enable proactive planning and resource optimization, reduce the risk of material shortages or recycling bottlenecks, and support a circular economy approach in managing the end-of-life phase of renewable energy systems.

4.2.5 Failure mode analysis

Failure Mode Analysis (FMA) is an integral part of Material Recovery Forecasting (MRF) that focuses on understanding and quantifying the premature failures of components within renewable energy systems. While many systems are designed to operate for specific lifespans, real-world conditions often lead to deviations from these expectations. FMA helps identify why, how, and under what conditions systems fail before their intended end-of-life, thus offering a more nuanced view of when and what types of materials may become available for recovery. Renewable energy technologies operate in diverse and often harsh environments. Wind turbines, for instance, are exposed to high mechanical stress, temperature fluctuations, and corrosive atmospheres. Common failure modes include blade delamination, gearbox fatigue, and generator malfunctions. Solar panels, particularly those deployed in desert or high-humidity regions, may suffer from microcracks, delamination of layers, hot spots, or corrosion of electrical contacts. Energy storage systems such as lithium-ion batteries can fail due to thermal runaway, electrolyte leakage, or electrode degradation.

Failure Mode Analysis involves collecting and analyzing data from operational systems, maintenance records, warranty claims, and field inspections. This data helps categorize different types of failures, their root causes, frequency, and the age at which they typically occur. Statistical models such as Weibull analysis or Monte Carlo simulations can then be used to estimate the probability of failure over time for each component. These models enhance the precision of MRF by accounting not only for standard retirements but also for unanticipated recoveries due to failures. Incorporating FMA into forecasting models allows stakeholders to prepare for early material inflows, which might otherwise be overlooked. For example, a significant number of battery packs may become available well before the average 8–10 year lifespan due to performance issues or accidents. Recognizing this possibility enables recyclers and refurbishers to scale their operations and develop protocols for handling components that are partially degraded but still usable.

Moreover, FMA contributes to improving the overall reliability and design of renewable energy technologies. Insights gained from failure data can feed back into the product development cycle, encouraging the use of more durable materials, better protection mechanisms, and enhanced quality control measures. This creates a virtuous cycle where forecasting not only anticipates material recovery but also informs strategies to extend product life and reduce waste. Another benefit of FMA is its role in shaping maintenance and service strategies. Predictive maintenance programs can use FMA data to schedule inspections and repairs more effectively, preventing unexpected breakdowns and optimizing asset utilization. For large-scale installations such as wind farms or solar power plants, this translates into higher efficiency, lower operational costs, and a more predictable material recovery timeline. Failure Mode Analysis enriches Material Recovery Forecasting by introducing a critical dimension of real-world performance. It ensures that forecasting models reflect both planned and unplanned material flows, enhancing their accuracy and utility. By anticipating failures and integrating their implications into recovery planning, FMA supports the development of robust, responsive, and circular energy systems.

4.2.6 Geographic and policy factors

Geographic and policy factors play a critical role in shaping the landscape of renewable energy adoption and recycling practices. These elements influence both the feasibility and success of clean energy projects and material reuse strategies, making them indispensable considerations in energy planning and sustainability modeling. Geographically, the availability of natural resources such as sunlight, wind, and biomass varies significantly across regions. For example, solar panel installations are highly efficient in equatorial and arid zones where sunshine is abundant, while wind energy thrives in coastal and high-altitude regions. This natural variability affects the distribution of renewable infrastructure and, consequently, the types and volumes of waste generated. For instance, areas with large-scale wind farms may generate higher volumes of turbine blade waste, necessitating region-specific recycling infrastructure and policies.

In parallel, transportation and logistics capabilities vary across geographies. In remote or landlocked areas, transporting end-of-life renewable components—like solar panels or batteries—to recycling facilities may be cost-prohibitive or logistically complex. This makes decentralized recycling and reuse strategies more viable in such regions. Additionally, geographic challenges such as terrain, population density, and climate resilience influence infrastructure development, site selection for renewable installations, and the ease of material recovery and reuse. Policy frameworks are equally pivotal. National and regional regulations on waste management, extended producer responsibility (EPR), and clean energy incentives define the operational boundaries for recycling industries and renewable energy providers. Countries with stringent e-waste and material recovery laws often see more advanced recycling ecosystems. For instance, the European Union's Waste Electrical and Electronic Equipment (WEEE) Directive has pushed manufacturers and recyclers to innovate in solar panel and battery recycling.

Policies also include subsidies, tariffs, carbon pricing, and green financing. Tax incentives for renewable installations can stimulate growth, but unless coupled with end-of-life management policies, they may lead to future waste surges. Moreover, policy stability and clarity are essential. Unpredictable shifts in environmental regulations or subsidy schemes can disrupt investment in both renewable energy and recycling infrastructure. Cross-border policy alignment is another critical dimension. As clean energy technologies are globally traded, international cooperation is necessary for managing the transboundary movement of renewable waste. Harmonized standards for product design, recyclability, and circularity enable more efficient recovery practices and reduce material incompatibility across markets. Finally, socio-political stability, public awareness, and administrative capacity also fall under the umbrella of policy-related factors. Public support for sustainability, robust institutional frameworks, and inclusive governance fosters innovation, industry compliance, and community engagement in clean energy transitions. In conclusion, geographic and policy factors collectively influence the life-cycle management of renewable energy systems. An integrated approach combining local resource assessment, logistical planning, and policy harmonization is essential for ensuring the environmental and economic viability of recycling and reuse within the clean energy transition.

4.2.7 Market dynamics and technological shifts

Market dynamics and technological shifts are key drivers that continuously reshape the recycling ecosystem and renewable energy landscape. These factors influence the supply demand equilibrium, cost structures, innovation rates, and competitiveness of sustainable solutions. From a market perspective, demand for renewable energy technologies such as solar panels, wind turbines, and batteries has surged due to climate goals, policy mandates, and corporate sustainability initiatives. This increasing demand has spurred mass production, which in turn leads to a growing volume of retired components. Consequently, the recycling market for these materials, such as silicon, rare earth metals, lithium, and composite plastics, is also expanding.

Price volatility in raw materials significantly impacts recycling decisions. When prices of virgin materials rise due to scarcity or geopolitical tensions, recycled materials become economically attractive. For example, lithium and cobalt prices have seen periodic spikes, prompting greater interest in battery recycling technologies. Conversely, if virgin material prices drop, recycling may become financially unsustainable without policy support or technological breakthroughs. Technological shifts often occur in tandem with market trends. Innovations in energy storage (like solid-state batteries), new solar cell materials (such as perovskites), or lightweight turbine blades demand new recycling methods. Traditional recycling systems may not be equipped to process these novel materials, creating a lag between technological advancement and recycling capability.

Moreover, as technologies evolve, so does product design. The push for design for disassembly and circular manufacturing is growing. Modular designs and standardized components can simplify end-of-life processing, improve recovery rates, and reduce costs. Industry leaders are investing in R&D for closed-loop systems where materials from decommissioned components are reintegrated into new product lines, creating a circular market dynamic.

Consumer behavior and investor interest also contribute to market evolution. As environmental consciousness rises, markets are shifting toward greener alternatives. Companies adopting circular strategies and sustainable technologies are often rewarded with better market positioning, customer loyalty, and access to green capital. Green bonds, ESG (Environmental, Social, Governance) metrics, and impact investing are reshaping capital flows toward circular economy ventures. Global competition adds another layer to market dynamics. Countries and companies compete to secure raw material access, dominate in clean energy manufacturing, and lead in recycling innovations. For instance, China's dominance in rare earth processing has pushed other nations to invest heavily in alternative supply chains and recycling. Furthermore, digital technologies like AI, IoT, and blockchain are transforming both the recycling and renewable sectors. Smart waste tracking, predictive maintenance of energy systems, and digital marketplaces for recycled materials are optimizing operations, improving transparency, and creating new business models.

4.2.7.1 Applications of material recovery forecasting

- *Planning urban mining operations*: Cities with aging solar installations can use MRF to prepare for large-scale recovery, establishing solar panel collection centers or recycling units.
- *Secondary raw material markets*: Industries can plan procurement from secondary sources based on predicted material availability, reducing cost and environmental impact.
- *Circular supply chain design*: Manufacturers can collaborate with recyclers to create closed-loop systems that feed recovered materials directly into new production.
- *Policy and investment guidance*: Governments can prioritize investment in recycling technologies for materials forecasted to be available in large volumes, like glass from solar panels or neodymium from wind turbines.

4.2.7.2 Challenges in material recovery forecasting

- *Data gaps*: Incomplete or proprietary data on product composition hinders accurate forecasting.
- *Technological uncertainty*: Rapid innovation can make long-term predictions less reliable.

- *Inconsistent product standards*: Varying designs make it difficult to standardize recovery models.
- *Limited EOL tracking*: Many renewable components are reused, sold second-hand, or abandoned, making tracking complex.

Material Recovery Forecasting is an essential tool for managing the sustainability of renewable energy systems. By anticipating material flows, stakeholders across the energy value chain, from manufacturers and recyclers to policymakers and investors, can make informed decisions that align with the principles of the circular economy. When paired with smart design, robust data collection, and supportive policy frameworks, MRF transforms waste into valuable resources, driving both economic and environmental gains.

4.2.8 Cradle to cradle thinking

Cradle to Cradle (C2C) Thinking is a transformative design philosophy that reimagines the lifecycle of products, systems, and infrastructure. Unlike the traditional "cradle to grave" model, which ends with disposal in landfills or incineration, Cradle to Cradle envisions a circular, regenerative system where materials are continuously reused, repurposed, or safely returned to nature. This concept has become foundational in advancing sustainable design, circular economy practices, and clean energy development. At its core, Cradle to Cradle is about designing products and systems from the outset so they can either be biologically decomposed or technologically recycled into new, high-value applications. Materials are thus categorized into two nutrient cycles:

Biological nutrients: Organic materials that can safely decompose and return to the environment, enriching ecosystems (e.g., biodegradable packaging, compostable polymers).

Technical nutrients: Synthetic or mineral-based materials that are designed to be recovered, refurbished, or remanufactured without degrading in quality (e.g., metals, plastics, composite components).

In the renewable energy sector, Cradle to Cradle Thinking is particularly relevant. Solar panels, wind turbines, and batteries contain rare and valuable materials such as lithium, cobalt, neodymium, and cadmium. Designing these products with disassembly and material recovery in mind ensures that their end-of-life value is not lost but reintegrated into new systems, creating a closed-loop supply chain. Take, for example, the case of solar photovoltaic (PV) modules. In a Cradle to Cradle framework, a PV module would be designed using materials that can be easily separated: nontoxic glass, recyclable aluminum frames, and standardized semiconductors. Upon reaching its end-of-life, the module wouldn't go to a landfill but would be sent to a specialized recycling facility where its components are extracted and reused in new modules or other industries. Cradle to Cradle also emphasizes energy efficiency and the use of renewable power throughout the product lifecycle, from material extraction and manufacturing to transportation and recycling.

This complements clean energy goals by minimizing carbon footprints and reducing reliance on fossil fuels during production and disposal stages. Furthermore, this approach extends to building materials, electronic devices, and consumer products, advocating for "materials as a service" rather than ownership. For instance, a wind turbine manufacturer might retain ownership of the turbines and offer energy as a service while taking responsibility for reclaiming and recycling the components, ensuring sustainable material stewardship. Certification systems like the Cradle to Cradle Certified Product Standard evaluate products on five performance categories: material health, material reutilization, renewable energy use, water stewardship, and social fairness. Companies across industries are increasingly seeking this certification to demonstrate their commitment to sustainability and circularity.

By embedding cradle to cradle thinking into policy, industrial design, and supply chain management, we move closer to zero-waste systems that not only minimize environmental harm but also regenerate resources and value. It shifts the mindset from minimizing damage to maximizing positive impact, fundamentally changing how we think about product design, consumption, and end-of-life recovery.

4.3 Reuse strategies for renewable energy assets

Reuse strategies for renewable energy assets focus on extending the life and value of components such as solar panels, wind turbines, batteries, and related infrastructure before they reach the end of their lifecycle. These strategies are essential to advancing the circular economy, reducing waste, and minimizing the need for virgin resource extraction. Table 4.5 illustrate reuse strategies for renewable energy assets.

4.3.1 Second-life photovoltaics

As the global deployment of solar photovoltaic (PV) systems accelerates, a growing number of panels are reaching the end of their primary operational life, typically 25 to 30 years. While these panels may no longer meet the efficiency standards required for grid-scale energy production, many still retain a significant portion of their energy-generating capacity. This opens the door to an innovative reuse strategy known as Second-Life PV, where decommissioned or downgraded panels are repurposed for alternative applications that do not demand peak performance. Second-life PV initiatives support the circular economy by extending the useful life of solar panels and deferring their entry into the waste stream. These systems are especially valuable in low-load, off-grid, or low-income settings where the initial cost of new PV technology is prohibitive. By diverting used panels from landfills and giving them a renewed function, second-life PV contributes to resource conservation, cost savings, and environmental protection. Here are some key applications where second-life solar panels can be effectively utilized:

TABLE 4.5 Reuse strategies for renewable energy assets.

Reuse strategy	Explanation (general)	Application in renewable energy systems
Component refurbishment	Repairing or restoring used components to extend their life.	Refurbishing inverters, turbines, gearboxes, and PV module frames to reintroduce them into operation.
Direct reuse of functional parts	Using fully functional components from decommissioned systems without major changes.	Reusing solar panels with >80% efficiency or wind turbine blades in secondary applications like lighting poles.
Repowering (upgrading systems)	Upgrading existing renewable assets with advanced, more efficient components.	Adding new turbine blades, replacing old cells in PV modules, or improving control electronics to boost output.
Second-life applications	Using old components in new, lower-demand applications.	EV and solar batteries repurposed for stationary energy storage or microgrids; older PV panels used for rural electrification.
Material reuse in manufacturing	Using recovered materials directly in new product manufacturing.	Reusing aluminum frames, glass from PV panels, copper wiring, and steel components in new renewable energy equipment.
Modular replacement strategy	Swapping only the defective or inefficient module instead of the entire system.	Replacing inverter modules, battery cells, or turbine control units to prolong the system lifecycle.
Functional reassignment	Using components for a different purpose than originally intended.	Wind turbine blades are converted into composite materials for furniture, bridges, or architectural structures.
Leasing and product-as-a-service models	Extending asset life through shared ownership and coordinated maintenance.	Companies leasing solar panels or batteries ensure controlled reuse and return for refurbishment.
Reverse logistics systems	Creating collection and transport networks for returned assets.	Helps recover solar panels, batteries, and turbine components for reuse, refurbishment, or repowering.
Service life extension programs	Systematic checks and upgrades to extend asset lifetime.	Operational monitoring and predictive maintenance to keep renewable energy assets functional longer.

Continued

TABLE 4.5 Reuse strategies for renewable energy assets.—cont'd

Reuse strategy	Explanation (general)	Application in renewable energy systems
Hybridization of used and new components	Combining old components with new technologies to optimize performance.	Mixing refurbished PV modules with new inverters or boosted battery cells to create hybrid systems.
Cascade reuse	Sequential reuse of components across different tiers of performance demand.	High-performance PV panels moved to industrial use → weaker ones to residential → final use in small devices.
Demonstration and training deployment	Using older systems for skill-building or demonstration purposes.	Retired solar panels, batteries, or turbines are used in academic labs and training centers.
Local community reuse programs	Allocating reusable components to local communities for electricity access.	Donated or low-cost reused PV modules or batteries for rural or low-income energy access projects.

Off-grid rural electrification: Second-life PV panels are particularly suited for providing basic electricity to rural or remote communities in developing countries. Even at 70% to 80% efficiency, these panels can power lights, small appliances, mobile charging stations, and essential medical devices in areas lacking access to centralized grids.

Street lighting and signage: Streetlights and public signage often require relatively low amounts of power and operate intermittently. Retired panels can be paired with energy-efficient LEDs and battery storage systems to power such installations in parks, public pathways, highways, and remote townships.

Irrigation and smart agriculture: Solar-powered water pumps and smart agricultural sensors for soil moisture, humidity, and temperature monitoring do not require high-capacity energy sources. Second-life PV panels offer an affordable, eco-friendly option to support precision farming and reduce reliance on fossil-fuel-based irrigation systems.

In addition to these applications, second-life PVs can also support:

- Educational labs and community projects, providing students with practical learning tools.
- Emergency response systems, offering backup power in disaster-prone areas.
- Green urban infrastructure, including solar-powered benches, kiosks, or mobile charging stations.

However, implementing second-life PV strategies requires careful assessment of panel condition, safe transportation, appropriate installation, and monitoring. It also necessitates developing legal and quality assurance frameworks to manage liability and ensure user safety. In conclusion, Second-Life Photovoltaics represent a practical, sustainable approach to maximizing the utility of solar energy technologies. By redirecting retired panels toward meaningful second-use applications, we not only reduce environmental impact but also foster energy equity and resilience in underserved communities.

4.3.2 Battery repurposing

As electric vehicles (EVs) and renewable energy systems gain global traction, an increasing volume of lithium-ion batteries is reaching the end of their first life. These batteries, though no longer suitable for high-performance applications like EVs due to capacity degradation, often retain 60% to 80% of their original storage ability. This remaining capacity can be repurposed for less demanding energy storage needs, a strategy known as battery repurposing or second-life battery application. Battery repurposing offers a critical pathway in the circular economy, reducing electronic waste and maximizing the life cycle of valuable resources such as lithium, cobalt, and nickel. It also addresses the rising demand for stationary energy storage solutions, which are essential for stabilizing renewable energy supply, enhancing energy access in remote regions, and ensuring backup during outages. Below are the key applications of repurposed batteries:

Home energy storage systems: Repurposed EV batteries are being increasingly used in residential energy storage units. When paired with rooftop solar installations, these second-life batteries allow homeowners to store excess solar power during the day and use it at night or during peak demand periods. This not only increases energy self-sufficiency but also reduces electricity bills and reliance on fossil-fuel-based grid systems. Companies like Nissan, BMW, and Tesla have piloted such systems using batteries retrieved from retired EVs. The relatively lower cost of repurposed batteries makes home energy storage more affordable and accessible.

Community-based microgrids: Second-life batteries are well-suited for deployment in community microgrids, especially in off-grid or underserved areas. These microgrids can integrate renewable energy sources like solar or wind, and use repurposed batteries to store electricity for communal use—powering homes, schools, clinics, and businesses. In many developing regions, where grid expansion is costly or impractical, this model ensures reliable electricity access at a fraction of the cost of new infrastructure. Such systems also promote energy democracy by allowing local control and resilience against grid failures.

Emergency backup solutions in remote locations: Remote and disaster-prone regions often struggle with interrupted power supply during emergencies. Repurposed batteries can be integrated into portable or fixed emergency backup power systems to support communication equipment, refrigeration for medicine, lighting, and basic appliances. These systems are particularly beneficial for hospitals, relief shelters, and mobile response units. Due to their portability and quick deployment, second-life batteries help ensure continuity of critical services during blackouts, natural disasters, or humanitarian crises.

4.3.2.1 Benefits of battery repurposing

- *Environmental sustainability*: Delays battery disposal and reduces mining pressure for new materials.
- *Cost efficiency*: Offers a cheaper alternative to new storage systems.
- *Waste minimization*: Keeps batteries out of landfills and reduces hazardous waste.
- *Energy access*: Expands clean energy storage to communities with limited financial resources.

4.3.2.2 Challenges and considerations

- *Battery health assessment*: Requires testing and diagnostics to evaluate capacity, safety, and usability.
- *Standardization*: Lack of uniform battery formats complicates repurposing at scale.
- *Safety*: Must address issues like thermal runaway, leakage, or short circuits.
- *Regulatory frameworks*: Clear guidelines are needed for transportation, repurposing, liability, and certification.

Battery repurposing transforms what would otherwise be considered waste into valuable energy storage assets. It plays a pivotal role in achieving sustainability in the energy and mobility sectors by bridging the gap between performance decay and complete disposal. When integrated thoughtfully into home systems, community microgrids, or emergency setups, second-life batteries not only reduce environmental impact but also contribute significantly to energy resilience and equity.

4.3.3 Wind turbine component reuse

As wind energy infrastructure ages, a substantial volume of turbine components such as gearboxes, nacelles, and blades is being retired. While some parts are recyclable, others, particularly the blades made from composite materials, are more challenging to dispose of sustainably. Instead of discarding these components, reuse strategies are emerging that breathe new life into wind turbine parts through innovative and practical applications. Reusing these components not only conserves resources but also supports the broader goals of the circular economy by extending the value chain of renewable energy systems.

Gearboxes and nacelles reused in training simulators: Gearboxes and nacelles, which house the turbine's drivetrain and generator, are among the most valuable and complex components of a wind turbine. Although they may no longer be suitable for commercial energy production after decommissioning, they still hold immense value for educational and training purposes. Technical institutions, engineering universities, and vocational training centers often lack real-world equipment for hands-on learning. Reusing gearboxes and nacelles in training simulators allows students and technicians to practice assembly, diagnostics, maintenance, and repair in a controlled, realistic environment.

This repurposing offers multiple benefits:

- It reduces training costs by providing real industrial hardware.
- It enhances workforce preparedness for the wind energy sector.
- It supports research on turbine performance, failure modes, and system upgrades.

In essence, the reuse of these components in educational settings ensures continued value extraction while building technical expertise and awareness around renewable energy technologies.

Blades repurposed into urban furniture, pedestrian bridges, and playgrounds: Wind turbine blades, typically made from fiberglass-reinforced composites, present a major disposal challenge due to their size, durability, and nonrecyclable nature. Instead of landfilling them, a growing number of architects, designers, and civil engineers are finding innovative ways to repurpose blades into functional and artistic structures in urban and rural environments.

Some notable reuse applications include:

Urban furniture: Sections of blades are transformed into stylish benches, bus shelters, bike racks, or public art installations. Their aerodynamic curves, strength, and weather resistance make them ideal for outdoor furniture that is both durable and visually striking.

Pedestrian bridges: Blade segments are being used as structural elements in small-scale pedestrian or cycle bridges, especially in parks or campuses. Their lightweight yet strong composition helps reduce construction costs and timelines while giving a second life to otherwise discarded materials.

Playgrounds: In some innovative projects, blades are being redesigned into playground equipment such as slides, tunnels, and climbing structures. This not only promotes sustainability but also sparks public interest in renewable energy through visible and interactive installations.

4.3.3.1 Benefits of wind turbine component reuse

- Reduces landfill burden and environmental impact.
- Extends product lifespan and resource efficiency.
- Encourages creativity and cross-disciplinary collaboration.

- Promotes public awareness and education on sustainability.
- Supports green infrastructure in cities and communities.

4.3.3.2 Challenges and considerations

- *Logistics*: Transporting large turbine parts can be complex and costly.
- *Design limitations*: Blades must be safely cut and adapted to new purposes.
- *Safety standards*: Repurposed components used in public spaces must meet structural and safety codes.
- *Awareness and demand*: Encouraging municipalities and designers to adopt these solutions requires advocacy and awareness.

The reuse of wind turbine components showcases the innovative potential of circular thinking in renewable energy. Whether in classrooms or cityscapes, giving a second life to gearboxes, nacelles, and blades transforms waste into value, demonstrating that sustainability is not only about energy production but also about how we design, use, and retire the technologies of tomorrow.

4.4 Repurposing—Innovation beyond energy

As renewable energy technologies reach their end-of-life, the focus has shifted from simply recycling components to repurposing them creatively and innovatively beyond their original energy-generation functions. This shift not only enhances sustainability but also opens up new industrial, architectural, and social applications that align with the principles of the circular economy. Table 4.6 illustrate attributes and properties of repurposing strategies for renewable energy assets.

The following sections explore three compelling areas where solar and wind energy components are being repurposed to bring value beyond their energy-producing years: solar panel glass as architectural material, wind blade composites for infrastructure, and repurposed inverter systems.

4.4.1 Solar panel glass as architectural material

Solar panels, particularly crystalline silicon-based models, contain high-quality tempered glass that is engineered to endure harsh environmental conditions for decades. When these panels are decommissioned due to reduced energy efficiency or upgrades, the glass layer remains physically intact and structurally sound. This durability makes it an ideal candidate for architectural reuse in a range of creative and functional applications.

One of the most promising uses is in green building design, where second-life solar panel glass is employed for façades, skylights, canopies, partitions, and balcony enclosures. This application provides both esthetic and environmental value. The glass typically has antireflective coatings and high light

TABLE 4.6 Attributes and properties of repurposing strategies for renewable energy assets.

Attribute	Property/ explanation	Examples/applications from attached content
1. Material durability	High structural strength and long-lasting materials enable repurposing beyond energy use.	Solar PV tempered glass used for façades, skylights; wind blade composites used for bridges, noise barriers, modular shelters.
2. Structural integrity	Decommissioned components retain enough physical integrity for secondary applications.	Wind blades are used as structural elements in pedestrian bridges; solar glass is used in building envelopes.
3. Environmental resilience	Components built to withstand harsh conditions continue performing in architectural and civil uses.	Weather-resistant blade materials used in urban furniture or sound barriers; anti-reflective solar glass for roofing.
4. Light transmission and optical properties	Transparent or semitransparent materials enable functional reuse in lighting and design.	Solar panel glass for skylights, atriums, indoor illumination, and UV-protective panels.
5. Acoustic properties	Components with inherent sound-dampening capabilities can be repurposed for noise reduction.	Wind blade segments used in roadside noise barriers.
6. Geometric versatility	Unique shapes allow innovative design integration in architecture or civic infrastructure.	Curved wind blade structures are used in playgrounds, benches, bus stops, amphitheater seating.
7. Safety and certification requirements	Repurposed components must undergo thorough testing before reuse.	Structural evaluation of wind blade sections; electrical testing and recertification of inverters.
8. Ease of disassembly and processing	Simplifies extraction, cleaning, modification, and repurposing.	Removing PV glass for construction use, cutting and transporting large wind blades.

Continued

TABLE 4.6 Attributes and properties of repurposing strategies for renewable energy assets.—cont'd

Attribute	Property/ explanation	Examples/applications from attached content
9. Modularity for repurposing	Components designed in modular form can be repurposed more easily.	Modular inverter systems reused in microgrids; modular blade segments used in shelters.
10. Cost efficiency	Repurposing reduces the cost of new materials while offering functional benefits.	Affordable housing using reused PV glass; low-cost microgrids using second-life inverters.
11. Social value creation	Repurposing supports community needs, education, and social development.	PV glass in off-grid schools; used inverters in academic labs and training centers.
12. Circular economy alignment	Repurposing minimizes waste and extends material life cycles.	Architectural reuse of solar glass; reuse of blades in civil engineering; inverters repurposed for industrial automation.
13. Educational and skill development potential	Provides hands-on experience with real technologies.	Repurposed inverters are used in engineering labs for training students.
14. Industrial innovation opportunities	Encourages new business models and product design strategies.	Startups using blade composites for furniture; factories using inverter components in robotics.
15. Aesthetic versatility	Components offer visually appealing design possibilities.	Decorative panels from solar glass; artistic installations from wind blades.
16. Technical Re-engineering potential	Electronic components can be extracted for new engineering uses.	Inverter IGBTs, capacitors, microcontrollers are reused in control systems and motor drivers.
17. Logistics and handling requirements	Large components require careful transport, cutting, and installation planning.	Blade transport challenges: cleaning and processing PV glass.
18. Policy and institutional support needs	Successful repurposing depends on guidelines, certifications, and incentives.	Grants for inverter repurposing; R&D for blade cutting tools; standards for reused solar glass.

TABLE 4.6 Attributes and properties of repurposing strategies for renewable energy assets.—cont'd

Attribute	Property/ explanation	Examples/applications from attached content
19. Cross-sector collaboration	Requires energy, construction, education, and design sectors to work together.	Joint projects between energy companies, architects, civil engineers, and educators.
20. Community awareness and engagement	Repurposed structures can act as public symbols of sustainability.	Wind blade playgrounds, bus stops, and benches that raise awareness of the renewable lifecycle.

transmission, making it ideal for indoor lighting solutions while offering UV protection and thermal insulation.

Key applications include:

- Skylights and atriums in commercial buildings or residential complexes, offering natural lighting and energy savings.
- Wall cladding or curtain walls in industrial spaces that value durability and sustainability.
- Decorative panels and furniture elements in eco-design projects.

By integrating used PV glass into construction, architects and builders can reduce the demand for new raw materials, lower project costs, and showcase a commitment to sustainable practices. Moreover, such initiatives enhance the visibility of renewable energy's life cycle and encourage wider adoption of circular design.

In developing regions, repurposed solar panel glass has also found use in affordable housing projects. It can replace conventional glass in windows or be used as translucent roofing material to provide light in off-grid shelters and schools. This not only addresses waste management but also supports social development.

However, challenges remain in terms of disassembly, cleaning, and standardization of solar glass for architectural purposes. To overcome these, future panel designs could incorporate modularity and ease of repurposing into their construction, supporting the philosophy of "design for circularity."

4.4.2 Wind blade composite for infrastructure

Wind turbine blades are constructed from fiberglass-reinforced composites, chosen for their strength-to-weight ratio and resistance to environmental wear.

However, these same properties make them difficult to recycle or dispose of once decommissioned. As the number of retired blades increases, attention has turned to repurposing wind blade materials in infrastructure and civil engineering projects.

One ground-breaking approach involves using blade segments as structural components in bridges, walkways, and sound barriers. Their curved, aerodynamic shape and structural integrity offer design flexibility and cost-effective alternatives to conventional materials like concrete and steel.

Notable applications include:

- *Pedestrian and cycle bridges*: Blade segments are used as load-bearing members or railings, as seen in several projects across Europe.
- *Roadside noise barriers*: Due to their acoustic properties and weather resistance, blade materials are ideal for highways and urban environments.
- *Modular housing or shelters*: Blade sections can be reconfigured as roofing structures or curved walls for emergency or low-income housing.

Moreover, repurposed blades have entered the realm of urban design and art installations. Public benches, amphitheater seating, playground structures, and even bus stops have been creatively crafted from wind turbine blades, combining functionality with environmental storytelling. These installations serve as tangible reminders of the energy transition and promote public engagement with sustainability goals. Cities that adopt such reuse practices benefit from reduced construction waste, resource conservation, and a unique esthetic that aligns with eco-conscious values. Additionally, this supports local circular economies by creating jobs in blade processing, logistics, and fabrication.

That said, safety remains a critical consideration. The structural integrity of repurposed blade segments must be thoroughly evaluated and certified. Moreover, logistics such as cutting, transportation, and installation require careful planning due to the size and shape of blades. To scale this innovation, more investment is needed in R&D for blade-friendly tools, policy support, and cross-sector collaboration among energy, construction, and design industries.

4.4.3 Repurposed inverter systems

Inverter systems are the technological heart of any renewable energy system, responsible for converting direct current (DC) from solar panels or wind turbines into usable alternating current (AC). As solar and wind systems age or upgrade, inverters are often decommissioned despite having functional electronic components. These decommissioned inverters hold substantial potential for reuse in secondary energy, educational, and automation applications. A primary area of reuse is in low-demand microgrid applications, where second-life inverters regulate power in off-grid or hybrid setups. For instance, a used inverter can serve as the core control unit for a community solar farm or battery backup system in a rural village. These applications do not require the inverter to operate at peak original efficiency but benefit significantly from the cost savings and sustainability gains.

Another important avenue is the use of repurposed inverters in academic labs and technical training centers. Educational institutions often lack access to advanced, real-world energy technology due to budget constraints. Repurposed inverters allow students to engage in practical learning related to power electronics, grid integration, and system diagnostics without the risk or expense associated with new units. Additionally, inverter electronics can be reengineered for industrial automation. Components such as IGBTs, capacitors, microcontrollers, and heat sinks can be extracted and repurposed in control panels, motor drivers, or robotics applications. This contributes to the broader goals of electronic waste reduction and circular manufacturing.

Benefits of repurposing inverters include:

- Cost savings in microgrids and educational setups.
- Skill development through hands-on learning tools.
- Reduced e-waste and recovery of valuable electronic parts.
- Acceleration of innovation in industrial control systems.

To facilitate repurposing, inverter manufacturers can support modular designs with documentation and diagnostic tools that ease the transition into second-life use. Governments and industry stakeholders can further encourage adoption by offering grants, technical guidelines, and refurbishment centers. The success of inverter reuse hinges on rigorous testing, recertification, and quality control to ensure safety and reliability in their new functions. With the right infrastructure, these systems can extend their utility well beyond their original deployment, reinforcing the interconnectedness of clean energy, education, and technology innovation.

Repurposing components from solar and wind energy systems exemplifies how innovation can flourish at the intersection of sustainability, design, and functionality. Whether it's solar panel glass illuminating green buildings, wind turbine blades becoming the backbone of pedestrian infrastructure, or retired inverters breathing life into new energy systems and classrooms, these strategies demonstrate that renewable energy's contribution does not end with electricity generation. By embedding circular design thinking into the lifecycle of renewable technologies and fostering collaboration across industries, we can unlock new forms of value, reduce environmental impact, and promote inclusive growth. As the energy transition progresses, repurposing will play a vital role not just in waste reduction but in reimagining what clean energy assets can become.

4.5 Technologies empowering circularity

As the global economy transitions toward sustainability and net-zero goals, the concept of circularity, which involves eliminating waste and reusing materials across a product's life cycle, has gained unprecedented attention. While circularity has traditionally been associated with design and manufacturing innovations, the emergence of advanced technologies is accelerating its potential. In the realm of renewable energy, particularly solar and wind power, technological advancements such as digital twins, artificial intelligence (AI), blockchain, and 3D printing are playing a crucial role in closing the loop of resource use. These technologies not only optimize existing infrastructure but also help manage waste, extend asset lifetimes, and create innovative pathways for material reuse. Fig. 4.5 illustrates technologies empowering circularity.

This chapter explores three transformative technology enablers that are fostering circularity in renewable energy systems:

Digital twins and artificial intelligence (AI)

Blockchain for traceability

3D Printing with recovered materials

4.5.1 Digital twins and AI

A Digital Twin is a real-time virtual representation of a physical object, system, or process that uses data from sensors embedded in the real-world counterpart. In renewable energy, digital twins are used to replicate wind turbines, solar panels, inverters, and even entire energy systems. They provide deep insights into equipment performance, aging, and material stress by simulating environmental, mechanical, and operational parameters. Digital twins are central to circularity in two primary ways:

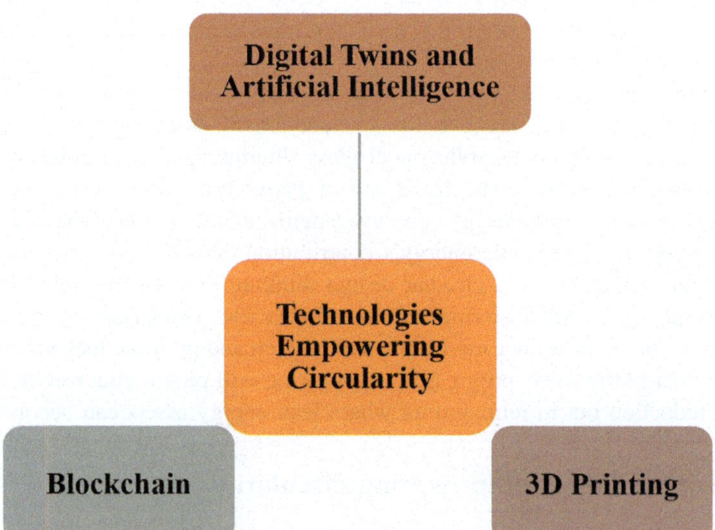

FIGURE 4.5 Technologies empowering circularity.

Lifecycle extension: By simulating and monitoring real-time wear and tear, digital twins allow operators to proactively schedule maintenance or parts replacement, thus extending the lifespan of components and reducing waste.

End-of-life intelligence: When a component is approaching the end of its useful life, the digital twin can help determine whether it should be repaired, reused in a different context (e.g., downcycled for less demanding use), or sent for material recovery.

For instance, the digital twin of a wind turbine gearbox could reveal patterns of microcracking or lubricant degradation well before they become critical. This data-driven understanding informs whether the gearbox could be repurposed for training simulators or industrial applications that demand lower performance thresholds.

Artificial Intelligence (AI) plays a pivotal role in promoting circular practices within renewable energy infrastructure through advanced tools like predictive maintenance systems, decision support systems (DSS), and automated sorting and disassembly mechanisms. These AI-enabled technologies not only enhance operational efficiency but also ensure that materials and components are retained within the value chain for as long as possible, a key tenet of the circular economy.

Predictive maintenance, one of the most impactful applications of AI in the renewable sector, utilizes machine learning algorithms to monitor the real-time performance and condition of assets such as wind turbine gearboxes, rotor blades, or photovoltaic panels. By analyzing vast datasets collected from sensors such as vibration, temperature, pressure, and electrical outputs, AI models can identify subtle patterns that precede equipment failure. This foresight allows operators to plan maintenance activities proactively, thus avoiding sudden breakdowns, extending the service life of components, and minimizing the resource-intensive emergency repairs typically required during unplanned downtimes. For example, instead of replacing an entire turbine gearbox when it fails unexpectedly, predictive analytics might identify early-stage bearing wear and recommend a targeted intervention week in advance, saving both material and labor costs while reducing environmental impact.

Complementing predictive maintenance is the use of Decision Support Systems (DSS) powered by AI, which aid in determining optimal end-of-life strategies for components. These systems integrate data on material health, degradation curves, and current market dynamics to suggest the most sustainable and economically viable pathways for component reuse, refurbishment, or recycling. Suppose a solar panel's efficiency has dropped below a viable threshold for commercial applications, but the DSS evaluates that the degradation is uniform and the physical structure remains intact. In that case, it may recommend redeploying the panel in off-grid or rural setups where high performance is not critical. Alternatively, if materials like silver, glass, or silicon wafers can be profitably extracted based on current commodity prices, the DSS may suggest material recovery instead of reuse. This intelligence enables asset managers and manufacturers to make informed, circular decisions rather than defaulting to disposal or premature recycling.

Furthermore, AI-powered automated sorting and disassembly systems are revolutionizing the way renewable energy components are processed at the end of their lifecycle. These systems, often integrated with computer vision and real-time analytics, are capable of identifying, sorting, and disassembling complex components such as wind turbine blades, nacelles, and solar modules with high precision and speed. Traditional manual disassembly is labor-intensive, error-prone, and often leads to material contamination, especially when dealing with mixed composite structures. AI-driven robotic arms, however, can distinguish between different materials such as resins, metals, and polymers on the basis of texture, reflectance, or embedded QR/RFID tags. They can then disassemble parts in the most efficient sequence, minimizing damage and maximizing the yield of recoverable materials. For instance, an AI-enabled robotic disassembler might first extract the valuable rare-earth magnets from a wind turbine generator before proceeding to remove copper windings and aluminum casing, ensuring that each component is handled optimally for reuse or recycling.

In essence, these AI-driven technologies, such as predictive maintenance, decision support systems, and automated disassembly, not only drive operational excellence but also establish a digital backbone for circularity in renewable energy systems. They help transition from a linear "use and discard" mind-set to a regenerative model where each component is tracked, maintained, repurposed, and eventually reintegrated into new products or applications, contributing to a truly sustainable energy future.

4.5.2 Blockchain for traceability

Blockchain is a decentralized, distributed ledger system that records transactions across many computers in a secure, tamper-proof way. It provides transparency, traceability, and immutability traits that are essential in validating sustainability and circularity claims.

In circular energy systems, blockchain can be used to:

- *Track material origins and movements*: Recording the source, material composition, manufacturing history, and maintenance records of components.
- *Verify recycled content*: Ensuring that materials claimed to be "recycled" or "upcycled" truly are, and identifying the exact recycling process used.
- *Monitor carbon footprint and environmental impact*: Smart contracts can track emissions and provide real-time ESG (Environmental, Social, and Governance) data related to the asset or component.

A wind turbine blade, for example, may have components manufactured in multiple countries. Over its 20- to 25-year lifecycle, it undergoes maintenance, retrofits, and possibly material fatigue. Blockchain provides a unified record of all these touchpoints:

- *Genesis block*: Records manufacturing date, material origin (e.g., composite resin source), and structural specifications.
- *Lifecycle blocks*: Store operational data (from digital twins), inspections, repairs, and transportation.
- *End-of-life blocks*: Document decisions about repurposing (e.g., converted into pedestrian bridges or furniture) or recycling steps taken.

Such traceability is especially valuable in Extended Producer Responsibility (EPR) frameworks, where manufacturers are held accountable for the postconsumer phase of their products.

Public and private tenders for solar or wind projects are increasingly requiring proof of circular sourcing. Blockchain supports such sustainable procurement by:

- Certifying vendors that supply reused or recycled materials.
- Automating compliance checks via smart contracts.
- Allowing for third-party verification by regulators or independent bodies.

Companies like Power Ledger and Energy Web Foundation are piloting blockchain solutions for traceable, decentralized energy trading. These platforms are extending their use cases into circularity by tracking component provenance and validating claims around recycled content and component reuse.

4.5.3 3D printing with recovered materials

3D printing, or additive manufacturing, has emerged as a transformative technology across industries due to its ability to fabricate complex, customized objects with minimal material waste. In recent years, the integration of recovered or recycled materials into 3D printing has opened new possibilities for promoting circular economy principles, reducing environmental impact, and making manufacturing more sustainable. "3D Printing with Recovered Materials" refers to the process of utilizing postconsumer waste, industrial scrap, or reclaimed natural resources as feedstock for producing printable filaments, resins, or powders used in additive manufacturing. This approach not only diverts waste from landfills but also reduces the dependency on virgin materials, thereby conserving natural resources and reducing the carbon footprint of production. One of the most common sources of recovered materials for 3D printing is plastic waste, especially thermoplastics such as PLA (polylactic acid), ABS (acrylonitrile butadiene styrene), and PET (polyethylene terephthalate). These plastics can be collected from discarded consumer products, packaging materials, and failed prints, then shredded, cleaned, melted, and extruded into new filaments. Companies and research institutions have developed desktop filament extruders that allow users to recycle plastic at a small scale, enabling a closed-loop production process. For example, recycled PET from plastic bottles can be transformed into durable 3D printing filament, suitable for making functional parts, tools, or even art pieces. Similarly, failed PLA prints can be repurposed into new raw material for future projects, reducing waste at the source. Beyond plastics, researchers are exploring innovative uses of industrial by-products and organic waste in 3D printing. Materials such as wood flour, sawdust, rice husk, and even coffee grounds can be mixed with bioplastics to create composite filaments that mimic the appearance and texture of natural materials. These composites not only make use of agro-industrial waste but also offer esthetic and functional benefits, making them popular for furniture, architectural models, and decorative items. Likewise, metal powders recovered from machining processes or e-waste can be used in powder-based 3D printing systems like Selective Laser Melting (SLM), enabling the reuse of valuable metals such as aluminum, copper, and titanium. Furthermore, construction-scale 3D printing is exploring the use of recycled concrete and demolition waste as base material for printing building components. By grinding down recovered concrete and combining it with binders, researchers have demonstrated the feasibility of printing structures that are both environmentally friendly and structurally sound. This not only reduces construction waste but also lowers the carbon intensity of infrastructure projects. Despite the promising potential, 3D printing with recovered materials faces challenges such as material consistency, contamination, and performance variability. Advances in material science, quality control technologies, and printer design are essential to overcome these issues and scale up the use of recycled inputs in additive manufacturing. Nonetheless, this approach aligns well with global sustainability goals, offering a path toward resource-efficient production, waste reduction, and localized manufacturing. As innovation continues, 3D printing with recovered materials is poised to redefine the future of eco-friendly fabrication and circular design. The following are the advantages of 3D Printing in Circular Systems:

- Material *efficiency*: Traditional manufacturing wastes up to 30% of materials through subtractive processes. 3D printing uses only what's necessary.
- *Customization and prototyping*: Engineers can rapidly prototype spare parts for old wind turbines or solar inverters without the need for mass production.
- *Localized manufacturing*: Reduces emissions and transportation costs by enabling on-site or near-site production from locally recycled materials.

In the evolving landscape of renewable energy, technological innovation is not only about efficiency and performance it is increasingly about sustainability and circularity. Digital twins, artificial intelligence, blockchain, and 3D printing are reshaping how we design, use, and retire components in solar and wind energy systems. These technologies empower stakeholders to make data-driven decisions that reduce waste, extend product lifecycles, and innovate reuse pathways. As these tools mature and integrate, they form a digital infrastructure for the Circular Economy 4.0 where every material has a history, a purpose, and a future. By investing in these enabling technologies and aligning them with policy support and business incentives, societies can leap toward a truly regenerative, resource responsible energy system. Table 4.7 illustrate properties of technology enablers supporting circularity in renewable energy systems.

4.6 Business models driving reuse and repurposing

In the face of escalating climate challenges, resource scarcity, and increasing consumer awareness, businesses are rapidly shifting toward more sustainable, circular approaches. Indeed, the linear "take-make-waste" model that has dominated industrial production is under increasing scrutiny. As governments enact stricter regulations and consumers place higher value on environmental stewardship, companies are redesigning their offerings—not just products, but entire business models. A few leading strategies stand out:

1. Product-as-a-Service (PaaS)
2. Circular Supply Chains
3. Buy-Back and Refurbishment Programs

Together, these models embody circular economy principles and open up new revenue streams while reducing waste, extending product life, and building stronger relationships between producers and users. Let's dive deeper into each one.

4.6.1 Product-as-a-service

In the PaaS model, manufacturers retain ownership of products—whether solar panels, wind turbines, appliances, or electronics—offering them to users as a service. Consumers pay for the utility (e.g., electricity generated, hours of usage, output) rather than ownership itself. Once the product reaches its end-of-life (EoL) or falls out of technological relevance, the company takes it back to refurbish, recycle components, or responsibly dispose of materials. PaaS is work into the following way:

Aligns value with utility: Companies are incentivized to manufacture longer-lasting, robust, and efficient products, since they bear the lifecycle cost and responsibility. Users benefit from lower upfront costs and predictable usage fees.

Boosts customer loyalty and recurring revenue: Service-based contracts ensure ongoing engagement with customers. Manufacturers gain continual insights into usage patterns, performance, and maintenance needs.

TABLE 4.7 Properties of technology enablers supporting circularity in renewable energy systems.

Technology enabler	Key property	Explanation/role in circularity
Digital twins and artificial intelligence (AI)	Real-time monitoring	Enables continuous tracking of asset performance, predicting failures, and optimizing lifespan.
	Predictive maintenance	AI forecasts component degradation, reducing downtime and extending system life.
	Design optimization	Digital twins simulate materials, loads, and wear to create circular, repair-friendly designs.
	Lifecycle impact analysis	AI models environmental footprints, supporting low-impact and recyclable system choices.
	End-of-life forecasting	Predicts when components will reach retirement, aiding repurposing and material recovery planning.
	Resource efficiency algorithms	Optimizes energy, material use, and system configurations to reduce waste.
	Virtual prototyping	Reduces physical prototyping waste by testing designs digitally before manufacturing.
	Circular supply chain coordination	Enhances coordination between manufacturers, recyclers, and refurbishers through predictive insights.
Blockchain for traceability	Immutable record keeping	Stores verifiable lifecycle data of components, enabling trustworthy circularity tracking.
	Material provenance verification	Ensures materials come from sustainable and certified circular sources.
	End-of-life compliance	Tracks proper dismantling, recycling, or repurposing to meet regulatory standards.
	Transparent supply chains	Provides stakeholders with visibility from raw material extraction to repurposing or recycling.

TABLE 4.7 Properties of technology enablers supporting circularity in renewable energy systems.—cont'd

Technology enabler	Key property	Explanation/role in circularity
	Anticounterfeiting	Prevents the reuse of unsafe or uncertified components by recording unique digital identities.
	Smart contracts	Automates circular processes such as take-back agreements, recycling incentives, and reverse logistics.
	Energy transaction tracking	Supports peer-to-peer energy trading using repurposed assets in microgrids.
	Carbon and waste tracking	Provides accurate, tamper-proof accounting for circularity metrics like carbon savings and recovered materials.
3D printing with recovered materials	Material circularity	Uses recycled polymers, metals, glass, or composites to create new functional components.
	Customization and modularity	Enables tailor-made, repairable parts that support modular and disassembly-friendly design.
	Localized manufacturing	Reduces transportation emissions and supports local circular economies.
	Waste minimization	Produces components with minimal leftover material compared to traditional manufacturing.
	Rapid prototyping	Quickly creates replacement parts from recovered materials, reducing downtime and extending asset life.
	Component refurbishment	Allows remanufacturing or strengthening of old parts using additive layering techniques.
	Design for repair	Creates simplified geometries or replacement parts that extend the life of renewable systems.
	Value recovery innovation	Turns waste streams (e.g., blade composites, PV glass powders) into new high-value products.

Encourages circular design: When companies retain ownership and face reuse responsibilities, they naturally design for disassembly, reparability, and upgradability from the outset.

4.6.1.1 Real-world examples

Energy-as-a-service (EaaS): Companies like Sunrun and Engie install and own solar panels on residential rooftops. Homeowners pay only for the electricity used, while the firm handles maintenance, monitoring, and eventual panel removal or recycling.

Procurement-as-a-service in electronics: Firms such as HP's Device-as-a-Service lease laptops and printers, providing regular hardware upgrades while taking responsibility for end-of-life device collection, data wiping, refurbishment, and remanufacturing.

4.6.1.2 Advantages

- Reduces capital expenditure for clients
- Promotes sustainable product lifecycle management
- Reinforces brand reputation for environmental responsibility

4.6.1.3 Challenges

- Requires sophisticated logistics and tracking infrastructure
- Must handle insurance, liability, and long-term asset management
- Profit margins depend on product durability and lifecycle control

4.6.2 Circular supply chains

A circular supply chain breaks away from the linear trajectory of resource use. Instead, it plugs waste streams back into production. In practice, this means integrating various actors from dismantlers and recyclers to refurbishers into a network where end-of-life components are captured, processed, and reintegrated into the manufacturing workflow. The following are the significance of circular supply chain:

Resource security: Companies can mitigate volatility in material prices (e.g., rare earth metals) by recovering and reusing valuable components.

Environmental impact: Keeps materials in circulation, reduces virgin extraction, lowers energy consumption, and minimizes emissions from production and transport.

Compliance readiness: Circular chains help firms meet increasingly strict regulations like the EU's Waste Electrical and Electronic Equipment (WEEE) and RoHS directives.

- Core components

Collection and dismantling networks: Strategically located local centers gather end-of-life goods, such as solar modules, printers, or batteries, and dismantle them for component retrieval.

Advanced recycling and reclamation facilities: These use mechanical, chemical, or thermal processes to extract pure materials, plastics, glass, and metals for reuse or resale.

Design for circularity: Products are engineered for nondestructive disassembly. Modular construction enables easy part replacement and refurbishment.

Information transparency: Material passports and blockchain tracking systems ensure traceability, demonstrating compliance and building trust.

4.6.2.1 Case examples

Automotive remanufacturing: Companies like Caterpillar and Volvo recover engine components, refurbish them to equivalent new quality, and resell them with warranties. This reduces raw material use by 50%–90% versus virgin production.

Electronics recycling ecosystems: Dell's closed-loop recycling program recovers plastics from old electronics to manufacture new equipment, reducing new plastic consumption by millions of pounds annually.

Solar panel recycling partnerships: Firms like First Solar collaborate with recyclers to recover glass, aluminum frames, and semiconductor material from decommissioned panels—reintegrating them into new module production.

4.6.2.2 Advantages

- Stabilizes supply chains
- Delivers environmental gains and economic value
- Fosters collaboration across industries

4.6.2.3 Challenges

- Requires major upfront network investment
- Tech maturation—especially for complex, composite materials—can be slow
- Regulatory, logistical, and standardization hurdles remain

4.6.3 Buy-back and refurbishment programs

Buy-back schemes entice customers to return used products in exchange for financial incentives like credits, discounts, or cash refunds. Returned items are then inspected, refurbished, and resold as "like-new," or broken down into components or raw materials. In the following way buy-back and refurbishment programs are going on:

Consumer incentive models: Consumer incentive models form the cornerstone of successful buy-back and refurbishment programs. These initiatives encourage customers to return used products like smartphones, laptops, or solar modules—by offering tangible rewards. Incentives often come in the form of monetary credits, discounts on new purchases, gift cards, or even cash payouts. A prominent example is Apple's Trade-In Program, where customers can return an old iPhone and receive credit toward the purchase of a newer model. This approach benefits both parties: consumers save money, and companies recover valuable products for reuse.

These models tap into behavioral economics, motivating users with immediate, visible rewards rather than abstract environmental benefits. Incentives reduce disposal in landfills, create a culture of conscious consumption, and extend product lifespans. Moreover, by retaining customer touchpoints during upgrades, brands strengthen loyalty and increase repeat purchases.

Designing effective incentive models requires balancing perceived product value, program cost, and ease of participation. Higher-value devices yield greater returns, while low-value items might only fetch nominal discounts. Therefore, customer awareness campaigns, digital tools to estimate trade-in value, and seamless return processes are vital for adoption. In sum, consumer incentive models foster sustainable behavior while keeping customers within a brand's ecosystem. They transform the concept of waste into a new value stream, financially and environmentally.

Logistics and collection: The success of any reuse or refurbishment program hinges on efficient logistics and collection systems. Once consumers are incentivized to return used products, companies must establish streamlined methods to retrieve those goods securely, quickly, and cost-effectively. Collection can happen through several channels: mail-in kits shipped to customers, drop-off points at retail stores or kiosks, and even pickup services for bulky items like appliances or solar panels.

These processes are part of reverse logistics; a specialized branch of supply chain management focused on moving goods from consumers back to manufacturers. Unlike traditional outbound logistics, reverse flows require additional handling, sorting, and coordination, especially since returned items vary in condition and may require individual assessment. Partnering with third-party logistics providers is often critical, particularly in large-scale or global operations. These partners specialize in warehousing, item sorting, secure data transport (especially for electronics), and preparing items for refurbishment or recycling. Transparency, tracking, and digital documentation are essential to ensure chain-of-custody compliance and user confidence.

Additionally, environmental compliance plays a role in logistics, with laws like the EU's WEEE directive requiring producers to offer free take-back systems. To meet such regulations and reduce emissions, companies are experimenting with low-carbon logistics, using electric vehicles or local processing centers. In conclusion, efficient logistics and collection systems are the physical backbone of circular business models. They enable the transition from linear consumption to closed-loop resource use, minimizing waste and maximizing reuse.

Grading and refurbishment: Once used products are collected, they undergo a grading and refurbishment process, which is essential to determine their condition and potential for reuse. Grading involves evaluating the returned items to classify them based on aesthetic, functional, and technological parameters. For instance, a returned smartphone might be graded A (like-new), B (minor wear), or C (damaged but repairable). This step helps companies decide whether a device should be refurbished, recycled, or cannibalized for parts. Refurbishment includes a series of activities: cleaning, hardware repair, component replacement, software updates, and battery testing. In electronics, data sanitization is crucial to ensure that all personal information is permanently wiped, protecting user privacy and complying with data protection laws like GDPR.

Technicians or specialized refurbishment partners follow quality assurance protocols to restore functionality and appearance to acceptable standards. Products may also undergo final inspections using automated tools and manual testing to ensure they meet performance benchmarks. Once refurbished, they are repackaged often with a new warranty and prepared for resale, usually at a discounted price. Companies like Dell and HP have robust refurbishment programs where returned laptops are professionally restored and sold through their certified outlet stores. Refurbishment not only gives a second life to products but also saves up to 80% of the energy and resources compared to making new ones. Ultimately, grading and refurbishment represent a sustainable business strategy that aligns environmental goals with profitability. By restoring value from used items, companies reduce e-waste, serve cost-conscious customers, and foster a more circular economy.

Resale or recovery: After grading and refurbishment, products enter the resale or recovery phase. This stage determines how a product is reintroduced into the market or broken down for material reclamation. For devices and equipment that meet quality standards postrefurbishment, the preferred path is resale. These products are sold via certified outlets, e-commerce platforms, or secondary markets at lower prices, thereby tapping into price-sensitive consumer segments such as students, small businesses, or customers in emerging markets. Resale of refurbished goods offers consumers access to branded, functional products at reduced costs while extending the lifecycle of valuable materials. Many firms back these sales with warranties, return policies, and certifications to assure quality and reliability, boosting consumer confidence.

When a product is beyond repair or economically nonviable for resale, it moves into recovery mode. Here, usable components, batteries, screens, chips, and motors are extracted and reintroduced into the supply chain as spare parts or raw materials. In electronics, rare metals like gold, cobalt, or lithium can be recovered using specialized processes, reducing dependence on virgin mining. Companies like Cisco and Lenovo run formal recovery programs where they reclaim components from EoL equipment for reuse in newer devices, helping reduce manufacturing costs and environmental impact. This dual-path model of resale or recovery ensures maximum value extraction from returned items. It reflects the circular economy philosophy, where waste is viewed not as an end, but as a beginning for new cycles of use. This not only increases overall material efficiency but also reinforces a brand's sustainability credentials in an increasingly eco-conscious marketplace.

Financial flows: The financial flows associated with buy-back and refurbishment programs form a critical foundation for their scalability and sustainability. Revenue generation in this model occurs through multiple channels, making it attractive from both ecological and economic perspectives. The most direct source of revenue is the sale of refurbished products. These units are sold at discounted rates, yet often yield high margins due to the relatively low refurbishment costs. For example, a company may refurbish a returned laptop for $50 and resell it for $300, capturing significant value while offering affordability.

Another key financial benefit is cost savings from reduced disposal and compliance expenditures. Instead of paying for landfill fees or hazardous waste handling, companies extract value from what would otherwise be a liability. In jurisdictions with Extended Producer Responsibility (EPR) regulations, this also helps brands meet legal recycling and take-back requirements more cost-effectively. Moreover, there are secondary financial gains in customer retention and loyalty. By offering trade-in credits and easy upgrade paths, companies keep customers within their product ecosystem. This leads to repeat purchases, increased brand affinity, and better cross-selling opportunities.

Additionally, the reuse of recovered components lowers procurement costs. For instance, reusing recovered hard drives or batteries in service replacements reduces dependence on expensive new parts. This also cushions the business from raw material price volatility. Financial success in such circular models requires investment in refurbishment infrastructure, skilled labor, and digital systems for tracking and analytics. However, the long-term returns, revenue diversification, enhanced customer relationships, and cost optimization make it a compelling approach. In summary, smart management of financial flows transforms environmental responsibility into a profitable business strategy. It ensures that sustainability efforts aren't just viable they'revalue-generating engines for modern enterprises.

4.6.3.1 Benefits

- *Extends product life*: Significantly delays end-of-life disposal and reduces demand for new units.
- *Consumer reach*: Affordable refurbished products can reach new markets, such as students, low-income buyers, and emerging economies.
- *Brand loyalty*: Trade-in incentives keep customers within the ecosystem and reduce the friction of upgrading.
- *Data recovery and privacy assurance*: Ensures users feel safe handing over old electronics.

4.6.3.2 Challenges

- *Operational complexity and costs*: Handling returns, maintaining refurbishment centers, and managing inventory control can be resource-intensive.
- *Quality assurance*: Ensuring refurbished items meet high standards is critical to maintain brand reputation.
- *Price perception*: Consumers may perceive refurbished units as inferior, even when quality is guaranteed. Table 4.8 illustrate integration with combined models in practice.

4.6.3.3 Key enablers and considerations

To succeed, these models often require:

- *Designing for circularity*: From day one, products must be modular, repairable, and built with traceable materials.
- *Infrastructure investment*: Reverse logistics, refurbishment facilities, and recycling plants are capital-intensive but critical.
- *Information systems*: Material tracking, usage monitoring, and predictive analytics are essential for lifecycle management.
- *Collaborative ecosystems*: Trust-based partnerships between manufacturers, logistics experts, refurbishers, and recyclers.
- *Regulatory alignment*: Supportive policies—landfill taxes, EPR laws, tax credits—can tilt profitability.
- *Customer activation*: Simple returns, clear incentives, and communication build program adoption.

The transition from a linear to a circular economy hinges on innovative business models that realign economic success with environmental stewardship. Product-as-a-Service, Circular Supply Chains, and Buy-Back and Refurbishment programs are frontline strategies in this transformation. They offer

TABLE 4.8 Integration—Combined models in practice.

Model	Key synergy
PaaS + circular supply chain	Manufacturers retain assets, collect them at EoL, dismantle and reuse components—closing the loop completely.
PaaS + buy-back programs	Long-term lease transitions to customer upgrades. Old units are taken back, refurbished, and deployed again.
Circular supply + buy-back	Companies recover mixed assets and refurbish them for resale while recycling nonrepairable items.

compelling economic advantages, reduce ecological burdens, and foster consumer trust, all while preserving profitability and competitive edge. To operationalize these models, businesses must think holistically: designing products for disassembly, investing in reverse logistics, deploying information systems, and fostering stakeholder collaboration. They must also align with evolving policy landscapes and shift consumer mindsets. Looking ahead, we can expect more industries, such as energy, electronics, healthcare, automotive, and construction, to adopt and innovate on these models. The result? A future in which waste becomes an input, ownership gives way to access, and value is preserved and amplified across entire product lifecycles. By embracing the circular business model framework outlined here, companies can turn sustainability from an obligation into an opportunity, paving the way for resilient, regenerative, and profitable growth in a resource-constrained world.

4.7 Conclusion

As the global renewable energy infrastructure continues to scale, embracing circular approaches is no longer optional it is essential. This chapter has demonstrated that circular design, reuse, and repurposing are not only environmentally beneficial but also economically and socially advantageous when supported by the right technologies and business models. The integration of smart systems like AI and robotics can significantly improve the efficiency and traceability of reuse operations, while business innovations such as Product-as-a-Service and circular partnerships ensure long-term stakeholder value. Furthermore, repurposing initiatives extend the utility of renewable components beyond their original energy applications, unlocking new opportunities in sectors like construction, transport, and urban design. The future of renewable energy lies in designing systems with their full lifecycle in mind, where every asset is seen not as waste at end-of-life, but as a valuable resource for the next cycle. A fully circular renewable energy ecosystem holds the promise of not only decarbonizing power generation but also redefining the sustainability of the materials and models that make it possible.

4.8 Case study

4.8.1 Case study 1

Repurposing decommissioned wind turbine blades into urban infrastructure.

4.8.1.1 Background

As wind farms expand globally, thousands of turbine blades reach end-of-life every year. These blades, made from composite materials, are difficult to recycle conventionally and often end up in landfills. Denmark's National Renewable Energy Innovation Center launched a pilot program in 2024 to explore circular reuse pathways.

4.8.1.2 Circular strategy applied

Instead of shredding or landfilling obsolete blades, the project repurposed the strong and lightweight blade structures into **pedestrian footbridges**, **bike shelters**, and **public benches**. Engineers redesigned selected blade sections, preserving their aerodynamic shape for structural strength while minimizing additional processing.

4.8.1.3 Outcomes

- **38 blades** were repurposed in the first year, diverting more than **400 tons of composite waste** from disposal.
- The city saved **40% construction cost** compared to building new public structures with virgin materials.
- Citizens appreciated the esthetic futuristic design, boosting public acceptance of renewable technologies.

4.8.1.4 Impact

The case became a model for European cities, proving that circular design in renewable energy can not only reduce waste but also strengthen community infrastructure. Danish authorities are now evaluating scaling the practice across all retiring wind sites by 2030.

4.8.2 Case study 2

Second-life applications of solar PV panels for rural microgrids in India**.

4.8.2.1 Background

Solar PV panels typically last 25 to 30 years, but many commercial solar farms replace them much earlier when efficiency drops below peak performance. By 2025, India faced rising solar waste, prompting a circular economy intervention.

4.8.2.2 Circular strategy applied

A collaboration between the Ministry of New and Renewable Energy (MNRE) and a private solar recycler initiated a program to **test, repair, and redeploy** partially degraded solar panels. Panels with **70% to 85% remaining efficiency** were refurbished and repurposed for **rural microgrid projects** in Rajasthan and Madhya Pradesh.

4.8.2.3 Outcomes

- Over **12,000 second-life PV panels** were deployed in 18 villages, powering schools, water pumps, and community centers.

- This reduced new panel procurement costs by **55%** for rural electrification programs.
- The initiative prevented nearly **1500 tons of solar waste** and reduced the carbon footprint associated with manufacturing new panels.

4.8.2.4 Impact

The second-life PV program provided **reliable electricity to 23,500 people**, extended the functional life of PV modules by 8 to 10 years, and created new local repair-and-refurbishment jobs. The project demonstrated how repurposing renewable energy components contributes to both sustainability and inclusive development.

4.9 Exercise

1. Explain the core principles of circular design in renewable energy systems. How do these differ from traditional design approaches?
2. Discuss how modularity and standardization support circularity in solar and wind energy technologies.
3. Describe three major reuse strategies employed in the wind turbine industry and analyze their environmental impact.
4. Evaluate the challenges involved in reusing solar photovoltaic modules after their first lifecycle.
5. How can artificial intelligence (AI) and Internet of Things (IoT) enable predictive maintenance and efficient reuse in renewable energy systems?
6. Analyze how wind turbine blades are being repurposed into nonenergy applications. Provide at least three examples.
7. What is the role of extended producer responsibility (EPR) in circular supply chains for renewable energy systems?
8. Design a conceptual Product-as-a-Service (PaaS) business model for solar panel deployment in rural communities.
9. Discuss the economic and environmental benefits of implementing buy-back and refurbishment programs for renewable energy components.
10. Examine how computer vision and robotics can be applied to automate the disassembly of renewable energy components.
11. What are the key differences between reuse, refurbishing, and repurposing in the context of renewable energy assets?
12. Suggest policies or incentives that governments can implement to promote circular design in clean energy industries.
13. Compare and contrast the circular business models applied in solar energy and wind energy systems.
14. How can recovered materials from decommissioned renewable systems contribute to other industrial applications?
15. Develop a case study to illustrate the role of circular economy principles in extending the lifecycle of a renewable energy project.

4.10 Objective Type

1. Which of the following is NOT a principle of circular design?
   a) Design for longevity
   b) Design for disassembly
   c) Design for single-use
   d) Design for upgradability
2. True or False: Reuse of wind turbine gearboxes is an example of a repurposing strategy.
3. What does PaaS stand for in circular business models?
   a) Platform as a solution
   b) Product as a service
   c) Process after sustainability
   d) Power as a source
4. Which component is most commonly repurposed from solar panels?
   a) Semiconductor
   b) Glass
   c) Plastic backsheet
   d) Wiring harness
5. Fill in the Blank:__________ supply chains facilitate the continuous circulation of components, materials, and products in the clean energy industry.
6. True or False: Refurbishment always involves changing the original function of the product.
7. Which of these technologies enhances predictive maintenance in renewable energy systems?
   a) Blockchain
   b) Augmented reality
   c) Artificial intelligence
   d) 3D Printing
8. Which renewable component is often repurposed into playground equipment or pedestrian bridges?
   a) Solar cells
   b) Battery packs
   c) Wind turbine blades
   d) Inverter units
9. Buy-back programs typically offer:
   a) Free replacements
   b) Usage tracking
   c) Financial incentives for returning old products
   d) Lifetime warranty

10. True or False: Circular business models reduce waste but increase dependency on virgin materials.

11. Which business model allows users to pay for performance rather than ownership?

a) Linear model
b) Refurbishment model
c) PaaS
d) Subscription model

12. Fill in the Blank: Repurposing involves finding a__________ function for a component or material after its original use ends.

13. Which of the following is a challenge in reusing solar panels?

a) High energy efficiency
b) Structural rigidity
c) Performance degradation
d) Lightweight design

14. True or False: Robotics plays a significant role in enabling automated disassembly of wind turbines.

15. Circular design in renewable energy systems aims to:

a) Maximize energy output only
b) Minimize the cost of installation
c) Extend product lifecycle and reduce waste
d) Reduce warranty period

Chapter 5

Circular economy with carbon credit and carbon sequestration

5.1 Introduction

The Circular Economy represents a fundamental shift in how societies understand growth, resource utilization, and environmental responsibility. Unlike the conventional linear economic model of "take, make, dispose," the Circular Economy seeks to design out waste, keep materials circulating within productive systems, and regenerate natural resources that have long been depleted by industrial activity. This framework has evolved over decades rooted in principles of ecology, industrial symbiosis, biomimicry, and sustainable engineering and is now recognized as a cornerstone for global sustainability strategies. At its core, the Circular Economy envisions materials not as disposable commodities but as valuable assets that should remain in continuous use through rethinking product design, enabling reparability, promoting reuse, supporting remanufacturing, and ensuring that end-of-life materials return safely to the environment or re-enter the industrial loop. As climate change accelerates and resource pressures intensify, this regenerative model offers an essential pathway to harmonize economic development with planetary boundaries.

Within this larger sustainability landscape, two concepts have gained immense importance: carbon credit and carbon sequestration. Carbon credits operate as market-based instruments that place measurable, tradable value on reducing or removing greenhouse gas emissions. Through regulated and voluntary carbon markets, organizations earn credits by implementing projects that lower emissions such as renewable energy adoption, improved energy efficiency, reforestation, or waste-to-energy initiatives and these credits can then be traded or sold to entities seeking to offset their own carbon footprints. Carbon sequestration, closely linked to these mechanisms, involves capturing atmospheric carbon dioxide and storing it through natural or engineered processes. Biological sequestration occurs in forests, soils, grasslands, and oceans, while technological methods include carbon capture and storage (CCS), mineralization, and emerging negative-emission technologies.

The Circular Economy and their Management. https://doi.org/10.1016/B978-0-443-44617-7.00009-7

Together, carbon credits and carbon sequestration provide a structured and accountable framework for large-scale climate action, enabling both environmental protection and economic incentives.

The Circular Economy's emphasis on reducing waste, improving resource efficiency, and regenerating ecosystems aligns naturally with global carbon-reduction goals. Many circular interventions directly lower carbon emissions by minimizing extraction of virgin materials, reducing energy use in manufacturing, and extending product lifecycles, all of which contribute to smaller carbon footprints. For instance, recycling aluminum saves up to 95% of the energy required to produce it from raw bauxite, while designing modular renewable energy systems can significantly reduce construction and decommissioning emissions. Circular strategies in the renewable energy sector such as repurposing discarded solar panels, remanufacturing wind turbine components, or innovating bio-based materials for batteries reinforce climate commitments by lowering embodied carbon and preventing landfill accumulation. Additionally, regenerative approaches like composting, regenerative agriculture, and afforestation directly support natural carbon sequestration, enhancing soil health and increasing long-term carbon storage. By integrating these circular actions with measurable carbon accounting, policymakers and industries can strengthen their ability to meet national and international climate targets.

The relationship between the Circular Economy and carbon mitigation becomes even more powerful when viewed through the lens of climate finance, particularly carbon markets. The convergence of carbon markets with material circularity creates new economic opportunities where emissions reduction and resource efficiency become mutually reinforcing priorities. As carbon prices rise globally and compliance markets become stricter, businesses are increasingly incentivized to adopt circular practices not only for environmental reasons but also for financial gain. Projects such as biochar production, recycling infrastructure, green building materials, and nature-based solutions can generate carbon credits while simultaneously supporting circular resource flows. This link enhances the financial viability of circular innovations, attracts private investment, and accelerates technological development. Moreover, carbon markets encourage standardized measurements of climate impact, which helps quantify the benefits of circular initiatives and integrate them into broader sustainability reporting frameworks.

The synergy also extends to national and international climate policies, where governments are recognizing the combined value of circular strategies and carbon mitigation. Many countries now include circular economy roadmaps within their climate action plans, emphasizing extended producer responsibility (EPR), eco-design regulations, and mandatory recycling targets. These policy tools not only reduce emissions across value chains but also create conditions for scalable carbon sequestration projects. As nations work toward achieving carbon neutrality, the integration of circular economy principles with carbon credit mechanisms offers a holistic model that addresses both resource scarcity and greenhouse gas reduction. The result is a future where environmental integrity, economic resilience, and social well-being intersect.

In essence, the Circular Economy, carbon credit systems, and carbon sequestration are not separate strands of sustainability but interconnected pillars that collectively shape a regenerative and low-carbon world. Their convergence marks a transformative moment in global climate action where waste becomes value, emissions become financial assets, and natural systems are restored through intentional design and innovation. This chapter explores this intersection in depth, highlighting how thoughtful integration of these principles can power the next generation of sustainable development.

5.2 Circular economy with carbon credit

As the global community races toward achieving climate neutrality and sustainable development, the intersection of two transformative concepts carbon credits and the circular economy has emerged as a vital area of innovation and policy. The circular economy, which emphasizes reducing waste, reusing resources, and regenerating natural systems, offers a regenerative framework that aligns perfectly with carbon mitigation efforts. In this evolving landscape, carbon credits serve as both a financial incentive and a measurement tool for quantifying and rewarding emission reductions, removals, and sustainable practices. Carbon credits, each representing one metric ton of carbon dioxide equivalent (tCO_2e) reduced or sequestered, are no longer limited to renewable energy or forestry projects. Increasingly, circular economy initiatives such as product reuse, material recycling, waste-to-energy systems, biochar production, and composting are being recognized for their potential to generate carbon credits. These initiatives not only divert waste from landfills but also reduce greenhouse gas emissions associated with production, transportation, and disposal of materials. Understanding carbon credits in the context of circularity also means recognizing the role of carbon markets both voluntary and compliance-based as mechanisms that support investments in low-carbon, circular systems. Circular business models that extend product lifespans or reduce virgin material use can now tap into carbon finance to fund scaling efforts. Moreover, advancements in monitoring, reporting, and verification (MRV) technologies, including blockchain and AI, are enhancing the credibility and traceability of circular carbon projects. This topic invites us to explore how carbon credit frameworks can evolve to include and promote circular practices, enabling a transition that is not only low-carbon but resource-efficient, inclusive, and economically viable. It sets the foundation for a future where decarbonization and circularity are two sides of the same sustainable coin.

5.2.1 What are carbon credits?

Carbon credits are a market-based mechanism designed to reduce greenhouse gas (GHG) emissions by putting a monetary value on carbon emissions. One carbon credit represents the reduction or removal of one metric ton of carbon dioxide (CO_2) or an equivalent amount of another GHG (such as methane or nitrous oxide). These credits serve both as incentives and accounting tools in the fight against climate change, allowing individuals, companies, or governments to offset their emissions by funding projects that reduce or remove GHGs elsewhere. Fig. 5.1 shows general image of carbon credit.

The concept of carbon credits is rooted in the understanding that while some emissions are unavoidable, they can be balanced by reductions made elsewhere. This is based on the idea of the "carbon offset." For instance, if a company emits 1000 tons of CO_2 annually, it can purchase 1000 carbon credits from a verified offset project such as reforestation, renewable energy, or methane capture which proves it has removed or prevented the release of 1000 tons of CO_2. In this way, the company is considered "carbon neutral." Carbon credits are broadly categorized into two main types:

Compliance carbon credits: Compliance carbon credits are an integral part of government-regulated carbon markets, established to meet legally binding emission reduction targets. These credits are typically generated, distributed,

FIGURE 5.1 Carbon credit.

and traded within a cap-and-trade system or emissions trading scheme (ETS), such as those operated in the European Union (EU ETS), California, or New Zealand. Under these programs, governments impose a cap on the total amount of greenhouse gases (GHGs) that can be emitted by entities like power plants, factories, or airlines. This cap is divided into a set number of carbon allowances or credits, each representing the right to emit one metric ton of carbon dioxide (CO_2) or its equivalent in other greenhouse gases. Companies receive or purchase these allowances and are required to surrender a number of credits equal to their emissions over a defined period. If a company reduces its emissions and has surplus credits, it may sell them to another company that is exceeding its cap. This creates a market-based incentive for industries to innovate and reduce their carbon footprint. Compliance credits are subject to stringent regulations, verified methodologies, and third-party audits to ensure integrity and prevent fraud. The credibility of these credits is ensured by robust monitoring, reporting, and verification (MRV) protocols. These programs often evolve with national and international climate policy frameworks, particularly aligning with goals of the Kyoto Protocol, Paris Agreement, and respective national determined contributions (NDCs). Notably, compliance markets drive large-scale investment in cleaner technologies and contribute significantly to a country's emission reduction targets. However, criticisms exist, including concerns over allowance over-allocation, market manipulation, or the lack of strict penalties for non-compliance. Nonetheless, the compliance carbon market plays a vital role in institutionalizing climate action by making carbon pricing a fundamental economic factor. In essence, it shifts the cost of pollution back to the polluter under the "polluter pays" principle, compelling industries to internalize environmental costs. Fig. 5.2 illustrate comparison between compliance and voluntary carbon credit.

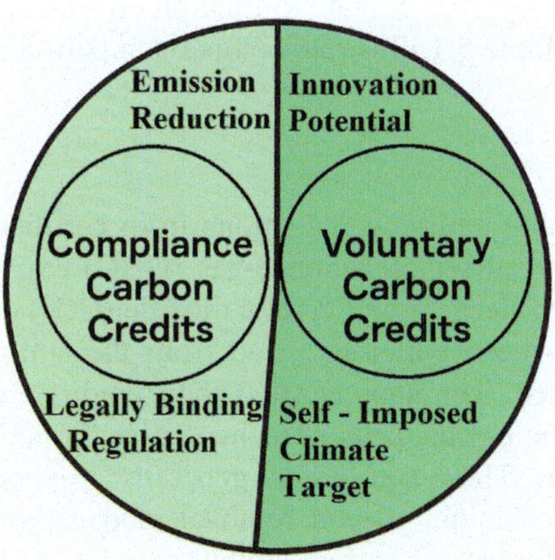

FIGURE 5.2 Comparison between compliance and voluntary carbon credit.

Voluntary carbon credits: Voluntary carbon credits operate outside of government mandates and are purchased by individuals, corporations, or organizations seeking to voluntarily offset their carbon footprint. These credits are typically generated through projects that actively reduce, remove, or avoid greenhouse gas emissions. Examples include afforestation and reforestation, renewable energy development, methane capture from landfills, and energy efficiency programs in developing regions. Each credit represents the avoidance or removal of one metric ton of CO_2 or its equivalent. Unlike compliance credits, participation in the voluntary market is not mandatory, which makes it a flexible and market-driven approach to climate responsibility.

Voluntary credits are issued by independent standards such as the Verified Carbon Standard (VCS), Gold Standard, or Climate Action Reserve, which assess project validity, additionality (proof that the carbon reduction wouldn't have occurred without the project), permanence, and social co-benefits. These standards maintain project registries and ensure transparency, verification, and third-party auditing. Corporations often use voluntary credits to achieve carbon neutrality, fulfill environmental, social, and governance (ESG) goals, or demonstrate leadership in sustainability. For instance, tech firms, airlines, or consumer goods companies might purchase voluntary credits to offset emissions from travel, logistics, or product lifecycles. The voluntary market has grown rapidly as climate consciousness increases and businesses face pressure from investors, consumers, and regulators to decarbonize. It plays a crucial role in financing climate solutions, particularly in regions where regulation is weak or absent. However, challenges persist, including inconsistent standards, risks of greenwashing, and concerns over the true additionally and permanence of some projects. To address this, initiatives like the Integrity Council for the Voluntary Carbon Market (ICVCM) aim to improve quality assurance and governance. In summary, voluntary carbon credits enable proactive climate action by offering a way to support emissions reductions beyond legal obligations, making them a valuable tool for global climate mitigation efforts and corporate environmental stewardship. Table 5.1 illustrate comparison between compliance and voluntary carbon credit.

5.2.1.1 Sources of carbon credits

Carbon credits are measurable, verifiable emission reductions from certified climate action projects that are used to offset greenhouse gas (GHG) emissions. Each carbon credit represents one metric ton of carbon dioxide (CO_2) or its equivalent in other GHGs that has been either removed from the atmosphere or prevented from being emitted. The sources of carbon credits are diverse and are categorized based on the mechanisms or technologies through which they achieve emission reductions. These sources are generally grouped into three broad categories: carbon reduction, carbon removal, and carbon avoidance projects. Fig. 5.3 illustrate sources of carbon credit.

TABLE 5.1 Compliance versus voluntary carbon credits.

Criteria	Compliance carbon credits	Voluntary carbon credits
Purpose	To meet legally mandated emission reduction requirements	To offset emissions voluntarily for sustainability goals or CSR
Market type	Regulated markets (government-mandated)	Unregulated, voluntary markets
Participants	Corporations, industries, and entities bound by law	Businesses, individuals, NGOs, and investors
Regulatory authority	Overseen by government or international bodies (e.g., EU ETS, California cap-and-trade)	Managed by independent standards (e.g., Verra, Gold standard)
Verification and standards	Strict, legally enforced measurement and verification	Third-party verification but more flexible
Project types	Often sector-specific (energy, industrial reductions)	Wider variety (forestry, renewables, community projects)
Credit price	Generally higher and more stable	More variable depending on project type and demand
Market drivers	Government regulation, mandatory caps	Corporate sustainability, branding, ESG, climate commitments
Penalties for non-compliance	Legal or financial penalties	No penalties participation is voluntary
Use cases	Meeting emission caps or trading allocated allowances	Achieving net-zero goals, enhancing brand image, offsetting travel/operations

Renewable energy projects: One of the most significant sources of carbon credits comes from renewable energy projects such as wind, solar, hydro, and biomass. These projects generate electricity without emitting CO_2, replacing fossil fuel-based power generation. For instance, a wind farm that displaces electricity that would otherwise be generated by a coal-fired plant results in a measurable reduction in emissions. The avoided emissions are quantified and translated into carbon credits. These projects are particularly valuable in developing countries where renewable alternatives replace more carbon-intensive sources.

Energy efficiency projects: Energy efficiency initiatives aim to reduce energy consumption and associated emissions without compromising output or performance. Examples include retrofitting buildings with energy-efficient lighting, improving industrial processes, or upgrading equipment in

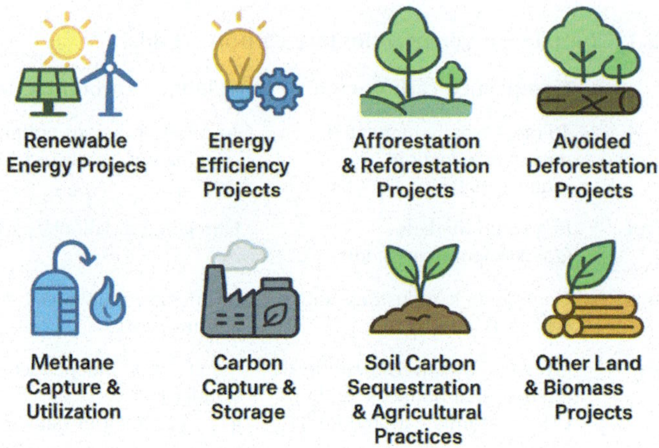

FIGURE 5.3 Sources of carbon credit.

manufacturing. The reduction in energy use translates into a corresponding decrease in CO_2 emissions, which can be calculated and issued as carbon credits. These projects not only generate credits but often also lead to cost savings for users.

Afforestation and reforestation projects: Forestry projects are prominent in carbon credit markets, particularly in the voluntary sector. Afforestation refers to planting trees on land that has not previously been forested, while reforestation involves replanting trees in deforested areas. Trees absorb CO_2 through the process of photosynthesis, effectively removing it from the atmosphere. These projects are not only critical for carbon sequestration but also offer biodiversity, soil conservation, and community co-benefits. However, these projects must ensure permanence and protection from deforestation or degradation.

Avoided deforestation (REDD+) projects: REDD+ (Reducing Emissions from Deforestation and Forest Degradation) projects aim to prevent deforestation and forest degradation in areas at risk. By preserving forests that would otherwise be cut down, these projects avoid the release of the carbon stored in trees and soil. The avoided emissions can be quantified and issued as credits. These projects often work in partnership with local communities and governments to ensure long-term conservation.

Methane capture and utilization: Methane (CH_4) is a potent greenhouse gas, with a global warming potential more than 25 times that of CO_2 over a 100-year period. Projects that capture methane emissions from landfills, wastewater treatment plants, coal mines, and agricultural operations (e.g., livestock manure) are significant sources of carbon credits. The captured methane can be flared (converted to CO_2) or used as energy, reducing its overall climate impact.

Carbon capture and storage (CCS): CCS technologies involve capturing CO_2 emissions at the source such as from power plants or industrial processes and storing it underground in geological formations. Though still emerging and expensive, CCS is seen as a crucial technology in decarbonizing hard-to-abate sectors like cement, steel, and chemicals. Carbon credits are issued for the amount of CO_2 captured and securely stored.

Soil carbon sequestration and agricultural practices: Improved agricultural practices, such as no-till farming, cover cropping, agroforestry, and optimized fertilizer use, can enhance soil's ability to store carbon. These practices increase the organic carbon content of soil, reduce emissions from synthetic inputs, and may generate carbon credits. Measurement and verification of soil carbon are advancing with technology and satellite data, enhancing credibility. Table 5.2 illustrate sources and their property of carbon credit.

Carbon credits originate from a variety of sources that contribute to mitigating climate change either by reducing, avoiding, or removing emissions. These sources offer not only environmental benefits but often social, economic, and biodiversity co-benefits, making them critical tools in the global transition toward a low-carbon economy. Carbon credits are a critical tool in the global climate strategy, enabling emission reductions and removals in a cost-effective and scalable way. When properly designed and governed, they create powerful incentives for innovation, sustainability, and circular economy practices, helping industries and nations progress toward net-zero emissions. As climate policies tighten and carbon markets expand, understanding and utilizing carbon credits will become even more important in achieving environmental, economic, and social sustainability goals.

5.2.2 Integrated circular–carbon value framework

The Integrated Circular–Carbon Value Framework (ICCVF) presents a transformative approach to sustainability by combining the principles of circular economy with advanced carbon management and sequestration strategies. This unified framework bridges material circularity, renewable energy integration, carbon markets, and ecological regeneration to create a future-ready model for low-carbon development. It moves beyond linear economic structures and establishes a holistic architecture where every material and energy cycle contributes to carbon reduction and long-term environmental resilience. Through its six interconnected layers, ICCVF enables industries, policymakers, institutions, and communities to transition toward carbon neutrality and ultimately carbon negativity. By integrating technological innovation, governance mechanisms, and market incentives, this framework positions circularity not only as an environmental necessity but also as a viable economic strategy driven by carbon value creation.

TABLE 5.2 Sources of carbon credits.

Source of carbon credits	How credits are generated	Key benefits	Common project examples	Scalability	Verification complexity
Renewable energy projects	By replacing fossil-fuel energy with clean energy (solar, wind, hydro, biomass)	Reduces large-scale CO_2 emissions; supports clean grid transition	Solar farms, wind farms, hydro mini-grids, biogas power	High (utility-scale possible)	Low–Moderate
Energy efficiency projects	By lowering energy consumption compared to a baseline	Cost-saving + direct emission reduction	Efficient lighting, efficient motors, HVAC upgrades, fuel-efficient cookstoves	Moderate–High	Moderate
Afforestation and reforestation (A/R)	Planting new forests or restoring degraded forests to absorb CO_2	Long-term carbon removal; biodiversity improvement	Reforestation of degraded land, mangrove restoration	Moderate (needs land)	High (long-term monitoring)
Avoided deforestation (REDD+)	Preventing deforestation that would have released CO_2	Protects ecosystems; prevents massive emissions	Forest conservation, protected areas, community forest management	High (large land areas)	High (baseline proof required)
Methane capture and utilization	Capturing methane (CH_4) and converting it to energy or flaring it	Removes high-impact greenhouse gases (CH_4 is 28–34× stronger than CO_2)	Landfill gas capture, livestock manure digesters, coal mine methane	Moderate	Moderate–High

Carbon capture and storage (CCS)	Capturing CO_2 at source and storing it underground	Large-scale emission reduction for heavy industries	Industrial CCS plants, direct air capture (DAC) + geological storage	High (industrial-scale)	Very high
Soil carbon sequestration and agricultural practices	Increasing soil organic carbon through improved farming	Enhances soil health + stores atmospheric CO_2	No-till farming, cover crops, biochar, regenerative agriculture	High (broad agricultural adoption)	Moderate–High
Other land and biomass-based Projects *(Optional category)*	Using biomass or land-use change to store or reduce emissions	Flexible approaches; community involvement	Agroforestry, sustainable grazing, biomass briquette production	Moderate–High	Moderate

At the foundation lies the System Layer, defined as the Circular Carbon Architecture (CCA). This layer re-engineers how materials, energy, and carbon flow across the economic system. Its first pillar, the Regenerative Material Loop (RML), emphasizes cradle-to-cradle design, ensuring that all resources re-enter ecosystems or industrial loops without generating waste. RML focuses on modularity, design for disassembly, and the continuous reuse, repair, remanufacture, and recycling of products. A ground breaking feature of this loop is the integration of carbon coefficients that quantify the carbon savings associated with every material flow, turning circular actions into measurable carbon benefits. The second component, Carbon Flow Synchronization (CFS), tracks carbon emissions across the product lifecycle from production through usage to end-of-life. CFS enables organizations to calculate avoided emissions, captured emissions, and net-negative contributions, allowing carbon performance to be embedded into product design. The third element, the Closed-Loop Energy Nexus (CLEN), integrates renewable energy, such as solar, wind, and bioenergy, into circular systems, ensuring that circular activities themselves are powered sustainably. Notably, CLEN allows renewable energy projects to automatically function as carbon credit–generating assets, creating a direct link between clean energy adoption and carbon revenue. Building upon the system foundation, the Technology Layer introduces Smart Carbon-Circularity Infrastructure. The Digital Twin for Circular Carbon (DTCC) creates a real-time virtual replica of a product, mapping its material loops, carbon flows, and potential environmental impacts. This enables industries to simulate and optimize circularity and carbon performance before physical production, thus maximizing carbon credit potential. The Blockchain Carbon Ledger (BCL) ensures immutable and transparent tracking of material flows and carbon transactions, reducing fraud and improving global market credibility. Meanwhile, the AI-Driven Sequestration Optimizer (AISO) uses machine learning to identify optimal sites for afforestation, soil carbon enhancement, and other sequestration strategies. It also helps predict carbon credit prices by analyzing global trends, enabling smarter investment decisions. Finally, the Additive Manufacturing Circular Hub (AMCH) leverages 3D printing with recycled materials, reducing virgin resource demand and enabling localized, low-emission production. This technology not only minimizes waste but also establishes micro-circular economies that support decentralized manufacturing. Fig. 5.4 illustrate integrated circular carbon value framework.

The Operational Layer connects circularity and technology to financial incentives by establishing robust carbon market mechanisms. At its core is the concept of Carbon Circularity Credits (CCC), a next-generation credit system that evaluates a product or process based on Material Circularity Savings (MCS), Energy Circularity Savings (ECS), and Sequestration Units (SU). This tri-dimensional approach broadens the scope of carbon markets, making them more inclusive and representative of true sustainability. The Circular-Carbon

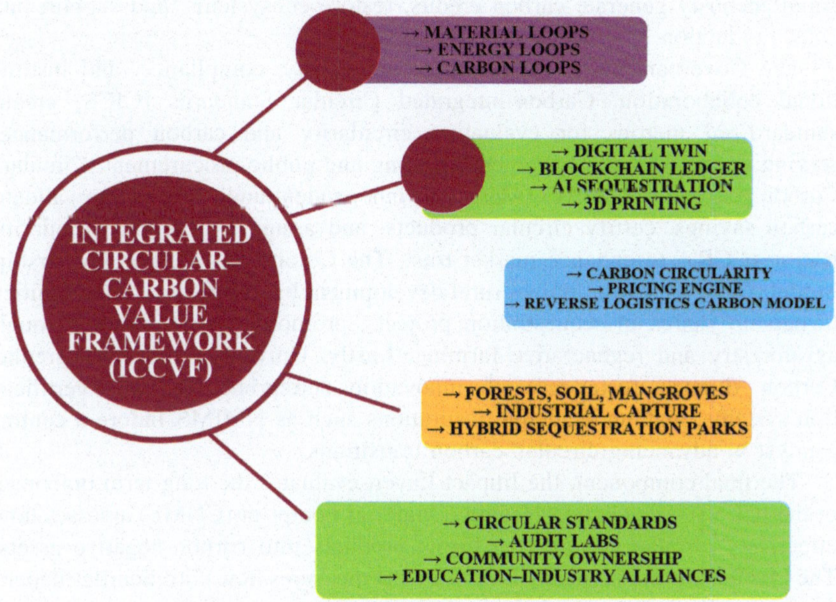

FIGURE 5.4 Integrated circular carbon value framework.

Pricing Engine (CCPE) uses AI-driven models to dynamically adjust carbon credit pricing by considering market volatility, sequestration rates, and recovery efficiencies. This provides industries with the ability to forecast carbon revenue and plan long-term investments. The Reverse Logistics–Carbon Synergy Model (RLSM) redesigns product return and recycling systems to be carbon-efficient, rewarding recycling centers and local collection hubs with carbon incentives. This model strengthens local circular ecosystems while reducing transportation emissions and improving material recovery.

The Environmental Layer operationalizes sequestration through integrated ecological and industrial systems. Nature-Based Circular Sinks (NBCS) include afforestation, reforestation, mangrove plantation, bio char-enriched soils, and wetland rehabilitation. These actions are linked to circular practices such as biomass reuse and agro-waste repurposing, ensuring environmental regeneration while creating secondary material loops. Complementing natural systems, Industrial Carbon Circular Sinks (ICCS) utilize engineered solutions such as carbon capture from factories, CO_2 mineralization into construction materials, and algae-based carbon capture integrated with biomass-to-product pathways. The third component, the Hybrid Circular Sequestration Parks (HCSP), combines nature-based and industrial strategies within multifunctional hubs integrating agriculture, renewable energy, manufacturing, and carbon sequestration. These parks are designed to simultaneously generate carbon credits, restore ecosystems, and support circular production.

The Governance Layer ensures accountability, compliance, and institutional collaboration. Carbon-Integrated Circular Standards (CICS) create standardized metrics for evaluating circularity and carbon performance, serving as prerequisites for green financing and public procurement. Circular-Carbon Assurance Labs (CCAL) act as independent audit bodies that validate carbon savings, certify circular products, and assign a "Carbon Circularity Rating (CCR)" to enhance market trust. The Community Carbon Ownership Model (CCOM) revolutionizes rural development by giving local communities ownership shares in sequestration projects, promoting participation through agroforestry and regenerative farming. Lastly, University–Industry Circular Carbon Alliances foster research, innovation, entrepreneurship, and certification systems, making academic institutions such as NMIMS Indore a central catalyst in advancing circular–carbon transitions.

The final component, the Impact Layer, evaluates the long-term outcomes of the ICCVF. The Carbon-Negative Material Footprint (CNMF) assesses how effectively circular design transforms products into carbon-negative assets. The Local Carbon Prosperity Index (LCPI) measures how communities benefit from circular–carbon initiatives through income generation, employment creation, and improved climate resilience. At the global scale, the Global Circular Carbon Score (GCCS) serves as a benchmarking tool that allows nations, corporations, and regions to compare their progress in integrating circularity with carbon sequestration. This score helps policymakers and international bodies identify leaders, encourage improvement, and design more effective cross-border climate strategies.

The Integrated Circular–Carbon Value Framework (ICCVF) serves as a pioneering blueprint for building regenerative, climate-positive, and economically viable systems. By seamlessly linking circularity, technology, carbon markets, ecological regeneration, and community empowerment, it provides a roadmap for transitioning from carbon-intensive linear models to dynamic, circular, and carbon-negative economies. This framework empowers industries, governments, institutions, and communities to collaboratively shape a sustainable future where carbon becomes a value-generating resource and circularity becomes the foundation of a resilient global economy.

5.2.3 The carbon credit value and governance framework

The global shift toward sustainable development has placed the **carbon credit ecosystem** at the center of environmental, economic, and societal transformation. As nations, businesses, and communities strive to achieve net-zero goals, a structured, transparent, and scalable framework becomes necessary to ensure that carbon markets serve both ecological restoration and economic progress. The **Carbon Credit Value and Governance Framework** brings together five critical layers *Emission Source Layer, Measurement and Verification Layer, Sequestration and Reduction Layer, Market and Monetization Layer, and Governance and Integrity Layer.* Each layer plays a distinct role yet remains interconnected, forming a unified architecture for generating, managing, and trading carbon credits responsibly. Every carbon credit begins with a clear understanding of where emissions originate. The Emission Source Layer identifies and categorizes the activities, processes, or assets that release greenhouse gases into the atmosphere. These sources range from industrial operations such as manufacturing plants and power stations to sectors like agriculture, transportation, infrastructure development, and buildings. By mapping emissions across Scope 1 (direct), Scope 2 (energy-related), and Scope 3 (supply-chain related), organizations gain a holistic view of their carbon footprint. This mapping serves two purposes first, it highlights hotspots where emissions can be reduced; second, it provides baseline data for carbon credit generation through mitigation or sequestration actions. Without accurate emissions identification, the credibility of carbon credits collapses. Therefore, this layer acts as the foundation on which the entire framework stands. Fig. 5.5 illustrate carbon quantification and baseline layer framework. Fig. 5.6 illustrate carbon reduction and sequestration layer.

Once emission sources are identified, the next essential step is quantification. The Measurement and Verification (M&V) Layer ensures that emissions are scientifically calculated, monitored, and independently verified. This layer integrates several advanced technologies:

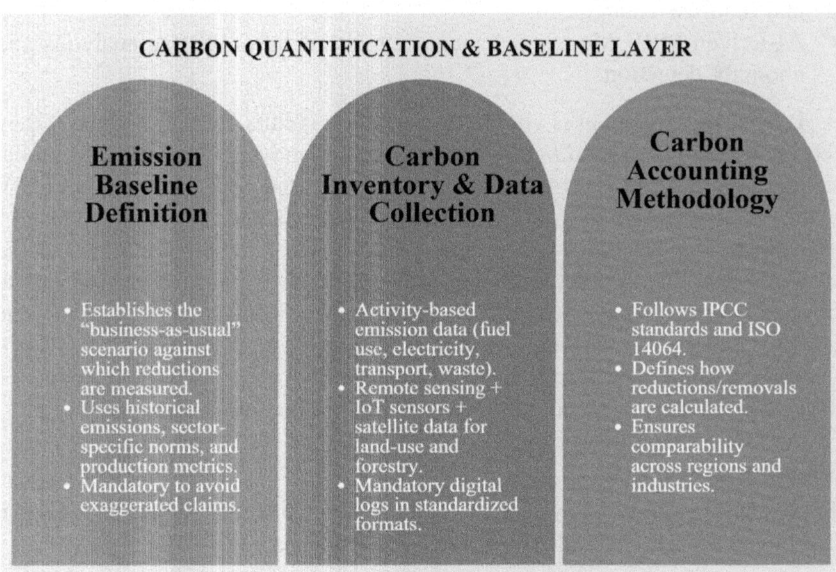

FIGURE 5.5 Carbon quantification and baseline layer.

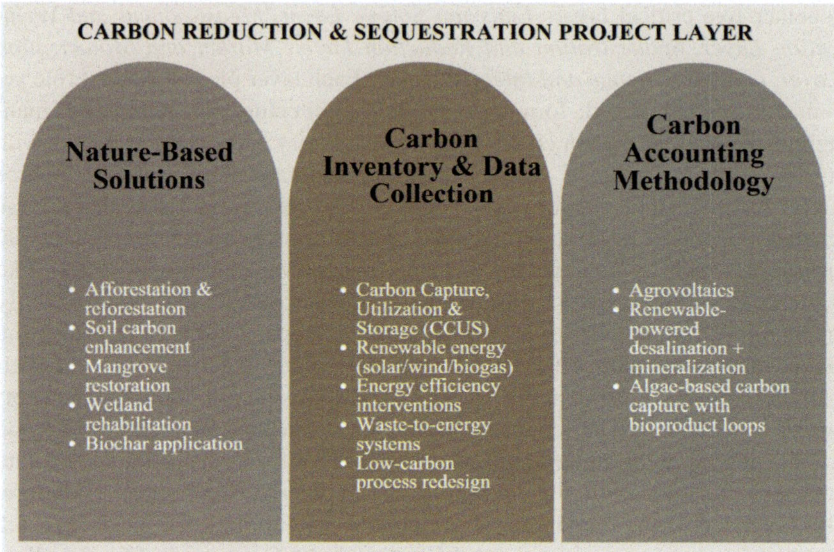

FIGURE 5.6 Carbon reduction and sequestration layer.

- Digital Twins replicate physical systems digitally to model real-time emissions.
- Smart Sensors and IoT devices continuously track carbon output.
- Satellite Monitoring provides macro-level verification for forests, oceans, and land-use changes.
- AI-Driven MRV Models enhance accuracy using predictive analytics and anomaly detection.

Precise measurement is vital because carbon credits only have value if they represent genuine, quantifiable reductions or removals. This is why global standards such as ISO 14064, GHG Protocol, and emerging digital MRV platforms play a critical role. Verification by accredited third parties ensures transparency and prevents fraudulent activities like double counting, inflated carbon reductions, or issuing credits for activities that are not truly additional. In short, this layer builds trust, the most valuable currency in carbon markets. Fig. 5.7 illustrate monitoring, reporting and verification layer.

Once carbon credits are generated and verified, they enter the *Market and Monetization Layer*. The modern carbon market includes two major types.

5.2.3.1 Compliance markets

These are regulated by governments. Companies must meet emission caps and can buy credits if they fall short. Examples include.

- EU Emission Trading Scheme (EU-ETS)

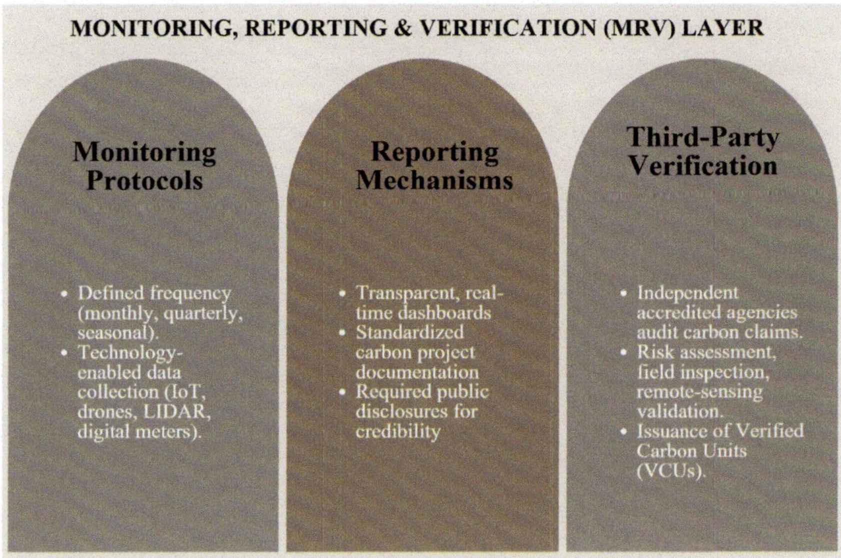

FIGURE 5.7 Monitoring, reporting and verification layer.

- California Cap-and-Trade Program
- Chinese National ETS

5.2.3.2 Voluntary carbon markets (VCM)

Here, companies and individuals purchase credits voluntarily to achieve climate commitments, brand goals, or sustainability targets.

This layer also includes:

- **Carbon trading platforms**: digital exchanges where credits are listed.
- **Pricing engines**: algorithms determine credit value based on quality, permanency, and co-benefits.
- **Tokenization and blockchain-based trading**: enhances transparency and prevents duplication.
- **Reverse logistics for carbon**: circular models that capture and reuse carbon in industrial cycles.

Credits become a financial asset that can be bought, sold, hedged, or used for offsetting emissions. This transforms sustainability into a revenue-driving opportunity for businesses, farmers, communities, and innovators who participate in sequestration or reduction projects. Fig. 5.8 illustrate carbon credit issuance and registry layer.

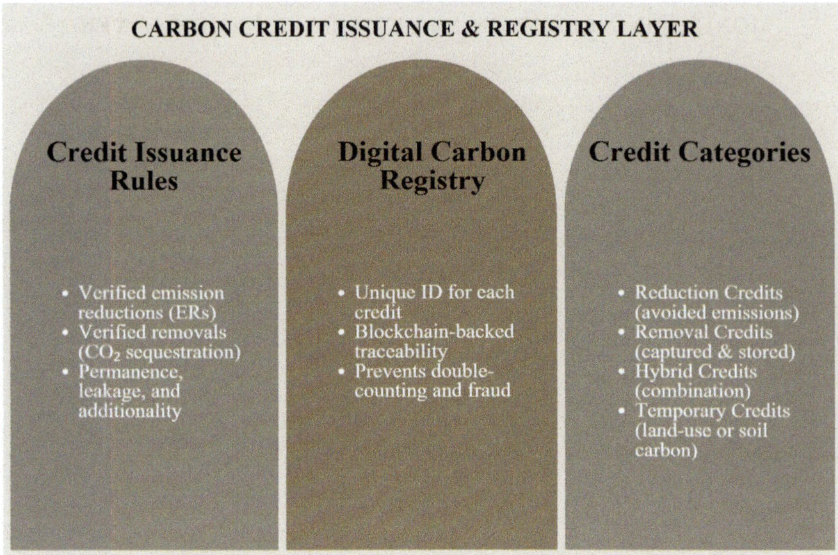

FIGURE 5.8 Carbon credit issuance and registry layer.

A carbon credit system is only credible if governed properly. The Governance and Integrity Layer ensures accountability, fairness, and long-term reliability. It includes:

- Policy Frameworks and National Registries to align with global climate goals.
- Standards and Certification Bodies such as Verra, Gold Standard, ISO.
- Audit Mechanisms to assess quality, risk, and compliance.
- Community Stakeholder Engagement to ensure local participation and benefit-sharing.
- Transparency Controls using blockchain-ledgers.
- Dispute Resolution Systems for ethical and legal protection.

Because carbon markets are rapidly growing, robust governance prevents greenwashing, protects consumers, and ensures that every credit truly represents a unit of climate impact. The integrity layer also ensures that indigenous communities, farmers, and forest dwellers receive fair compensation for their participation. Fig. 5.9 illustrate carbon market trading layer. Fig. 5.10 illustrate governance, ethics and policy layer. Fig. 5.11 illustrate community and socio—economic integration layer. Fig. 5.12 illustrate impact and transformation layer.

The Carbon Credit Value and Governance Framework is more than a structural model it is a roadmap for the future of climate action. By integrating scientific accuracy, technological innovation, financial incentives, and ethical

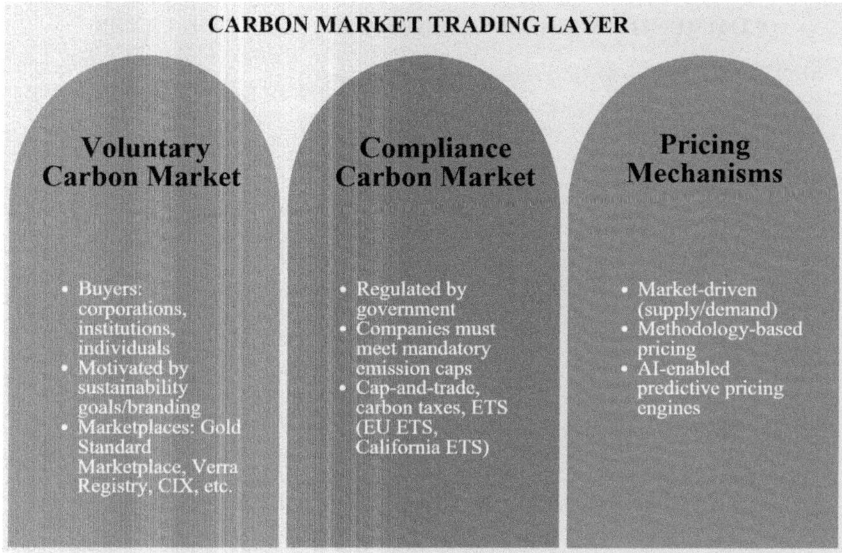

FIGURE 5.9 Carbon market trading layer.

governance, the framework ensures that carbon credits become a reliable tool for global decarbonization. This layered structure creates a seamless pipeline from emission detection to climate intervention, from quantification to monetization, and from governance to long-term sustainability. Through this

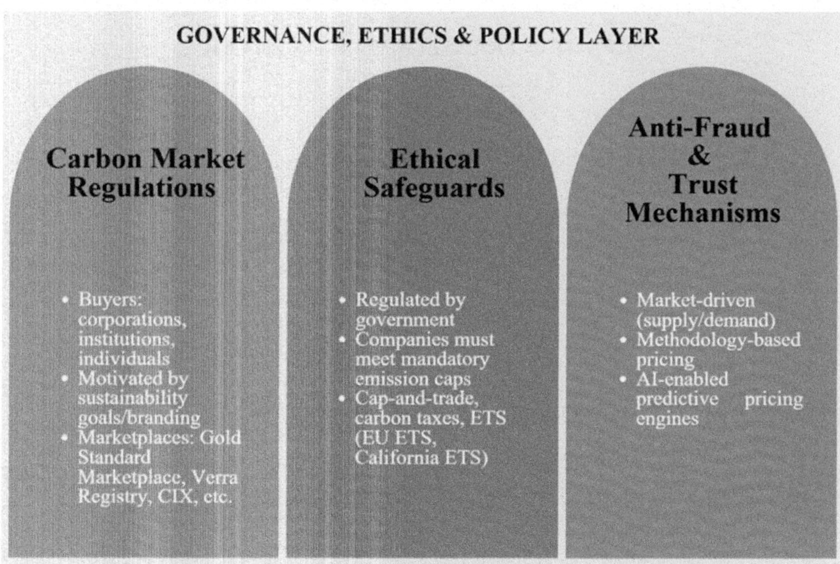

FIGURE 5.10 Governance, ethics and policy layer.

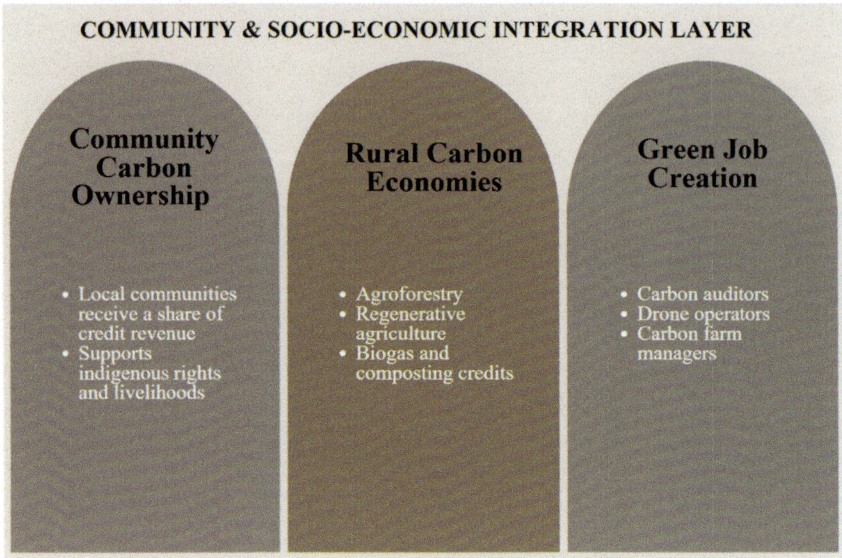

FIGURE 5.11 Community and socio—economic integration layer.

model, circular economy principles and carbon sequestration not only protect the planet but generate continuous economic value. Ultimately, the framework represents a bold vision: *a world where restoring the Earth becomes profitable, traceable, and universally accessible*.

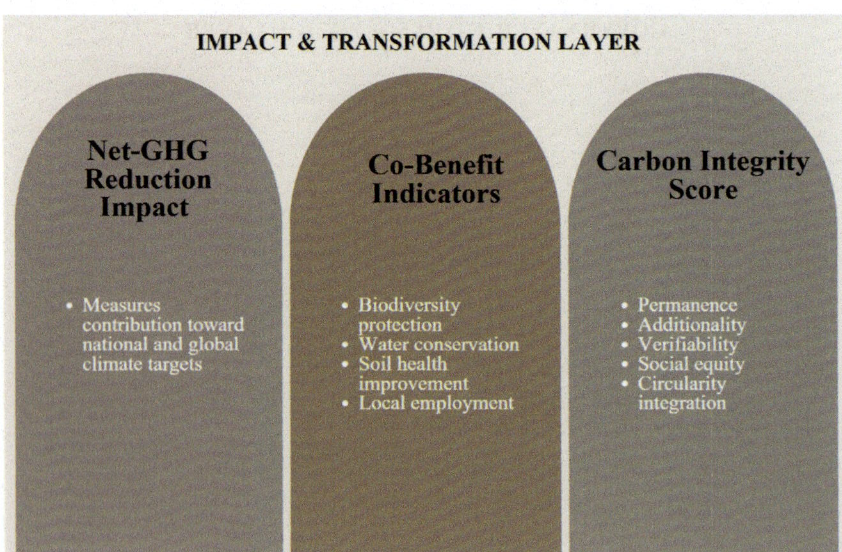

FIGURE 5.12 Impact and transformation layer.

5.2.4 Circular activities that earn carbon credits

In the context of growing global concerns about climate change, carbon emissions, and the need to transition toward a sustainable economy, the concept of a circular economy has emerged as a vital framework. Unlike the traditional linear model of "take, make, dispose," the circular economy emphasizes extending the lifecycle of products, minimizing waste, and making the most of resources. Within this framework, various circular activities not only support environmental goals but also have the potential to generate carbon credits by reducing, avoiding, or removing greenhouse gas (GHG) emissions. Carbon credits are tradable certificates that represent the reduction of one metric ton of carbon dioxide (CO_2) or its equivalent in other GHGs. Circular activities that contribute to emission reductions can be verified and quantified, thereby becoming eligible to earn carbon credits under recognized standards. This detailed explanation explores key circular activities that generate carbon credits, including composting, recycling, anaerobic digestion, biomass energy, sustainable product design, and avoided emissions from reuse and refurbishment.

Composting, recycling, anaerobic digestion, biomass energy: Circular waste management activities such as composting, recycling, anaerobic digestion, and biomass energy play a crucial role in reducing carbon emissions and supporting the generation of carbon credits. Composting is the process of converting organic waste like food scraps, garden trimmings, and agricultural residues into nutrient-rich compost through biological decomposition. Organic waste sent to landfills decomposes anaerobically, producing methane (CH_4), a greenhouse gas approximately 28 times more potent than CO_2 over a 100-year period. Composting diverts this waste from landfills, thus significantly reducing methane emissions.

- Composting reduces methane emissions from organic waste decomposition in landfills.
- It enhances soil health, sequesters carbon in the soil, and reduces the need for synthetic fertilizers.
- Composting projects can be monitored and verified to earn carbon credits under standards like Verra or Gold Standard.

Recycling involves the processing of used materials like paper, plastic, glass, and metals into new products, thereby avoiding the energy-intensive processes associated with extracting and manufacturing virgin materials. For example, recycling aluminium saves up to 95% of the energy required to produce new aluminium from raw ore.

- Reduces emissions associated with raw material extraction and manufacturing.
- Prevents landfill usage and the associated emissions.

- Projects supporting industrial-scale recycling programs are eligible for carbon credit generation.

Anaerobic digestion (AD) is a process where microorganisms break down organic materials in the absence of oxygen, producing biogas (primarily methane and CO_2) and digestate (a nutrient-rich residue). This biogas can be used as a renewable energy source, displacing fossil fuels.

- AD systems capture methane that would otherwise be released uncontrolled.
- Reduces reliance on fossil fuels by providing clean energy.
- The process reduces GHG emissions from organic waste and earns carbon credits for avoided emissions.

Biomass energy utilizes organic materials such as agricultural residues, forestry by-products, or dedicated energy crops to produce electricity, heat, or biofuels. When managed sustainably, biomass is considered carbon-neutral since the CO2 emitted during combustion is offset by the CO_2 absorbed during the growth of the biomass.

- Provides a renewable alternative to fossil fuel-based energy.
- Utilizes waste materials that would otherwise decompose and release methane.
- Eligible for carbon credits when displacing fossil fuel usage in power generation.

Together, these activities represent foundational circular economy strategies that reduce emissions and support regenerative environmental practices. Their role in waste-to-resource conversion creates measurable climate benefits, justifying their inclusion in carbon offset schemes.

Sustainable product design reducing lifecycle emissions: Sustainable product design is another powerful circular activity that helps reduce carbon emissions across the entire lifecycle of a product. This includes everything from the choice of materials to production methods, usage efficiency, and end-of-life recyclability. By intentionally designing products to minimize their environmental footprint, companies can significantly lower associated carbon emissions and potentially generate carbon credits when these reductions are validated.

Sustainable design approaches typically focus on:

- Material selection: Using recycled, renewable, or low-emission materials.
- Energy-efficient manufacturing: Reducing emissions during production.
- Durability and reparability: Ensuring products have a long lifespan and can be repaired instead of replaced.
- Modularity: Designing components that can be replaced or upgraded.
- Design for disassembly and recycling: Facilitating the recycling process at the end of product life.

These strategies contribute to significant emission reductions by avoiding the need for raw material extraction, lowering energy consumption, and reducing landfill contributions. Products designed with circularity in mind typically require fewer resources and produce less waste over time. For example, a modular smartphone that allows component replacement extends its usable life and reduces the need for producing entirely new devices.

- Reduction in energy and material use across the lifecycle contributes to carbon savings.
- Avoided emissions can be quantified through lifecycle assessment (LCA) methodologies.
- Certified reductions in emissions from sustainable design practices may be credited under certain voluntary carbon markets.

Furthermore, companies can use the outcomes of sustainable design to meet internal emissions targets, pursue carbon neutrality, and contribute to science-based climate goals. As consumers and regulators increasingly demand climate-conscious products, sustainable design not only reduces emissions but also enhances market competitiveness and corporate reputation.

Avoided emissions from reuse and refurbishment: One of the most direct and impactful circular economy strategies is the reuse and refurbishment of products, which contributes to avoided emissions. This approach extends the functional life of products and materials, delaying or preventing the need for new production and reducing the associated GHG emissions. In many cases, the production of goods especially electronics, appliances, and construction materials accounts for a substantial portion of their total lifecycle emissions. By opting to reuse or refurbish existing items, businesses and consumers effectively reduce demand for new manufacturing and the emissions it entails. Reuse refers to the repeated use of a product in its original form, with minimal or no alteration. Common examples include reusable packaging, second-hand clothing, or used furniture. Reuse is considered one of the most energy-efficient and low-carbon strategies available. Refurbishment involves repairing, restoring, or upgrading a product to bring it back to a usable condition. This is common in sectors like electronics, machinery, automotive, and consumer appliances. Refurbished products typically use a fraction of the energy and materials required for new manufacturing.

- Reuse and refurbishment avoid emissions from the production and transportation of new goods.
- These practices reduce waste generation and landfill usage.
- Projects promoting large-scale reuse and refurbishment, such as electronics take-back schemes or furniture remanufacturing programs, can be documented and verified to generate carbon credits.

Incorporating reuse and refurbishment into business models is becoming increasingly viable through digital platforms and reverse logistics. Companies like IKEA and Apple have launched programs to take back and refurbish used products, highlighting the scalability and business potential of these circular activities. Moreover, the environmental benefits of reuse and refurbishment are amplified when paired with renewable energy, sustainable logistics, and eco-design. In this way, avoided emissions not only earn carbon credits but also contribute to broader goals of sustainable development and corporate social responsibility.

- Reduces material extraction and production emissions.
- Enables circular business models that prioritize service over ownership.
- Aligns with consumer trends favoring sustainability and product longevity.

Circular economy practices are no longer just tools for resource conservation they are now central to climate strategy. Composting, recycling, anaerobic digestion, biomass energy, sustainable product design, and reuse/refurbishment all represent tangible actions that reduce greenhouse gas emissions. These reductions, when measured and verified, can generate carbon credits under voluntary and compliance carbon markets. Such credits provide financial incentives for businesses and communities to invest in sustainable practices, making climate action economically attractive. Moreover, carbon credits from circular activities support co-benefits like job creation, biodiversity preservation, and public health improvements. As global carbon markets evolve and mature, the integration of circular economy principles will become increasingly vital in achieving net-zero goals. Ultimately, by aligning economic activities with environmental limits through circular strategies, we can create systems that not only sustain the planet but also regenerate it transforming waste into wealth, emissions into opportunities, and products into circular assets.

5.2.4.1 Solved Example 1

Project summary (assumptions):

- Project type: **Reforestation/Afforestation**
- Area: **100 ha**
- Crediting period: **20 years (Year 1 → Year 20)**
- Sequestration (gross CO_2 uptake per ha per year): a realistic growth curve
- Year 1: **1.0 tCO_2/ha**
- Year 2: **2.5 tCO_2/ha**
- Year 3: **5.0 tCO_2/ha**
- Year 4: **7.5 tCO_2/ha**
- Year 5: **9.0 tCO_2/ha**
- Years 6–20: **10.0 tCO_2/ha** (steady mature growth)
- Deductions/adjustments (conservative, common in VCM):
- **Leakage**: 7% (emissions caused outside project because of project activity)

- **Risk buffer (permanence)**: 10% (pool deduction to insure against future losses)
- **Measurement uncertainty discount**: 2%
- Price per carbon credit: **₹1000/tCO_2** (example VCM price)
- Discount rate for NPV: **8%**

Method (formula, applied each year)

1. **Gross CO_2 sequestered (tCO_2)** = Area (ha) × Sequestration rate (tCO_2/ha)
2. **After leakage** = Gross × (1 − leakage)
3. **After risk buffer** = After leakage × (1 − buffer)
4. **After uncertainty** = After risk buffer × (1 − uncertainty)
5. **Credits issued that year** = After uncertainty (1 credit = 1 tCO_2)
6. **Annual revenue** = Credits × Price per credit
7. **PV of revenue** = Annual revenue/(1 + discount_rate)^year

We apply these steps for every year 1.20 and then sum.

Below are the numeric results for Years 1–10 (rounded to 3 decimals for clarity). The same method continues through Year 20 (Table 5.3).

Computation example for Year 1 (explicit):

- Gross = 100 ha × 1.0 = 100 tCO_2
- After leakage (7%) = 100 × 0.93 = 93.0
- After buffer (10%) = 93.0 × 0.90 = 83.7

TABLE 5.3 Dataset of reforestation project.

Year	Gross (tCO_2)	After leakage-buffer-uncertainty → credits (tCO_2)	Annual revenue (₹)
1	100.0	82.026	82,026
2	250.0	205.065	205,065
3	500.0	410.130	410,130
4	750.0	615.195	615,195
5	900.0	738.234	738,234
6	1000.0	820.260	820,260
7	1000.0	820.260	820,260
8	1000.0	820.260	820,260
9	1000.0	820.260	820,260
10	1000.0	820.260	820,260

- After uncertainty (2%) = 83.7 × 0.98 = 82.026 → 82.026 credits
- Revenue = 82.026 × ₹1000 = ₹82,026

Totals over 20 years

- Total gross sequestration (sum of gross for years 1–20) = sum of (area × rates) = 15,625 tCO_2 (This is the cumulative raw uptake before deductions.)
- Total net credits issued (after leakage, buffer, uncertainty) = 14,354.55 credits (tCO_2) over 20 years.
- Total nominal revenue at ₹1000/credit (no discounting): = 14,354.55 × ₹1000 = ₹14,354,550

We assume credits are sold in the same year they are issued. Discounting annual revenues at 8%:

- **NPV of future carbon revenues** = **₹6,310,324** (approx.)

This shows that, because revenue is spread over 20 years, the present value is substantially below the sum of nominal revenues.

5.2.4.2 Solved Example 2

Project summary:

- **Project type:** Solar PV (avoided emissions)
- **Capacity:** 250 kW (0.25 MW)
- **Capacity factor (average):** 20% (0.20)—typical for good solar sites
- **Hours per year:** 8760
- **Emission factor displaced:** 0.70 tCO_2 per MWh (assumed grid/diesel mix)
- **Deductions/conservative adjustments:**
- Leakage = **2%** (0.02)
- Permanence/buffer = **5%** (0.05)
- Measurement uncertainty discount = **2%** (0.02)
- **Credit price (VCM example):** ₹1200 per tCO_2 (per credit)
- **Crediting horizon used here:** 20 years (same annual generation each year)
- **Discount rate for NPV:** 8% (0.08)

Step 1: Annual energy generation (MWh)

Annual energy (MWh) = Capacity (kW) × Capacity factor × 8760 ÷ 1000

Annual energy (MWh) = Capacity (kW) × Capacity factor × 8760 ÷ 1000

Capacity = 250 kW
Capacity factor = 0.20
Hours/year = 8760

Annual energy $= 250 \times 0.20 \times 8760 \div 1000$

Annual energy $= 250 \times 0.20 \times 8760 \div 1000 = 250 \times 0.20 \times 8760/1000$

$= 250 \times 0.20 \times 8760/1000$

$= 50 \times 8760/\ 1000$

$= 50 \times 8760/1000$

$= 438.0$ MWh/year

Step 2: Gross avoided CO_2 emissions (tCO_2/year)

Gross avoided CO_2 = Annual energy (MWh) × Emission factor (tCO_2/MWh)

Gross avoided CO_2 = Annual energy (MWh) × Emission factor (tCO_2/MWh)

Annual energy = 438.0 MWh

Emission factor = 0.70 tCO_2/MWh

Gross avoided $CO_2 = 438.0 \times 0.70 = 306.60$ tCO_2/year

Step 3: Apply leakage deduction (2%)

Leakage reduces credited savings by emissions shifted elsewhere.

After leakage = Gross × (1 −leakage)

After leakage = Gross × (1−leakage) $= 306.60 \times (1\text{–}0.02) = 306.60 \times 0.98 = 300.468 tCO_2$

$= 306.60 \times (1\text{–}0.02) = 306.60 \times 0.98 = 300.468$ tCO_2

Step 4: Apply Permanence/Buffer Deduction (5%)

A buffer pool is commonly deducted to insure against future loss.

After buffer $= 300.468 \times (1 - 0.05) = 300.468 \times 0.95 = 285.4446$ tCO_2

After buffer $= 300.468 \times (1\text{–}0.05) = 300.468 \times 0.95 = 285.4446$ tCO_2

Step 5—Apply measurement uncertainty discount (2%)

After uncertainty $= 285.4446 \times (1 - 0.02) = 285.4446 \times 0.98 = 279.735 tCO_2$

After uncertainty $= 285.4446 \times (1\text{–}0.02) = 285.4446 \times 0.98 = 279.7357 tCO_2$

Round reasonably → 279.736 credits/year (1 credit = 1 tCO_2)

So Annual credits issued ≈ 279.736 tCO_2 (credits)

5.3 Carbon sequestration: natural and technological approaches

Carbon sequestration refers to the process of capturing and storing atmospheric carbon dioxide (CO_2) to mitigate the adverse impacts of climate change. It plays a critical role in achieving global net-zero emissions by offsetting residual emissions from hard-to-abate sectors like aviation, steel, and cement. There are two primary approaches to carbon sequestration: natural and technological. Each method offers distinct benefits, limitations, and implementation challenges, and both are essential to a balanced climate strategy. Natural carbon sequestration involves the use of ecosystems such as forests, grasslands, wetlands, and oceans to absorb and store carbon dioxide from the atmosphere. Forests act as powerful carbon sinks through the process of photosynthesis, where trees absorb CO_2 and store it in their biomass and soil. Afforestation, reforestation, and forest conservation are widely recognized natural carbon sequestration strategies that also provide co-benefits such as biodiversity conservation, improved water cycles, and rural livelihoods. Similarly, mangroves, peatlands, and grasslands store large quantities of organic carbon and are critical to maintaining long-term ecological balance. Agricultural practices like conservation tillage, agroforestry, and cover cropping can also enhance soil carbon sequestration. These methods are low-cost and nature-friendly but can be susceptible to reversals caused by land-use change, deforestation, or climate-induced disturbances like wildfires. Fig. 5.13 illustrate carbon sequestration framework.

On the other hand, technological approaches to carbon sequestration aim to capture CO_2 from industrial processes or directly from the air and store it underground or convert it into useful products. Carbon Capture and Storage (CCS) involves capturing CO_2 emissions from sources like power plants or

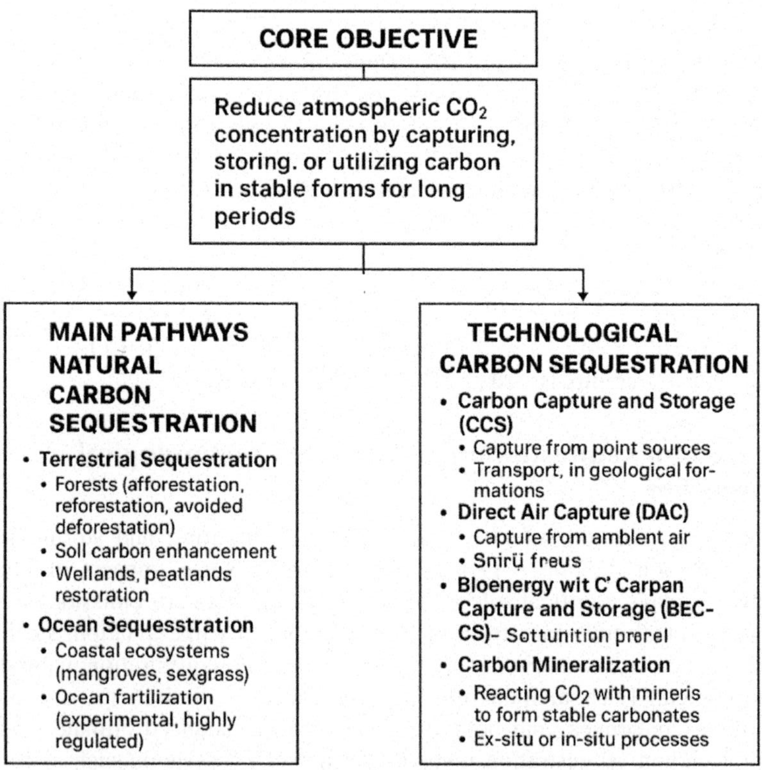

FIGURE 5.13 Carbon sequestration framework.

factories and injecting them into deep geological formations such as saline aquifers or depleted oil and gas reservoirs.

Direct Air Capture (DAC) is an emerging technology that extracts CO_2 directly from the atmosphere using chemical filters, and while energy-intensive and costly, it holds promise for removing legacy emissions. Another category bioenergy with carbon capture and storage (BECCS) combines biomass energy production with CCS to deliver negative emissions. Additionally, mineral carbonation and ocean-based techniques, such as artificial alkalinity enhancement, are being explored as long-term solutions. Despite their potential, technological approaches face barriers including high upfront costs, energy requirements, regulatory uncertainty, and public acceptance. However, as innovation advances and carbon markets mature, these solutions are becoming increasingly viable and critical, especially in sectors where decarbonization is difficult. Both natural and technological carbon sequestration approaches are indispensable in the global fight against climate change. While nature-based solutions offer immediate, low-cost mitigation with added ecological benefits, technological methods provide scalable, durable solutions for long-term carbon removal. A hybrid approach that integrates both strategies supported by science, policy, and finance can significantly accelerate global efforts to stabilize the climate and transition to a carbon-neutral future.

5.3.1 Natural sequestration

Natural carbon sinks play a foundational role in stabilizing the Earth's climate by capturing and storing atmospheric carbon dioxide (CO_2), a key greenhouse gas responsible for global warming. These biological and ecological systems have evolved over millions of years to absorb carbon from the atmosphere through natural processes such as photosynthesis, microbial activity, and sedimentation. Unlike technological solutions, natural sinks are low-cost, self-regenerating, and provide a host of co-benefits including biodiversity preservation, improved water cycles, and enhanced soil fertility. Among the most significant natural carbon sinks are forests, wetlands, and soils rich in organic carbon. Additionally, emerging nature-based solutions like bio char created from organic waste offer a long-term, stable form of carbon sequestration. Together, these systems are not only crucial for climate mitigation but also for supporting global food security, water regulation, and ecosystem health.

Forests are perhaps the most visible and important terrestrial carbon sinks. Through the process of photosynthesis, trees absorb CO_2 from the atmosphere and store it in their trunks, branches, leaves, and roots. Forests also enhance soil carbon by dropping leaves and other organic matter that decompose and enrich the forest floor. Globally, forests store over 80% of all terrestrial aboveground carbon and more than 70% of soil organic carbon. Tropical rainforests like the Amazon and Congo Basin are particularly valuable because of their year-round productivity and dense biomass. Temperate and boreal forests, though slower-growing, also play a significant sequestration role due to their extensive land cover. Reforestation (planting trees in deforested areas) and afforestation (planting trees in areas that were not previously forested) are key strategies for enhancing carbon storage. Forest conservation efforts such as REDD+ (Reducing Emissions from Deforestation and Forest Degradation) have also emerged as market-based mechanisms to protect existing carbon sinks by assigning economic value to the carbon stored in trees. However, the carbon sequestered in forests is vulnerable to disturbances such as wildfires, pest outbreaks, and illegal logging all of which are becoming more frequent due to climate change. Therefore, forest-based carbon sequestration must be complemented with robust protection policies and long-term ecosystem management. Fig. 5.14 illustrate framework of natural sequestration.

Wetlands, including marshes, peatlands, mangroves, and swamps, are among the most efficient yet often overlooked carbon sinks. They function differently from forests in that they sequester large amounts of carbon in their waterlogged soils, where anaerobic conditions slow down the decomposition of organic matter. This results in the accumulation of peat, a carbon-rich layer that can store carbon for thousands of years. Peatlands, which cover only about 3% of the Earth's land surface, hold nearly 30% of all soil carbon twice as much as all the world's forests combined. Mangrove forests are another vital wetland ecosystem that sequesters carbon both above ground in their woody biomass and below ground in sediments. In fact, mangroves, seagrasses, and salt marshes are collectively known as "blue carbon" ecosystems due to their exceptional ability to store carbon in coastal environments. Despite their value,

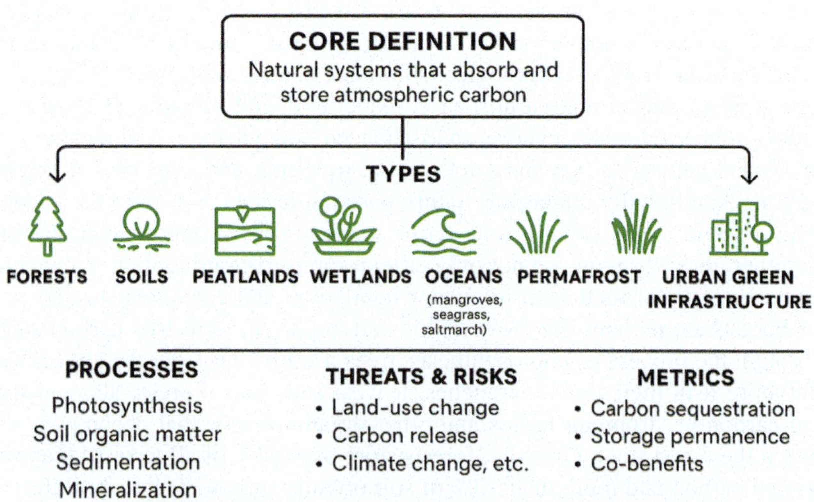

FIGURE 5.14 Framework of natural sequestration.

wetlands are among the most degraded ecosystems globally, with over 35% lost in the past 50 years due to urbanization, agriculture, and infrastructure development. Restoring wetlands through rehydration, native vegetation planting, and pollution control not only enhances carbon sequestration but also improves flood control, water purification, and wildlife habitats. Soil organic carbon (SOC) is another crucial and vast natural sink. Soils contain more carbon than the atmosphere and all terrestrial vegetation combined. This carbon is primarily stored in the form of decomposed plant and animal material, microbial biomass, and root exudates. Practices such as deforestation, intensive agriculture, overgrazing, and erosion can deplete SOC levels and release large amounts of CO_2 back into the atmosphere. Conversely, sustainable land-use practices can help increase SOC levels and turn degraded soils into effective carbon sinks. These practices include conservation tillage (minimal or no soil disturbance), cover cropping, crop rotation, organic manure application, and agroforestry (integrating trees into farmland). Such regenerative agricultural techniques not only improve soil health and fertility but also enhance water retention and crop productivity. Efforts like the "4 per 1000" Initiative, launched during the 2015 Paris Climate Conference, advocate for increasing global soil carbon stocks by 0.4% per year, which could significantly offset anthropogenic CO_2 emissions. Nevertheless, measuring and verifying SOC changes remains challenging due to variations in soil type, climate, and land management history. In addition to naturally occurring sinks, biochar is gaining attention as a highly stable, engineered method of carbon sequestration derived from natural sources. Biochar is a carbon-rich material produced by the pyrolysis (thermal decomposition in the absence of oxygen) of organic biomass such as agricultural waste, forestry residues, or animal manure. When applied to soil, biochar can remain stable for hundreds to thousands of years, locking carbon away that would otherwise return to the atmosphere through decomposition. Moreover, biochar improves soil properties by increasing nutrient retention, water-holding capacity, and microbial activity. Its porous structure helps reduce leaching of fertilizers and enhances root development, making it particularly useful in degraded or sandy soils. Biochar production can be decentralized, low-cost, and adaptable to rural communities, especially in regions where agricultural waste is abundant and underutilized. By converting waste into a value-added product, biochar contributes to circular economy models while simultaneously supporting climate resilience in agriculture. However, the net climate benefits of biochar depend on several factors including feedstock type, production temperature, transportation emissions, and application rate. More research is needed to optimize biochar systems and standardize protocols for carbon accounting and certification. Natural carbon sinks forests, wetlands, soils, and biochar are indispensable tools in the fight against climate change. They offer scalable, cost-effective, and multi-functional benefits that extend far beyond carbon mitigation, including biodiversity enhancement, ecosystem resilience, and rural livelihood improvement. However, the effectiveness of these sinks is increasingly threatened by human activity and climate-induced stressors. To harness their full potential, a coordinated global effort is required to conserve, restore, and manage these ecosystems. This includes integrating nature-based solutions into national climate policies, incentivizing sustainable land use through carbon markets, investing in scientific research, and promoting community-based stewardship. When combined with technological solutions and emission reduction strategies, natural carbon sinks form a powerful triad for achieving a sustainable, low-carbon future. As the world accelerates toward net-zero goals, prioritizing the protection and enhancement of nature's carbon reservoirs is not only a climate imperative but also a moral responsibility to future generations. Table 5.4 illustrate Natural versus Technological Carbon Sequestration.

5.3.2 Technological sequestration

Technological sequestration refers to engineered processes that capture carbon dioxide (CO_2) emissions from the atmosphere or industrial sources and securely store or repurpose them to prevent their release into the environment. These methods are crucial complements to natural carbon sinks, particularly for sectors where emissions are hard to eliminate, such as heavy industry, transportation, and waste management. Three prominent approaches within this domain include Direct Air Capture (DAC) with mineralization, Carbon Capture and Storage (CCS) from Waste-to-Energy (WtE) plants, and the integration of sequestration into circular design principles. Each of these solutions addresses a specific aspect of the carbon cycle, aiming to reduce the net load of CO_2 in the atmosphere. Fig. 5.15 illustrate technological sequestration framework.

Direct air capture (DAC) with mineralization: Direct Air Capture is a cutting-edge carbon removal technology that extracts CO_2 directly from ambient air using chemical processes. Specialized machines use filters or liquid solvents to absorb CO_2, which is then isolated, compressed, and either utilized or stored. One of the most promising methods for permanent storage is mineralization a natural reaction where CO_2 reacts with minerals like magnesium or calcium-rich rocks (e.g., basalt) to form stable carbonates. This process locks the carbon in solid form for thousands of years, effectively removing it from the carbon cycle. Companies like Climeworks in Iceland are pioneering this technology by combining DAC with underground mineral storage. Though energy-intensive and currently expensive, DAC with mineralization holds immense potential for offsetting unavoidable emissions and even drawing down historical carbon from the atmosphere. It is particularly useful in regions with abundant renewable energy and suitable geology.

Carbon capture and storage (CCS) from waste-to-energy (WtE) plants: Carbon Capture and Storage is a more established technology that captures

TABLE 5.4 Natural versus technological carbon sequestration.

Category	Natural carbon sequestration	Technological carbon sequestration
Definition	The process by which ecosystems absorb and store CO_2 through biological and geological pathways.	Human-engineered methods designed to capture, store, or utilize CO_2 from air or industrial sources.
Primary mechanism	Photosynthesis, soil carbon accumulation, ocean absorption, mineral weathering.	Direct air capture (DAC), carbon capture and storage (CCS), bioenergy with CCS (BECCS), mineralization, engineered sinks.
Key systems	Forests, wetlands, soil, oceans, peatlands.	Industrial plants, capture facilities, pipelines, injection wells, synthetic mineral processes.
Storage duration	Ranges from short-term (plants) to millennia (soils, oceans, geological formations).	Often long-term (geological storage), depending on technology used.
Scalability	Limited by land availability, ecosystem health, biodiversity constraints.	Potentially high but dependent on energy input, cost, and infrastructure.
Cost	Generally low but varies with restoration needs.	Currently high; DAC ranges from $100–$600 per ton CO_2 removed.
Energy requirements	Mostly solar energy through natural processes.	High energy demand, often requiring low-carbon energy to be effective.
Advantages	Supports biodiversity, improves soil, enhances water cycles, co-benefits for ecosystems.	High precision, measurable removal, scalable to industrial needs, permanent geological storage.
Limitations	Vulnerable to disturbances (fires, logging, climate change), slow sequestration rate.	High cost, energy-intensive, requires infrastructure, potential leakage concerns.
Permanence risk	Moderate to high (forest fires, deforestation, land-use change).	Low with proper geological storage; risks depend on site integrity.
Monitoring difficulty	Complex due to ecosystem variability.	Easier due to controlled industrial processes and measurement tools.
Time horizon	Slower uptake; depends on ecological regeneration.	Can remove large amounts relatively quickly once deployed.
Role in climate strategy	Essential baseline for global carbon balance; nature-based solutions.	Critical as a complementary solution when deep decarbonization isn't enough.

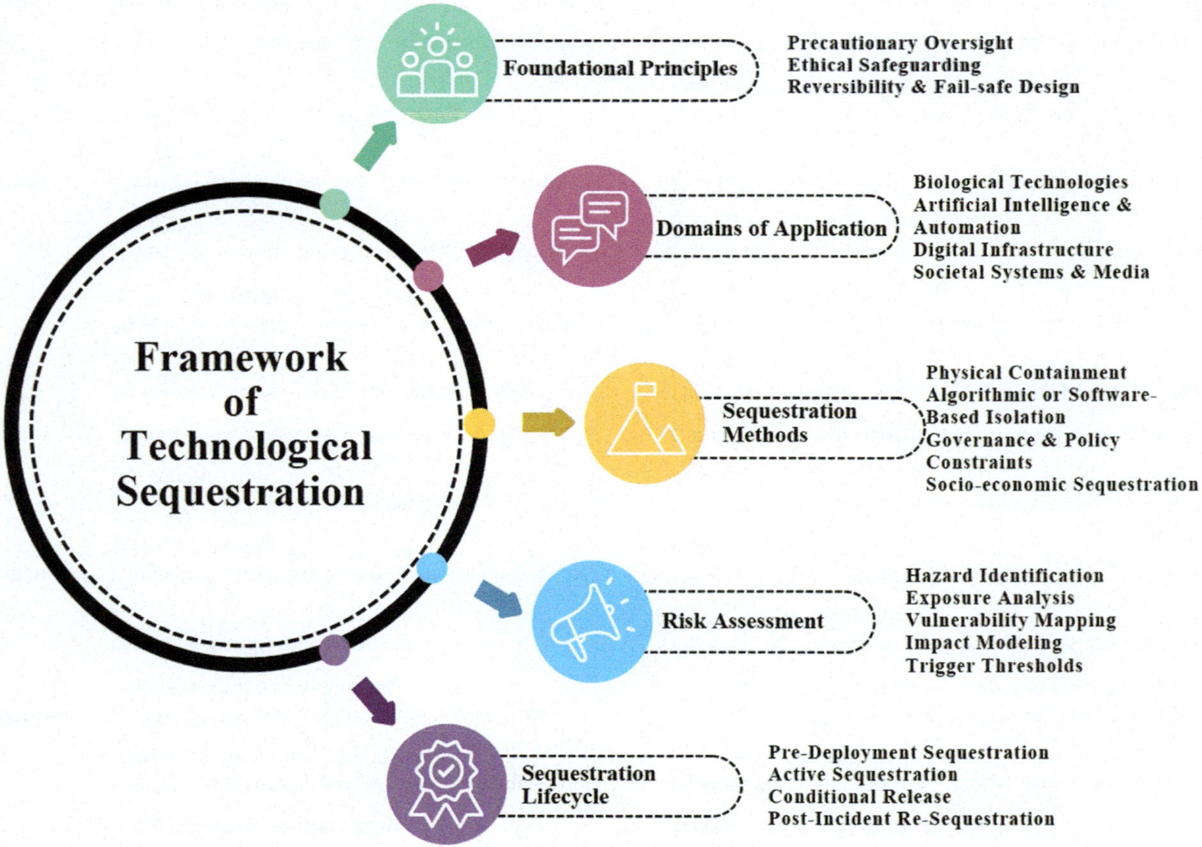

FIGURE 5.15 Technological sequestration framework.

CO_2 emissions from point sources like power plants, cement factories, or Waste-to-Energy (WtE) facilities. In the case of WtE plants, which burn municipal solid waste to generate electricity, integrating CCS systems allows for the capture of CO_2 before it is released into the atmosphere. This is critical because waste incineration while reducing landfill burden produces substantial carbon emissions, especially when fossil-based plastics and materials are burned. By capturing and compressing this CO_2, it can be injected into deep geological formations, such as depleted oil and gas fields or saline aquifers, where it remains safely sequestered. Some CCS setups are also exploring enhanced oil recovery (EOR), where CO_2 is injected into oil wells to boost extraction while simultaneously storing the gas. Applying CCS to WtE aligns with circular economy goals by ensuring that energy recovery from waste is not undermined by carbon leakage, and it provides a viable decarbonization pathway for urban waste management.

Integration of sequestration into circular design: Integrating carbon sequestration into circular design principles represents a transformative approach to reducing emissions across product life cycles. Circular design involves designing materials, products, and systems to minimize waste, extend usability, and regenerate natural systems. When sequestration is embedded into this framework, materials are chosen not only for recyclability or durability but also for their ability to store or absorb carbon. For example, construction materials such as carbon-negative concrete, which incorporates CO_2 during curing, or bio-based plastics that store atmospheric carbon, are being developed and scaled. Similarly, manufacturing processes are being redesigned to incorporate captured CO_2 as feedstock for new products ranging from fuels and chemicals to textiles and packaging. By closing material loops and turning carbon into a resource, circular design coupled with sequestration helps shift industries from linear "take-make-dispose" models to regenerative systems that actively remove carbon from the atmosphere while creating value.

Technological sequestration through DAC, CCS, and circular integration is becoming an indispensable pillar of climate action. While challenges remain in terms of cost, scalability, and infrastructure, these innovations represent a proactive means to achieve net-zero goals and transform carbon from a liability into an asset.

5.4 Circular economy strategies for carbon reduction

Circular economy (CE) strategies have become central to global efforts for carbon reduction, offering a systemic shift away from the traditional linear economic model of "take-make-dispose" toward a regenerative system designed to minimize waste, preserve natural resources, and decouple economic growth from carbon emissions. These strategies aim to keep materials, products, and components at their highest utility and value for as long as possible, reducing both direct and embedded carbon emissions throughout the product lifecycle. Among the most impactful approaches within this framework are resource efficiency and decarbonization, product life extension, and waste-to-energy and biochar systems. Together, these interconnected strategies support climate mitigation while promoting sustainable industry, innovation, and inclusive economic development. Fig. 5.16 illustrate circular economy strategies for carbon reduction.

The circular economy offers a transformative pathway for reducing carbon emissions by rethinking how materials, products, and energy systems are designed, used, and regenerated. At its foundation, the circular economy challenges the traditional linear model of "take, make, dispose" and replaces it with a system that preserves value, reduces waste, and minimizes the extraction of virgin resources. The first foundational principle decoupling growth from resource use is essential because economic growth has historically been tied to increased material consumption, which in turn drives carbon emissions from mining, manufacturing, transportation, and disposal. In a circular system, value is created through smarter design, shared use, extended product lifetimes, and efficient recovery, meaning growth no longer depends on consuming more physical materials. This decoupling enables societies to expand prosperity while staying within planetary carbon limits.

The second foundational principle, systemic resource efficiency, emphasizes designing products and systems in a way that maximizes their useful life. When products are designed for durability, easy repair, modular replacement, and eventual disassembly, their carbon footprint is drastically reduced. This is because the carbon embedded in producing a new item whether a smartphone, vehicle, or building component can be "distributed" across a longer lifespan or avoided entirely through refurbishing or remanufacturing. Much of today's carbon emissions come from the rapid turnover of products and materials that were never designed for reuse. Circularity reverses this logic by designing waste out of the system from the beginning. The third foundational principle, renewable and low-carbon inputs, complements these efforts by prioritizing materials and energy sources that inherently emit less carbon. For example, replacing fossil-based plastics with bio-based polymers or integrating renewable electricity into manufacturing reduces emissions both at the production stage and across a product's lifecycle.

The framework's first strategic pillar design for circularity is arguably the most impactful because up to 80% of a product's environmental impact is determined at the design stage. Circular design emphasizes modular components that can be replaced rather than discarded entirely, lightweight materials that reduce both production and transportation emissions, and standardized parts that simplify repair and refurbishment. Designing for disassembly ensures that once a product reaches the end of its life, its components can be easily separated, sorted, and reintegrated into new production cycles without degrading quality. When applied across industries, this approach significantly reduces the carbon embedded in new material extraction and manufacturing.

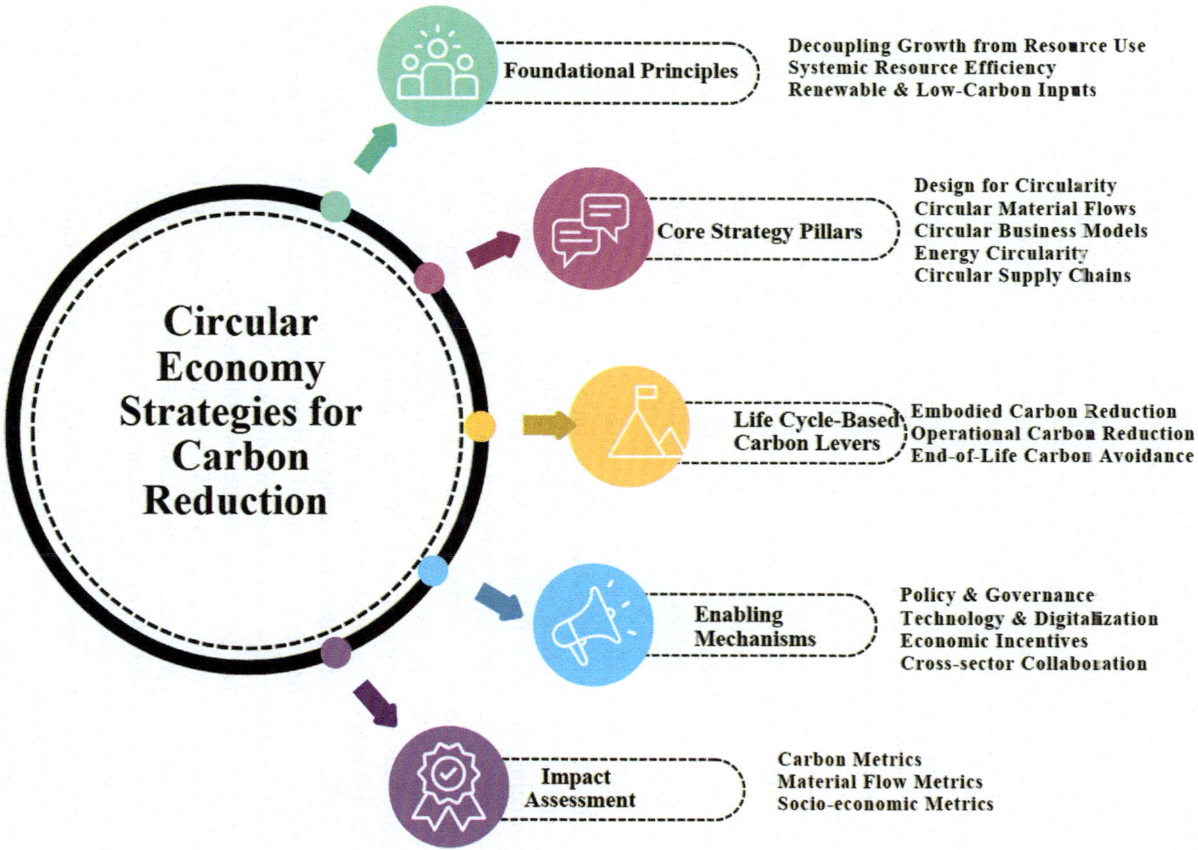

FIGURE 5.16 Circular economy strategies for carbon reduction.

The second pillar, circular material flows, focuses on how materials circulate within industrial and societal systems. Reuse, refurbishment, and high-quality recycling loops prevent materials from ending up in landfills, where they can emit methane or require carbon-intensive replacement. Industrial symbiosis where the waste from one industry becomes the input for another is a powerful method for reducing carbon emissions by treating emissions and by-products as resources. For example, waste heat from one factory can warm nearby buildings, or organic waste from food production can be transformed into bioenergy or compost. These interconnected loops reduce energy demand and avoid emissions associated with producing virgin materials.

A third strategic pillar, circular business models, reshapes how products reach consumers. Models such as Product-as-a-Service, leasing, sharing platforms, and take-back programs extend product lifespans and reduce the need for new production. By shifting incentives away from selling high volumes of products toward maintaining long-lasting ones, companies reduce emissions from manufacturing while generating value from services, repairs, and upgrades. For instance, a company that leases machinery has a strong incentive to design equipment that lasts longer and consumes less energy because it remains responsible for performance throughout the product's life cycle.

The fourth pillar, energy circularity, complements material circularity by addressing carbon emissions from operational energy use. Waste-heat recovery systems capture heat generated from industrial processes and repurpose it, reducing the need for additional fossil-fuel heating. Closed-loop industrial energy systems integrate processes so that one facility's energy output supports another's, improving overall efficiency. At the same time, the integration of renewable energy solar, wind, geothermal reduces operational carbon footprints and ensures that circular material flows do not rely on carbon-intensive energy sources.

The fifth pillar, circular supply chains, focuses on minimizing emissions across global transportation and logistics. Regionalized production can drastically reduce emissions from long-distance shipping. Reverse logistics systems ensure that products at the end of their life cycle can be collected efficiently for reuse, repair, or remanufacturing. Transparent tracking systems, enabled by digital tools, allow companies to measure carbon flows across supply chains and identify opportunities to reduce emissions, eliminate waste, and optimize resource use.

To measure progress, the framework includes life-cycle-based carbon levers. Embodied carbon reduction occurs through material substitution, increased recycling rates, and extending product longevity, all of which reduce the need for carbon-intensive manufacturing. Operational carbon reduction arises from energy efficiency improvements, renewable energy integration, and closed-loop energy systems. End-of-life carbon avoidance targets methane emissions from landfills, wasted materials, and inefficient disposal practices by redirecting products and materials back into the productive system.

Enabling mechanisms are critical to scaling circular strategies. Effective policy and governance tools such as Extended Producer Responsibility, carbon pricing, green procurement, and digital product passports create accountability and align incentives with carbon reduction goals. Technology and digitalization amplify circularity by improving material tracking, enabling predictive maintenance through digital twins, and optimizing recycling or sorting using AI. Economic incentives, such as subsidies for recycled materials or reduced taxes for repair services, help shift market dynamics toward low-carbon circular systems. Cross-sector collaboration among industries, governments, and community's fosters innovation and shared resource systems that significantly reduce emissions.

Finally, the implementation roadmap outlines how organizations or governments can phase in circular strategies. In the short term, conducting audits of material and carbon flows helps identify hotspots where circular interventions will have the biggest impact. Mid-term efforts focus on redesigning products, creating reverse logistics systems, and building industrial symbiosis networks. Long-term actions aim to establish fully closed-loop supply chains and regenerative systems that operate with net-zero carbon emissions. In essence, a circular economy provides a systemic pathway for deep carbon reduction by addressing emissions embedded in materials, energy, and product lifecycles. It shifts value creation from extraction to regeneration, making it a cornerstone of any credible long-term climate strategy.

5.5 Technologies enabling circular carbon systems

The vision of a circular carbon economy where carbon emissions are minimized, reused, recycled, and ultimately balanced requires more than just strategic policies and innovative business models. It also demands a suite of powerful enabling technologies that can monitor, verify, optimize, and communicate complex data related to resource use, emissions, and circularity. As the global economy shifts from linear to circular models, particularly in the context of climate action, digital technologies are playing a central role in making carbon flows visible, traceable, and manageable. Among the most impactful are Artificial Intelligence (AI) for tracking emissions and resource usage, Blockchain for carbon credit verification and trade, Digital Product Passports (DPPs) and Internet of Things (IoT) sensors for carbon footprint auditing, and carbon accounting platforms for quantifying emissions associated with circular products. Together, these technologies form the backbone of circular carbon systems by enabling transparency, traceability, accountability, and efficiency in carbon management. Fig. 5.17 illustrate integrated circular carbon technology architecture.

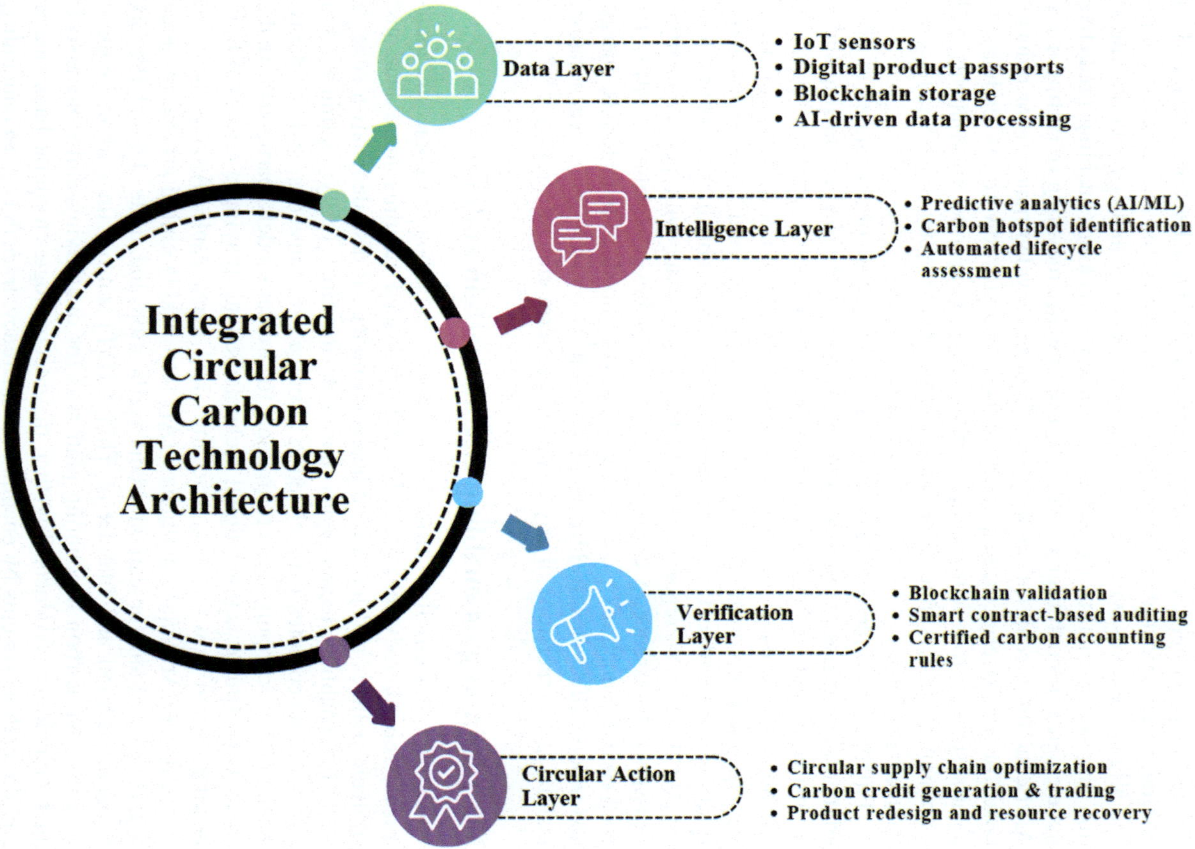

FIGURE 5.17 Integrated circular carbon technology architecture.

AI for tracking emissions and resource use: Artificial Intelligence (AI) has emerged as a powerful tool for environmental monitoring and decision-making, particularly in the realm of carbon management and circular economy operations. AI systems can process massive datasets in real time, detect patterns, and generate actionable insights for emission reductions and resource efficiency. In manufacturing and industrial settings, AI algorithms analyze energy consumption patterns, material flows, and production data to identify inefficiencies and optimize processes that minimize waste and carbon output. For instance, AI can help companies track Scope 1 (direct), Scope 2 (indirect from purchased energy), and Scope 3 (value-chain) emissions across supply chains by automating the collection and integration of disparate data sources. AI also supports predictive maintenance, which reduces unplanned downtimes and material waste by forecasting equipment failures before they occur. This not only improves productivity but also reduces the carbon footprint of maintenance logistics and replacement parts. Fig. 5.18 illustrate AI for tracking emissions and resources use.

In agriculture, AI-driven platforms can guide regenerative practices by analyzing satellite imagery, weather data, and soil health indicators to optimize crop cycles, fertilizer use, and carbon sequestration potential. Moreover, AI plays a key role in smart logistics routing delivery vehicles based on real-time traffic and emissions data to minimize fuel consumption. Combined with machine learning, AI models evolve over time, improving their accuracy and helping businesses and policymakers make more informed, low-carbon decisions. As carbon tracking becomes a mandatory aspect of corporate sustainability reporting, AI will be essential in automating data collection, benchmarking emissions, and recommending circular interventions.

Blockchain for carbon credit verification and trade: Blockchain technology is revolutionizing the way carbon credits are verified, issued, and traded. Traditionally, carbon markets especially voluntary ones have suffered from issues like double counting, lack of transparency, fraud, and inefficiency. Blockchain, as a decentralized and tamper-proof digital ledger, addresses these challenges by creating immutable records of carbon offset transactions and ownership. Every carbon credit tokenized on a blockchain represents a verified emission reduction or removal, and its lifecycle from issuance to retirement can be transparently tracked. This improves confidence in the credibility and integrity of carbon credits, especially in nature-based solutions like afforestation or mangrove restoration where on-the-ground verification is complex. For instance, when a credit is sold or retired, the blockchain ledger is updated in real time, ensuring no credit is used more than once. Blockchain also facilitates peer-to-peer carbon trading, reducing the need for intermediaries and lowering transaction costs. Fig. 5.19 illustrate blockchain for carbon credit verification and trade.

Startups and platforms like KlimaDAO, Toucan Protocol, and Verra-integrated solutions are already utilizing blockchain to democratize access

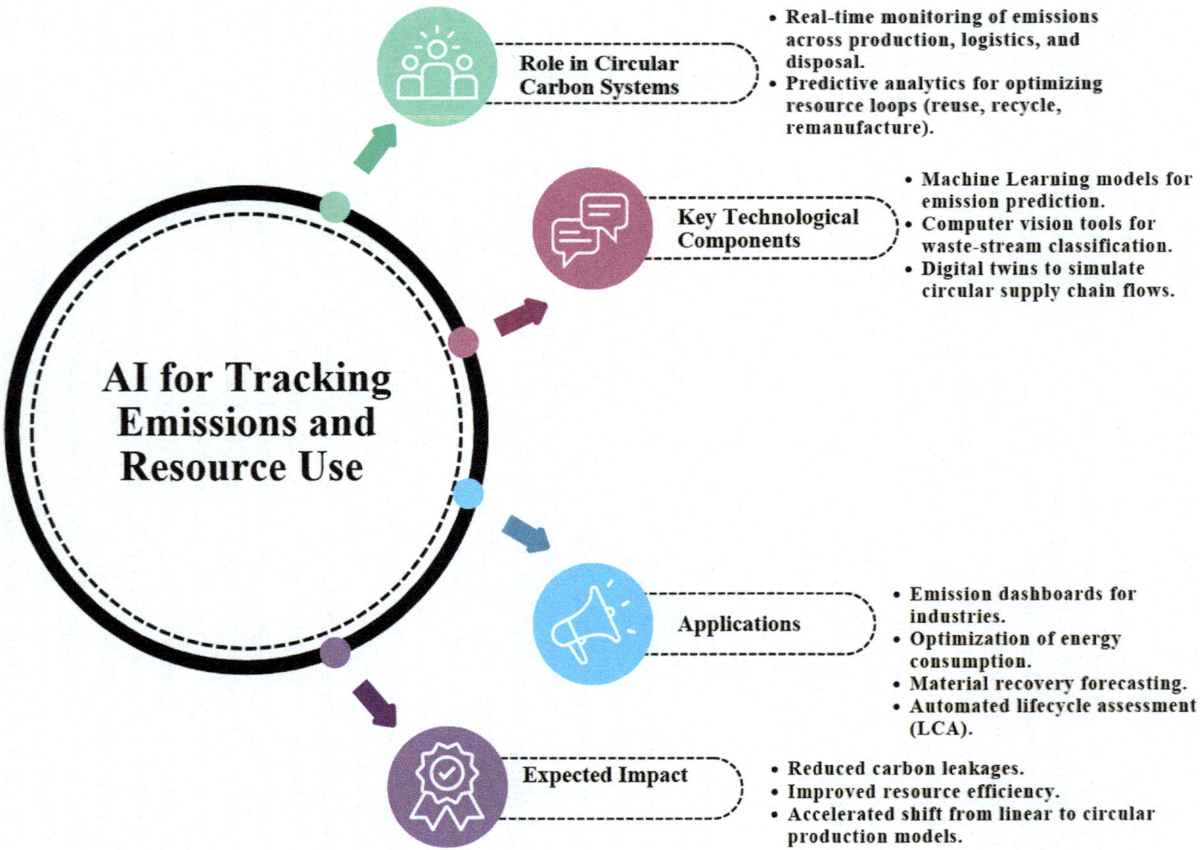

FIGURE 5.18 AI for tracking emissions and resources use.

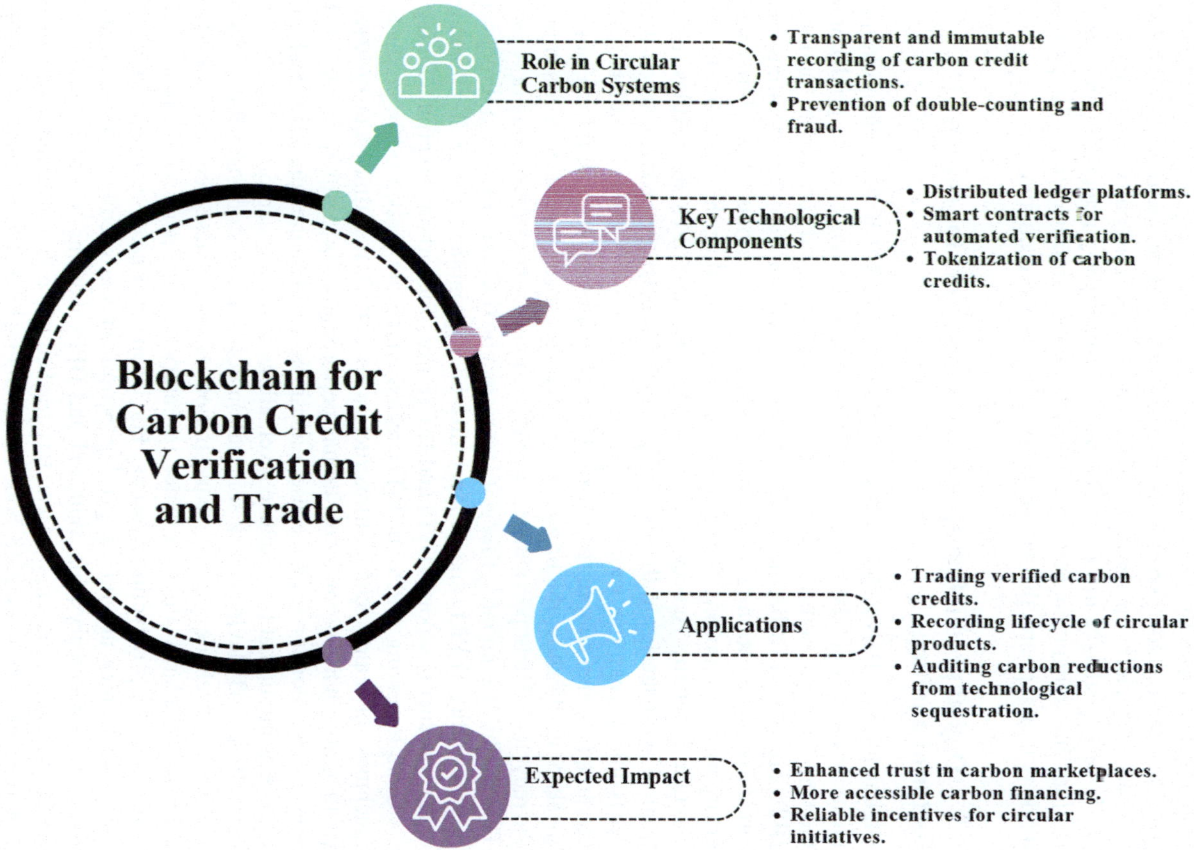

FIGURE 5.19 Blockchain for carbon credit verification and trade.

to carbon markets and create decentralized carbon registries. Smart contracts self-executing codes on blockchain automate processes such as credit issuance, verification triggers, or payments based on third-party sensor data, reducing administrative burdens. Importantly, blockchain can be integrated with IoT and remote sensing to bring near-real-time monitoring to offset projects, especially in developing regions where third-party verification is costly or logistically difficult. Through transparency, traceability, and automation, blockchain enables a trustworthy digital infrastructure for carbon markets that supports the scaling of circular carbon economies.

5.5.1 Digital product passports and IoT sensors for carbon footprint auditing

Digital Product Passports (DPPs) are digital records that store and communicate information about a product's lifecycle materials used, production methods, energy consumption, carbon emissions, and end-of-life options. DPPs are poised to become essential tools in carbon footprint auditing and circularity verification as regulations tighten around product traceability and environmental reporting. The European Union is leading this shift through its Circular Economy Action Plan, which will require DPPs for textiles, electronics, and batteries in the coming years. DPPs make embedded carbon data accessible across the supply chain, enabling manufacturers, retailers, and consumers to make more sustainable choices. For example, a DPP for a pair of sneakers might show the origin of its rubber, the energy used in its assembly, the recycled content percentage, and instructions for proper recycling or repair. This visibility supports lifecycle assessments (LCAs), Extended Producer Responsibility (EPR), and product take-back programs. Fig. 5.20 illustrate digital product passports (DPP) and IoT sensors for carbon footprint auditing.

Complementing DPPs are IoT sensors, which are embedded in products, machines, or supply chain infrastructure to collect real-time data on environmental conditions and operational performance. In a smart factory, IoT sensors can monitor resource use, emissions, temperature, pressure, and energy consumption at the machine level, allowing precise carbon accounting. In agriculture, IoT devices track soil carbon content, moisture, and fertilization metrics to evaluate carbon sequestration efforts. These technologies also enhance circularity through better inventory management, waste reduction, and closed-loop recycling systems. For instance, sensors can alert facilities when waste bins are full, triggering optimized waste collection routes and reducing unnecessary trips. With IoT and DPPs working together, the carbon footprint of products can be accurately tracked and communicated across their full lifecycle, strengthening both environmental performance and customer trust.

Carbon accounting platforms for circular products: Carbon accounting platforms serve as centralized systems for collecting, analyzing, reporting, and

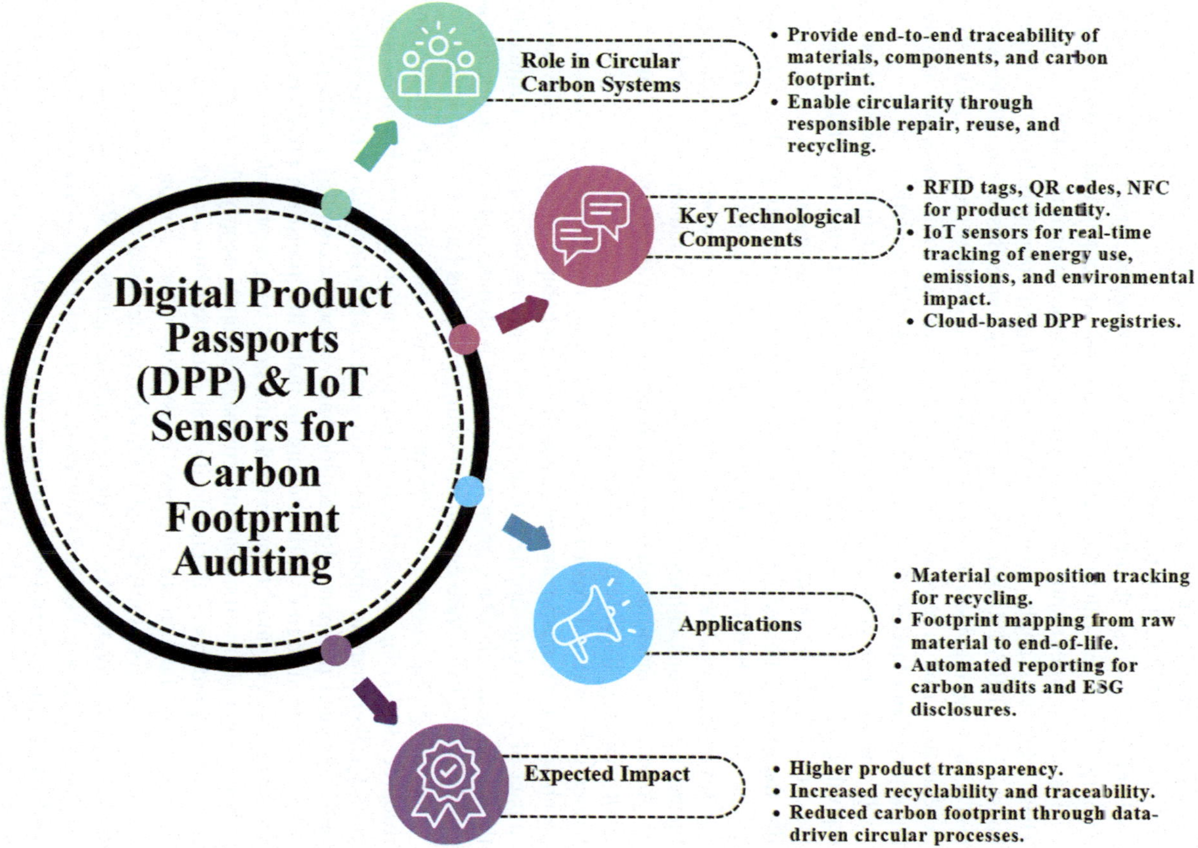

FIGURE 5.20 Digital product passports (DPP) and IoT sensors for carbon footprint auditing.

verifying greenhouse gas emissions associated with products, services, and organizational activities. As circular products become more mainstream those designed for reuse, repair, recycling, or remanufacturing there is a growing need to accurately quantify their carbon impact to justify sustainability claims and participate in carbon markets.

Platforms such as *Watershed*, *Normative*, *Persefoni*, and *CarbonChain* are helping businesses calculate emissions across all scopes and across supply chains, often integrating with enterprise resource planning (ERP) systems. These platforms apply internationally recognized standards like the Greenhouse Gas (GHG) Protocol, ISO 14064, and Science Based Targets initiative (SBTi) to ensure consistency and reliability. For circular products, they factor in reductions due to recycled content, energy savings from reuse, and avoided emissions from landfill diversion or virgin material extraction.

In the circular economy, carbon accounting is not just about emissions reporting—it becomes a tool for strategic decision-making. By understanding the carbon footprint of different product designs, sourcing options, and end-of-life treatments, businesses can prioritize circular innovations that offer the greatest climate benefit. These platforms also support emissions disclosure under environmental, social, and governance (ESG) frameworks and sustainable finance regulations, enhancing corporate accountability and investor confidence.

Advanced carbon accounting platforms are increasingly using AI and machine learning to automate data categorization and emissions estimation, reducing the time and expertise needed for accurate reporting. Some also feature dashboards and scenario planning tools that allow companies to model the carbon impact of switching suppliers, materials, or distribution strategies. Ultimately, these tools make circular carbon management both measurable and actionable at scale. Fig. 5.21 illustrate carbon accounting platforms for circular products.

As the global economy confronts the twin crises of climate change and resource depletion, the shift toward circular carbon systems becomes a necessity rather than an option. Technological enablers such as AI, blockchain, digital product passports, IoT sensors, and carbon accounting platforms provide the critical digital infrastructure to make this transition possible. These tools empower organizations to track emissions with unprecedented accuracy, verify carbon claims transparently, manage complex supply chains efficiently, and design products with circularity and decarbonization at their core.

Together, they create a robust ecosystem where carbon can be traced, valued, minimized, reused, and ultimately sequestered all within a regenerative economic model. As carbon regulations tighten, ESG expectations rise, and carbon pricing becomes universal, early adoption of these technologies will provide a competitive advantage and prepare businesses for the future of sustainable commerce. Therefore, investing in these digital enablers today is not only an environmental imperative but also a strategic one for organizations aiming to thrive in a carbon-conscious, circular world.

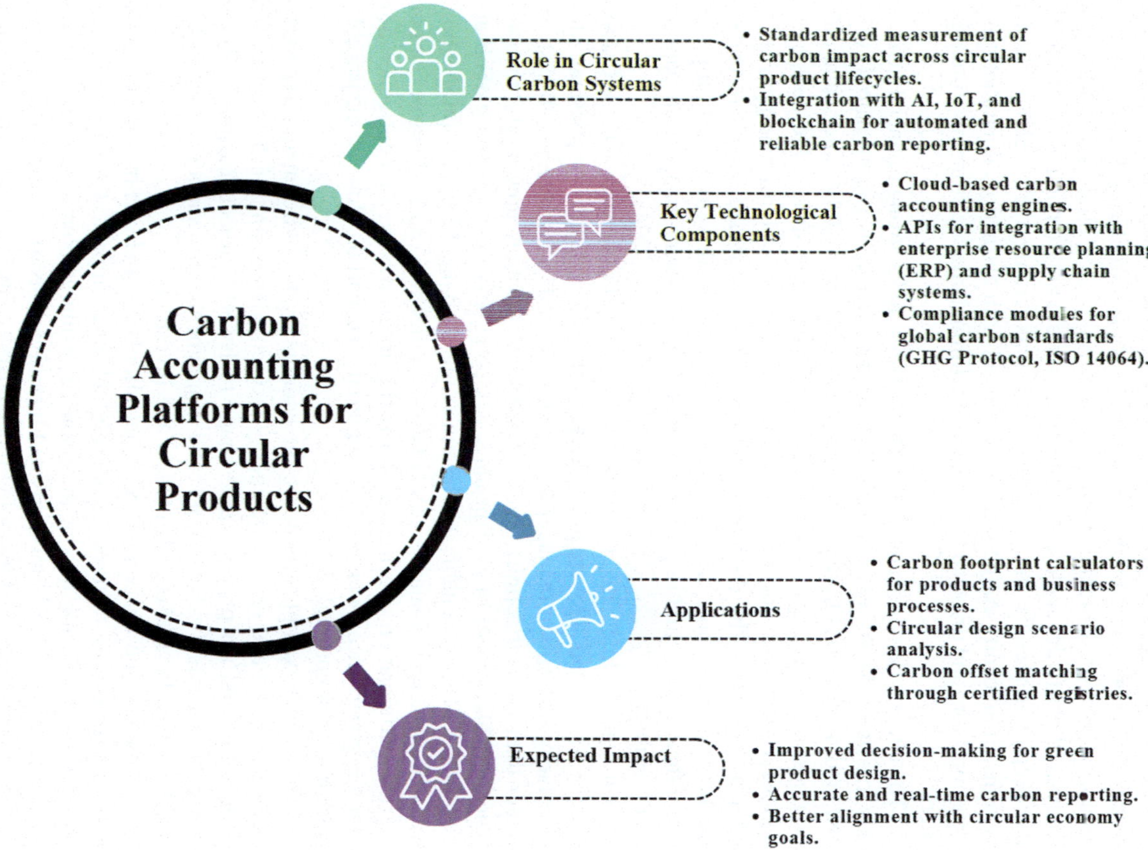

FIGURE 5.21 Carbon accounting platforms for circular products.

5.6 Conclusion

This chapter concludes that the integration of Circular Economy principles with carbon credits and carbon sequestration forms one of the most powerful and future-ready pathways for combating climate change while supporting sustainable economic growth. A key insight is that circularity does not merely reduce waste it fundamentally redesigns material cycles, product lifetimes, and resource flows to decouple economic activity from environmental degradation. By keeping materials in circulation and maximizing their value, circular systems significantly cut emissions associated with extraction, manufacturing, transportation, and disposal areas where traditional climate strategies often fall short.

The chapter demonstrates that carbon credits add a strong financial layer to this transition. When emissions reductions and removals are measured, verified, and monetized, circular activities gain economic viability and scalability. Composting, recycling, bioenergy, refurbishment, and sustainable design are no longer just environmentally beneficial they also become revenue-generating climate solutions. The distinction between compliance and voluntary carbon markets reveals how policy-driven and market-driven forces together strengthen climate ambition, while robust MRV systems ensure transparency and credibility.

Through the Integrated Circular–Carbon Value Framework and the Carbon Credit Value and Governance Framework, the chapter emphasizes that circularity and carbon management must be treated as interconnected systems rather than isolated actions. These frameworks illustrate how digital technologies AI, blockchain, digital product passports, IoT sensors, and carbon accounting platforms—enable precise emissions tracking, lifecycle optimization, transparent credit issuance, automated verification, and effective carbon market participation. Such tools form the infrastructure of next-generation climate governance, where every material and carbon flow is measurable, traceable, and optimizable.

The solved examples reinforce the economic and environmental outcomes of sequestration and renewable energy projects, demonstrating how crediting, deductions, and net present value shape real-world project feasibility. These practical insights bridge theory and implementation, supporting decision-making for policymakers, industries, and investors.

Ultimately, the chapter affirms that circularity and carbon markets must evolve together to achieve meaningful climate impact. Circular economy strategies reduce emissions at their source, while carbon credits and sequestration address residual emissions and incentivize deeper action. Technologies ensure transparency, and financing mechanisms enable scale. When combined, these elements form a regenerative economic model where carbon becomes a measurable asset, waste becomes a resource, and sustainability becomes a driver of innovation and prosperity.

The conclusion is clear: the intersection of circular economy, carbon credits, and carbon sequestration offers a holistic, technologically empowered, and financially viable pathway for creating a carbon-neutral and inclusive global future.

5.7 Case study

5.7.1 Case study 1. Circular economy with carbon credit

5.7.1.1 Background

XYZ Industries Ltd. is a mid-sized Indian FMCG and packaging manufacturer located in Pune. The company produces plastic bottles, food-grade pouches, and household packaging material for major brands across India.

For nearly a decade, the company operated on a linear production model:

- Raw materials → Production → Use → Disposal
- 78% of their products ended up in landfills
- Recycling rate was only 12%
- Annual carbon emissions: 58,000 t of CO_2e

In 2025, after increasing pressure from government policies and its corporate customers, GreenLoop committed to transform its operations using Circular Economy principles integrated with Carbon Credit mechanisms.

Their new strategy was based on three pillars:

1. **Material circularity**—redesign packaging to be recyclable and reusable
2. **Reverse logistics**—collect post-consumer packaging waste
3. **Carbon credit generation**—quantify emission reductions and sell the credits in voluntary carbon markets

5.7.1.2 The circular economy transformation

5.7.1.2.1 Product redesign

Green Loop redesigned 40% of its packaging portfolio:

Impact:

- 3500 t of virgin plastic reduced
- 9800 t CO_2 avoided annually

Old design	New circular design
Multi-layer plastic difficult to recycle	Mono-material recycle-friendly bottles
High-carbon virgin raw material	45% recycled plastic inclusion
No modularity	Detachable caps, refillable pouches

5.7.1.2.2 Establishing reverse logistics network

The company partnered with:

- 8 waste aggregators
- 3 municipal corporations
- 400 scrap dealers
- 65 collection kiosks in malls

They launched a "Return Your Bottle" campaign with ₹5 cashback per bottle.

Collected waste:

- Pre-project: **1200 t/year**
- After reverse logistics: **9600 t/year**

Approximately **68%** was recycled back into new bottles.

5.7.1.2.3 Renewable energy transition

Green loop installed:

- 2.5 MW solar rooftop
- 1 MW wind energy PPA

Energy shift results:

Annual reduction: 11,200 t of CO_2

Source	Before	After
Fossil fuel	82%	32%
Renewable	18%	68%

Challenge	Description	Solution
High initial investment	~₹40 crores	Blended finance + green bonds
Lack of segregation at source	Poor household waste sorting	Consumer awareness + aggregator incentives
Verifying carbon credits	Complex certification	Blockchain tracking + third-party auditors
Ensuring community participation	Farmers initially hesitant	Profit-sharing model for carbon credits

Energy source	Before	After
Coal and diesel	76%	18%
Renewable	24%	82%

5.7.1.2.4 Nature-based carbon sequestration project

The company invested in a 300-acre agroforestry + bamboo plantation in Satara under a community partnership.

- Expected sequestration: 7500 t CO_2e/year
- The community receives 40% of carbon revenue

5.7.1.3 Carbon credit generation mechanism

The company registered its project under the Verified Carbon Standard (VCS).
Projects Verified.

1. **Plastic waste recycling project**
- Formula: CO_2e saved per tonne of recycled plastic = 2.6 t CO_2e
    - Annual recycling: 9600 t

Carbon credits: 9600 × 2.6 = 24,960 credits.

2. **Renewable Energy Transition Project**
- Emission savings = 11,200 credits

3. **Agroforestry sequestration**
- Sequestration = 7500 credits

Total Carbon Credits Earned Annually **Total:** 43,660 credits.

5.7.1.4 Financial outcomes (carbon + circularity revenue)

Revenue from carbon credits.
Market rate (voluntary market): $7 per credit Conversion: ₹85 per USD
Annual carbon revenue: 43,660 × 7 × 85 = ₹25.9 Crores
Operational savings from circularity

- Virgin plastic reduction: ₹6.8 crores saved
- Energy cost reduction: ₹4.1 crores saved
- Waste disposal savings: ₹1.4 crores saved

Total circularity savings: ₹12.3 crores/year.
Marketing and brand value impact

- 3 major clients signed long-term contracts
- Product labelled as "Circular and Carbon Neutral Packaging"
- 35% increase in customer trust score
- Awarded India's "Green Manufacturing Excellence Award 2026"

Source	Savings
Raw material substitution	₹8.3 crores
Waste disposal elimination	₹1.2 crores
Energy savings	₹2.9 crores
Reverse logistics optimization	₹1.1 crores
Total circular savings	**₹13.5 crores**

5.7.1.5 *Challenges faced*

Challenge	Description	Solution
Seasonal fluctuations in crop residue	Inconsistent supply to factory	Multi-crop sourcing model
High cost of technology	Blockchain, IoT, AI expensive	Green innovation grant + CSR partnerships
Farmer awareness gap	Lack of knowledge on benefits	Capacity-building workshops
Ensuring credibility of carbon credits	MRV complexity	Third-party verification + digital infrastructure

5.7.1.6 *Results after 2 years*

Environmental impact

- 39% total carbon reduction
- 68% packaging circularity achieved
- 300-acre landscape restoration

Economic impact

- Total financial gain: ₹38.2 crores/year
- ROI achieved in 2.3 years

Social impact

- 1200 livelihood opportunities created
- Community carbon share: ₹6.2 crores
- Increased rural resilience and income stability

5.7.1.7 *Strategic lessons learned*

1. Circular economy + carbon credits = maximum synergy

A circular system automatically reduces carbon intensity, making it easier to generate credits.

2. Reverse logistics is the backbone of circularity

Strong collection networks ensure material availability and reduce emissions.

3. Technology is a critical enabler

- Digital twins track carbon flow
  - Blockchain ensures transparent credit issuance
  - AI optimizes waste routing and carbon forecasting

4. **Community participation ensures long-term success**

   Shared ownership creates motivation and sustainability.

5. **Carbon credits create a new revenue model**

   Circularity is no longer a cost—it becomes a profit center.

5.7.2 Case study 2: Circular economy with carbon credits

5.7.2.1 Background

XYZW is a fast-growing manufacturer of biodegradable and compostable packaging materials. Their key clients include organic food brands, dairy processors, and e-commerce companies transitioning toward sustainable packaging.

However, despite producing eco-friendly packaging, the company's operations and supply chain were not circular.

- Agricultural residues (raw materials) came from scattered farms with high transport emissions.
- Production waste (bio-fillers, starch residues) was dumped.
- Energy use depended 76% on fossil fuels.
- No mechanism existed to track carbon footprint or circularity.

5.7.2.2 Drivers for transformation

External drivers

- New Indian EPR regulations for packaging
- International buyers asking for carbon-neutral packaging
- Investors demanding environmental performance report
- Opportunities for revenue through voluntary carbon markets

Internal drivers

- Increasing raw material prices
- High waste disposal costs (₹1.2 crores annually)
- Long-term sustainability vision of founders

5.7.2.3 Circular economy strategy implemented

XYZW adopted a closed-loop value chain based on three circular pillars.

5.7.2.3.1 Biomass circularity: Turning waste into raw material

AgroPack's packaging is made from:

- Wheat straw
- Sugarcane bagasse

- Corn husk
- Paddy residues

Earlier these residues were burnt, generating CO_2 and air pollution. XYZW partnered with 3500 farmers across different cities.

Circular Model:

Farm Waste → Aggregation → Bio-Pulp → Packaging Sheets → Compostable products → Soil Enrichment.

Outcomes:

- 42,000 t biomass rescued from burning
- Avoided emissions = 87,360 t of CO_2e/year
- Farmers earn ₹4500–₹5200 per tonne of biomass
- Soil enriched using composted packaging waste

5.7.2.3.2 Zero-waste manufacturing system

Earlier:

- 18% of raw material turned into rejects
- 6 t/day waste dumped

Circular shift included:

- Onsite bio-digesters
- Reuse of sludge as filler
- Internal recycling loop for starch waste
- Zero-waste certification under ISO 14051

Impact:

- Waste reduced by 95%
- 3200 t of internal material looped back
- 6800 t CO_2e saved annually

5.7.2.3.3 Renewable energy transition

AgroPack installed:

- 1.8 MW ground-mounted solar
- Biogas plant using biomass rejects

Energy profile changed dramatically:
Annual emission reduction from renewable energy: 12,400 t CO_2e/year.

5.7.2.3.4 Compost return program (CRP)

AgroPack launched a compost-back initiative:

- Consumers return used packaging
- Packaging is processed into nutrient-rich compost
- Compost returned to the same farms supplying biomass

Closed loop benefits:

- Supports regenerative farming
- Enhances soil carbon content
- Strengthens farmer-industry-community ecosystem

Soil carbon sequestration achieved: 5900 t CO_2e/year

5.7.2.4 Carbon credit generation mechanisms

AgroPack registered three project categories:

1. **Avoided burning of crop residue**
- Formula: 2.08 t CO_2e avoided per tonne biomass
- Biomass recovered: 42,000 t
- Carbon Credits: 87,360 credits/year

2. **Renewable energy transition**
- Carbon credits: 12,400 credits/year

3. **Soil carbon sequestration program**
- Sequestration: 5900 credits/year

Total credits per year = 105,660 Carbon Credits per Year
Market price = \$6.5 per credit Indian exchange rate = ₹83/USD
Carbon revenue:

$$105{,}660 \times 6.5 \times 83 = ₹57.1 \text{ Crores/year}$$

5.7.2.5 Financial impact of circular economy

A. Cost savings (Annual)
B. New revenue streams
- Carbon credits: ₹57.1 crores
- Compost sales: ₹3.8 crores
- Premium pricing (certified carbon-neutral packaging): ₹6.4 crores

Total new revenue: ₹67.3 crores/year.

5.7.2.6 Stakeholder impact

Farmers

- Income increased by 18%–25%
- No residue burning → reduced health issues
- Better soil fertility through compost return

Consumers

- Sustainable packaging with a traceable carbon footprint
- QR code label shows emissions avoided per product

Investors

- ESG score improved by 44%

- Made the company eligible for low-interest green funding

Community

- Village-level biomass aggregation centers created 2400 jobs
- Air quality improved due to avoided crop burning

5.7.2.7 Technology enablers

XYZW deployed.

Technology	Purpose
Blockchain ledger	Track biomass origin, waste loops, and carbon credits
AI logistics optimizer	Reduce fuel consumption in biomass transportation
IoT sensors	Monitor renewable energy production, compost quality
Digital twin for circular flow	Simulate carbon savings for design changes

5.7.2.8 Challenges faced

Challenge	Description	Solution
Seasonal fluctuations in crop residue	Inconsistent supply to factory	Multi-crop sourcing model
High cost of technology	Blockchain, IoT, AI expensive	Green innovation grant + CSR partnerships
Farmer awareness gap	Lack of knowledge on benefits	Capacity-building workshops
Ensuring credibility of carbon credits	MRV complexity	Third-party verification + digital infrastructure

5.7.2.9 Final outcomes after 2 years

Environmental

- 105,660 t CO_2e mitigated annually
- 100% zero-waste certified manufacturing
- 42,000 t of crop waste diverted from burning

Economic

- Net annual gains: ₹80.8 crores
- Payback period of investment: 1.7 years
- 4× increase in global export orders

Social

- 2400 rural jobs

- Improved soil fertility
- Women participation increased in composting units

5.8 Exercise

1. **Explain the relationship between the Circular Economy and Carbon Credit systems.**
   How do these two concepts support global decarbonization goals?
2. **Differentiate between Compliance Carbon Credits and Voluntary Carbon Credits.**
   Discuss their governance, verification standards, and market drivers.
3. **Describe the major sources of carbon credits such as renewable energy, methane capture, soil sequestration, and afforestation.**
   Explain how verification challenges differ across sources.
4. **What is the Integrated Circular–Carbon Value Framework (ICCVF)?** Describe its six layers and how they collectively support carbon-negative economic systems.
5. **Discuss the role of AI, IoT, Digital Product Passports, and Blockchain** in enabling circular carbon systems. How do these technologies improve monitoring, verification, and transparency?
6. **Explain Nature-Based Circular Sinks (NBCS)** and provide examples of how they contribute to long-term carbon storage.
7. **Describe Technological Sequestration approaches** such as Direct Air Capture (DAC), CCS from waste-to-energy plants, and CO_2 mineralization.
8. **Analyze the Carbon Credit Value and Governance Framework.** How do the Emission Source Layer, MRV Layer, and Governance Layer ensure credibility in carbon markets?
9. **Explain how composting, recycling, anaerobic digestion, and biomass energy generate carbon credits.** Include the benefits and limitations of each.
10. **Describe the significance of circular design principles** in reducing lifecycle emissions and enabling carbon credit generation.
11. Explain how leakage, buffer, and uncertainty deductions affect carbon credit issuance.
12. Explain the step-by-step calculation of annual avoided emissions and net carbon credits.
13. **Compare Natural and Technological Sequestration** across criteria such as cost, scalability, permanence, and monitoring complexity.
14. **Discuss the importance of reverse logistics and circular supply chains** in achieving carbon reduction under the circular economy framework.
15. **Explain how Circular Economy Strategies such as product life extension, resource efficiency, biochar systems, and renewable energy integration reduce carbon emissions.**

5.9 Objective

1. **One carbon credit represents the reduction of:**
   a) 1 kg CO_2
   b) 1 metric ton CO_2 equivalent
   c) 10 kg CO_2
   d) 100 kg CO_2

2. **Which standard certifies voluntary carbon credits?**
   a) ISO 9001
   b) Verra (VCS)
   c) GATT
   d) Basel III

3. **Which of the following is NOT a source of carbon credits?**
   a) Afforestation
   b) Solar energy
   c) Mining activities
   d) Methane capture

4. **What does DAC stand for?**
   a) Digital Access Control
   b) Direct Air Capture
   c) Distributed Algorithm Chain
   d) Direct Agricultural Carbon

5. **Methane has a global warming potential approximately:**
   a) 5 times CO_2
   b) 10 times CO_2
   c) 28–34 times CO_2
   d) 100 times CO_2

6. **In ICCVF, which technology ensures transparent carbon tracking?**
   a) Digital Passports
   b) Blockchain Carbon Ledger
   c) SAP ERP
   d) RFID

7. **Which ecosystem stores the highest soil carbon?**
   a) Grasslands
   b) Peatlands
   c) Deserts
   d) Coral reefs

8. Which circular activity avoids methane generation most directly?
a) Refurbishment
b) Composting
c) Solar power
d) Biochar production

9. Which of the following is an enabler of carbon credit integrity?
a) Community Benefit Sharing
b) Blockchain Audits
c) Third-party MRV
d) All of the above

10. CCS from Waste-to-Energy plants primarily captures CO_2 from:
a) Solar farms
b) Municipal waste incineration
c) Hydropower
d) Nuclear plants

11. Which of the following technologies is most associated with real-time emission tracking?
a) Blockchain
b) IoT Sensors
c) Biochar Kilns
d) Crop Rotation

12. Carbon mineralization converts CO_2 into:
a) Liquid fuels
b) Organic fertilizers
c) Stable carbonates
d) Hydrogen gas

13. Which circular activity directly contributes to soil carbon enhancement?
a) Product leasing
b) Composting
c) Shared mobility
d) Carbon trading

14. The primary purpose of the Governance and Integrity Layer in carbon markets is to:
a) Increase electricity production
b) Enforce transparency and prevent fraud
c) Promote fossil-fuel subsidies
d) Reduce community engagement

15. In the reforestation carbon credit example, which factor accounts for uncertainties in measurement tools and methodology?

a) Leakage

b) Permanence buffer

c) Uncertainty discount

d) Market adjustment rate

Further reading

Anand, A., Kumar, V., Kaushal, P., 2025. Techno-economic and emission impact evaluation of crop residues-biochar system for carbon sequestration. Bioresource Technology Reports 30.

Angelopoulou, T., Balafoutis, A.T., Chabrillat, S., 2025. Earth observation technologies for agricultural carbon credits: A review. Smart Agricultural Technology 12.

Arcusa, S.H., Lackner, K.S., 2025. Carbon sequestration ought to be permanent on climate-relevant timescales. Environmental Science & Policy 173.

Chen, W.-H., Biswas, P.P., Chang, J.-S., 2025. Achieving carbon credits through biomass torrefaction and hydrothermal carbonization: A review. Renewable and Sustainable Energy Reviews 208.

da Silva, J.C.G., Rodrigues, L.A.M., Mockaitis, G., 2025. From semi-arid biomass to carbon sequestration: Modeling the impact of pyrolysis atmosphere and soil temperature on the carbon removal potential of Agave-based biochar. Biomass and Bioenergy 203.

Giordano, M., Miosi, P.G., Maucieri, C., 2025. Short-term soil carbon accumulation following sustainable practices: A life Cycle Assessment and carbon credit analysis. Italian Journal of Agronomy 20.

Gorain, S., Dutta, S., Das, S., 2025. Harnessing green wealth: A two-decade global assessment of forest carbon sequestration and credits and the economic implications of sustainable forest management practices. Journal of Environmental Management 393.

Hitch, M., Li, J., 2025. Developing a verification framework for carbon sequestration through mineral carbonation of mine tailings: An Australian context. The Extractive Industries and Society 23.

Jia, X., Song, P., Lei, Y., 2025. Managing urban parks to unlock carbon sequestration potential: Active LiDAR and machine learning approach. Urban Forestry and Urban Greening 112.

Joshi, R.K., Garkoti, S.C., 2025. Quantifying ecosystem carbon storage, tree diversity, and carbon credit potential in the urban semi-arid forests of Delhi, India. Environmental Development 56.

Modolo, G.S., Silva, T.V.M., Ferreira, M.J., 2025. Enrichment planting to restore secondary forests in Central Amazon: A perspective on timber production and carbon sequestration. Forest Ecology and Management 594.

Ruhlman, E., Alam, S.S., Depcik, C., 2025. Estimation of atmospheric carbon dioxide sequestration by soybean crops for an updated pathway emissions model of soybean biodiesel. Journal of Cleaner Production 534.

Saikia, N., Pooniya, V., Goswami, K., 2026. Two decades of organic farming in rice-wheat agroecosystems: An impact assessment of environmental footprints reduction and carbon credit potential for climate resilience. European Journal of Agronomy 173.

Siddesh, R., Pathak, H., Reddy, K.S., 2026. Estimating carbon sequestration in orchards of semi-arid tropics: A modeling approach. Biomass and Bioenergy 206.

Subbiah, G., Singh, R.P., Kamakshi Priya, K., 2025. Continuous pyrolysis of rice husk for sustainable biochar production and carbon sequestration: Recent advances and techno-economic perspectives. Results in Engineering 27.

Waheed, M., Haq, S.M., Iqbal, M., 2026. Carbon sequestration and credit potential of native and non-native monoculture plantations in a subtropical irrigated forest. Biomass and Bioenergy 205.

Wheeler, C.E., Begliomini, F., Coomes, D., 2026. The path to robust evaluation of carbon credits generated by forest restoration and REDD+ projects. Remote Sensing of Environment 332.

Xu, W., Liu, W., Wu, Z., 2026. Rubber-based agroforestry systems enhancing soil carbon sequestration through improved soil aggregate stability. Soil and Tillage Research 257.

Yadav, V., Malanson, G.P., Lant, C., 2009. Modeling watershed-scale sequestration of soil organic carbon for carbon credit programs. Applied Geography 29.

Zhang, P., Fu, Y., Al-Hussein, M., 2025. Quantifying operational-phase carbon sequestration in urban concrete buildings: A spatiotemporal analysis of Qingdao City. Energy and Buildings 349.

Zhou, Q., Wu, D., Wu, W., 2025. Carbon sequestration potential of biochar-modified building materials: A critical review. Journal of Building Engineering 117.

Zhu, X., Long, Y., Zhang, H., 2025. Impact of carbon policies on forest carbon sequestration supply at intensive and extensive margins. Forest Policy and Economics 181.

Chapter 6

Policy, economics, and business models for circular clean energy

6.1 Policy frameworks supporting the circular economy

The circular economy has emerged as a transformative model capable of redefining how societies create, consume, and regenerate resources. Unlike the traditional linear "take–make–dispose" system, the circular economy promotes restorative and regenerative processes that keep materials circulating within the value chain for as long as possible. In the global transition toward sustainability, policy frameworks have become the backbone enabling nations, industries, and communities to shift from linear to circular systems. Effective policy mechanisms guide strategic implementation, offer incentives, standardize practices, and stimulate innovation, thereby accelerating the adoption of circular principles across markets. As climate change, resource scarcity, and waste mismanagement intensify, circular economy policies serve as essential tools for achieving environmental resilience, economic competitiveness, and social well-being. At the heart of circular economy policy frameworks lies the objective of reducing resource extraction and waste generation through systemic interventions. National-level policies typically begin by defining clear targets and regulatory mandates. For instance, many countries have adopted extended producer responsibility (EPR) regulations that oblige manufacturers to manage the end-of-life stage of their products. By holding producers accountable for waste management, EPR encourages product redesign, greater recyclability, modularity, and the use of secondary materials. Such policies not only reduce pressure on landfills but also incentivize businesses to integrate circularity into core operations. Waste hierarchy laws further strengthen this shift by prioritizing prevention, reuse, recycling, and recovery, ensuring that disposal becomes the least preferred option.

The circular economy represents a transformative shift from traditional linear economic practices toward systems designed to eliminate waste, regenerate natural ecosystems, and optimize resource cycles. To support this

The Circular Economy and their Management. https://doi.org/10.1016/B978-0-443-44617-7.00006-1

transition, policy frameworks play a foundational role by establishing regulations, incentives, standards, and collaborative structures that guide governments, industries, and citizens toward circular practices. The infographic illustrates this comprehensive ecosystem through ten interconnected pillars, each representing a dimension of policy support essential for a thriving circular economy. The following paragraphs provide a detailed explanation of each pillar and their interrelationships.

The first pillar, **Vision and strategic alignment**, highlights the importance of establishing a clear national circular economy (CE) vision. This vision acts as the backbone for all policy actions, ensuring coherence across government departments and alignment with global agendas such as the Sustainable Development Goals (SDGs). Countries adopting circularity embed these principles into climate action plans, industrial strategies, and resource-efficiency frameworks. Sector-wise development roadmaps further help industries such as manufacturing, agriculture, and construction to adopt specific circular principles tailored to their unique needs. This strategic alignment ensures that circularity is not viewed as a stand-alone goal but an integrated approach that uplifts economic growth, environmental protection, and social well-being. Fig. 6.1 illustrate policy framework for circular economy.

The second pillar, **Regulatory mechanisms**, serves as the most direct instrument for shaping circular behavior. Within this, Extended Producer Responsibility (EPR) forms a cornerstone of circular policy. EPR mandates that producers take responsibility for the entire lifecycle of their products, especially at the post-consumer stage. Through mandatory take-back schemes and recycling targets, industries are motivated to design products that are recyclable, durable, and modular. Waste hierarchy legislation further strengthens regulatory efforts by legally prioritizing waste prevention, reuse, and recycling over landfilling and incineration. Eco-design regulations, another critical component, require manufacturers to incorporate longevity, reparability, and material efficiency into product design. Sector-specific regulations such as plastic bans, e-waste directives, and textile recycling rules address the nuanced needs of different industries, ensuring comprehensive regulatory coverage.

The third pillar focuses on Economic Instruments and Incentives, which help create market conditions that reward circular practices. Governments often introduce fiscal incentives such as tax rebates for companies that use recycled materials, subsidies for recycling facilities, and reduced VAT on repair services to encourage sustainable choices. Market creation tools also play a vital role especially Green Public Procurement (GPP), where governments commit to purchasing goods that meet circular criteria like durability, reparability, and recycled content. This not only sets an example but also stimulates demand in the private sector. Financing mechanisms such as green bonds, concessional loans, and innovation grants further empower businesses especially small and medium enterprises (SMEs) to invest in circular infrastructure, research, and technology.

FIGURE 6.1 Policy framework for circular economy.

The fourth pillar emphasizes Digital and Technological Enablers, acknowledging that modern circular systems require advanced technological support. Digital Product Passports (DPPs) are emerging as powerful tools that track materials throughout a product's lifecycle, ensuring transparency about origin, composition, repair history, and recyclability. Blockchain enhances verification and creates trust in areas such as carbon credit trading and recycled material certification. Technologies like AI and IoT optimize waste sorting, monitor real-time resource use, and support predictive maintenance, all of which improve efficiency within circular systems. By embedding digital innovation into policy frameworks, governments enable transparency, accountability, and efficiency in material flows.

Next, the fifth pillar Infrastructure and System Support highlights the importance of building physical and systemic structures that facilitate circular operations. Repair hubs, resource recovery centers, and material exchange platforms make it easier for businesses and consumers to participate in reuse and recycling. Urban circular systems, such as decentralized composting units, recycling clusters, and community repair centers, embed circularity into everyday life. Industrial symbiosis zones allow waste from one company to serve as raw material for another, creating industrial ecosystems that minimize waste and maximize value. Policies that support such infrastructure ensure that circular practices are accessible, scalable, and regionally customized.

The sixth pillar, Capacity Building, Education, and Awareness, underscores the need for knowledge and cultural change. Even the strongest regulations are ineffective without awareness and skills. Governments often introduce training programs for industries and municipal officials to strengthen their capability to adopt circular solutions. Certification programs and academic courses build long-term expertise in circular design, waste management, and sustainable innovation. Public awareness campaigns encourage behavioural changes such as repair over replacement, segregating waste, and purchasing durable goods. Through education, circularity becomes a cultural norm rather than an imposed regulation.

The seventh pillar addresses Monitoring, Metrics, and Compliance, which form the backbone of accountability. To measure progress, governments must develop standardized indicators for material circularity, carbon reduction, waste minimization, and resource efficiency. Annual circular economy reporting, environmental audits, and third-party verification ensure that industries remain compliant and transparent. National databases and material flow analysis tools help track progress at regional and sectoral levels. This data-driven approach supports informed policymaking, ensures regulatory compliance, and guides future interventions.

The eighth pillar highlights Multi-Stakeholder Collaboration, recognizing that circular transformation requires the involvement of government bodies, private companies, civil society, academia, and consumers. Public-private partnerships accelerate innovations and bring together diverse capabilities.

Civil society involvement empowers communities, enhances social acceptance, and strengthens grassroots innovation. International cooperation supports harmonized standards, cross-border waste management, and global learning.

The ninth pillar, Innovation and Research Promotion, encourages the development of new technologies, materials, and business models. Governments invest in R&D centers, pilot projects, innovation labs, and start-up incubators to explore cutting-edge solutions such as bioplastics, regenerative materials, advanced recycling, and circular digital platforms. Innovation sandboxes allow companies to experiment with circular solutions in a controlled regulatory environment. These initiatives foster an ecosystem where creativity and experimentation thrive. Table 6.1 illustrate policy frameworks supporting the circular economy.

Finally, the 10th pillar International Linkages and Trade Policies ensures alignment with global standards and cross-border cooperation. Harmonized regulations for recycled materials, safe waste movement, and circular product certification help create international markets for circular goods. Trade incentives encourage companies to adopt circular models, while participation in international treaties strengthens environmental governance. Together, these ten pillars create a holistic, integrated, and forward-looking framework essential for nurturing a circular economy. By combining regulation, innovation, infrastructure, technology, and collaboration, policy frameworks shape a resilient and sustainable future grounded in efficiency, regeneration, and responsible stewardship of resources.

6.2 Business models for circular economy

The circular economy represents a paradigm shift from traditional linear economic systems toward regenerative, restorative, and resource-efficient models of production and consumption. Unlike the "take–make–dispose" cycle that dominates industrial activity, circularity redefines value creation through restorative design, prolonged product life, and closed-loop material flows. As global industries face increasing pressure to reduce environmental footprints, improve resource efficiency, and remain competitive in a volatile resource-constrained world, the development of innovative business models becomes not only an opportunity but a necessity. The framework presented here provides a comprehensive and multi-layered approach to understanding how businesses can successfully transition to circularity. It encompasses six core layers: value regeneration, digital intelligence, customer experience, stakeholder collaboration, financial enablement, and regenerative impact. Each layer contributes to the formation of a holistic, future-ready system in which environmental and economic value reinforce each other.

TABLE 6.1 Policy frameworks supporting the ircular economy.

Pillar/Component	Key elements	Purpose/Impact
1. Vision and strategic alignment	• National CE vision • Alignment with SDGs and climate goals • Sector-wise CE roadmaps	Establishes long-term direction; ensures national and sectoral coherence in CE adoption.
2. Regulatory mechanisms	Extended producer responsibility (EPR) • Take-back rules • Recycling targets waste hierarchy regulations • Prioritization: prevent → reuse → recycle → recover eco-design standards • Durability and reparability rules • Material efficiency norms sector-specific rules (plastic, e-waste, textiles, construction)	Creates legal obligations that enforce circular behavior, reduce waste, and promote responsible product design.
3. Economic instruments and incentives	• Tax rebates for recycling and repair • Subsidies for circular manufacturing • Reduced VAT on repair services • Green public procurement (GPP) • Green bonds and concessional loans	Encourages financial attractiveness of circular practices; boosts market demand for circular products.
4. Digital and technological enablers	• Digital product passports (DPP) • Blockchain for traceability • AI and IoT for resource monitoring • Automated waste sorting technologies	Improves transparency, traceability, accountability, and efficiency within the CE ecosystem.
5. Infrastructure and system support	• Resource recovery centers • Repair and refurbishment hubs • Reverse logistics networks • Industrial symbiosis parks • Urban circular systems (decentralized compost, recycling clusters)	Builds physical support systems that enable widespread adoption of circular activities.

6. Capacity building, education and awareness	• Training for industries and government • Circular economy curriculum • Consumer awareness campaigns • Certification programs	Strengthens skills, knowledge, and cultural acceptance of circular practices at all societal levels.
7. Monitoring, metrics and compliance	• Standardized circular indicators • Waste and material flow metrics • Mandatory CE reporting • Third-party audits • National CE databases	Ensures accountability, tracks progress, and provides data-driven insights for future policies.
8. Multi-stakeholder collaboration	• Public-private partnerships • Inter-ministerial coordination • Civil society and community involvement • Academia-industry collaboration	Promotes collective ownership and shared responsibility; accelerates innovation and implementation.
9. Innovation and research promotion	• R&D funding • Circular innovation labs • Start-up incubation • Pilot projects • Innovation sandboxes	Supports development of new materials, business models, and technologies that enable circularity.
10. International linkages and trade policies	• Harmonized global standards • Cross-border waste management rules • Trade benefits for circular products • Participation in international treaties	Ensures global cooperation. facilitates international circular markets, and standardizes CE norms.

6.2.1 Value regeneration layer (core circular pillars)

At the foundation of the circular economy lies the value regeneration layer, which describes how products, materials, and components circulate through multiple lifecycle loops. This layer integrates five fundamental models: product life extension, resource recovery, product-as-a-service, sharing economy, and circular supply chains. Together, these models ensure that products retain value far beyond the conventional linear lifecycle.

Product life extension model: The product life extension model focuses on prolonging the functional lifespan of products through processes such as repair, refurbishment, remanufacturing, and modular upgrades. Traditional products are designed for short lifespans, often leading to premature obsolescence. Circular design principles, however, emphasize durability, maintainability, and reparability. Companies integrating this model shift their business from selling more products to maintaining, upgrading, and extending the life of existing ones. Economically, this reduces the need for new raw materials and increases long-term customer relationships. Environmentally, it prevents excessive waste generation and decreases energy consumption associated with manufacturing.

Resource recovery model: Resource recovery involves reclaiming valuable materials at the end-of-life stage of a product. This model is particularly crucial in energy-intensive and raw-material-dependent sectors such as electronics, automotive, construction, and textiles. Through advanced recycling technologies, companies extract metals, plastics, minerals, and biomaterials, reintroducing them into the production cycle. Industrial symbiosis where waste from one industry becomes the input for another further enhances this model's efficacy. By adopting resource recovery, organizations reduce dependency on virgin materials, create new revenue through secondary materials, and minimize environmental impacts.

Product-as-a-service (PaaS) model: PaaS represents one of the most transformative models in the circular economy. Here, customers pay for the utility of a product rather than its ownership. Examples include leasing furniture, renting electronics, subscription-based mobility, and pay-per-use machinery. Through this model, companies maintain ownership and responsibility for the product throughout its lifecycle. This incentivizes businesses to design long-lasting, high-quality, easily maintainable products. Financially, PaaS offers predictable revenue streams, stronger customer retention, and reduced lifecycle costs.

Sharing and access economy model: The sharing economy optimizes underused resources by providing shared access instead of individual ownership. Digital platforms enable people to share vehicles, tools, equipment, spaces, and even household appliances. This model reduces overall consumption, lowers environmental impact, and increases social connectivity. Businesses benefit by turning idle assets into profitable ones and reducing the need for manufacturing new products.

Circular supply chain model: Circular supply chains integrate renewable, recyclable, or biodegradable materials to minimize resource extraction. They optimize logistics to reduce emissions and incorporate circular design principles across sourcing, production, distribution, and reverse logistics. By localizing supply chains, companies reduce transportation costs and strengthen resilience against global disruptions. Circular supply chains ensure that materials continuously circulate between production and re-production systems, ultimately creating a closed-loop ecosystem.

6.2.2 Digital intelligence layer (tech-enabled circularity)

Circular business models are increasingly powered by digital technologies that optimize resource flows, enhance traceability, and improve product lifecycle management. The digital intelligence layer includes digital product passports, AI systems, blockchain platforms, and IoT-enabled reverse logistics.

Digital product passports: A digital product passport is a data-rich identity profile embedded within a product. It contains information on the materials used, production process, sourcing, carbon footprint, repair history, and recyclability. DPPs enhance transparency, authenticity, and circularity by informing consumers, recyclers, and suppliers about how to maintain and recover value from products. They also play a pivotal role in enabling automated sorting and recycling systems, ensuring that products enter the correct recovery streams at end-of-life.

AI-driven material flow optimization: Artificial intelligence supports circular models by predicting maintenance needs, optimizing repair schedules, identifying material recovery opportunities, and forecasting demand for recycled materials. AI-powered analytics optimize reverse logistics through route planning, ensuring timely collection of returned or used products. In manufacturing, AI improves design efficiency and minimizes waste. It also helps identify new circular opportunities by analyzing consumption patterns and material flows across systems.

Blockchain for circular trust: Blockchain technology supports transparent, tamper-proof tracking of materials throughout their lifecycle. Through decentralized ledgers, companies can verify the origin, quality, and quantity of recycled materials. Blockchain enhances consumer trust in refurbished goods and supports regulatory compliance for materials like e-waste, minerals, and plastics. It also strengthens marketplaces for secondary raw materials by eliminating fraudulent claims and ensuring authenticity.

IoT-enabled smart reverse logistics: IoT technologies enable real-time monitoring of product usage, enabling predictive maintenance and automated return systems. Sensors in products or packaging support efficient collection and sorting. Smart bins equipped with IoT can detect the type of waste deposited and guide it to suitable recovery processes. IoT-based tracking for rental or shared products ensures reduced losses, optimal use, and efficient redistribution.

6.2.3 Customer experience and value creation layer

While circular models focus heavily on resource management, customer experience is equally essential. Businesses must ensure that circular products are not only environmentally beneficial but also appealing, accessible, and high-value to consumers. This layer includes loyalty ecosystems, transparency models, experience-based offerings, and premium refurbished markets.

Circular loyalty ecosystems: Loyalty programs reward consumers for adopting circular behaviours such as returning products for recycling, choosing refurbished items, participating in repair programs, or using rental and leasing systems. Such incentives build long-term customer engagement, increase circular participation rates, and create a positive brand image.

Green branding and transparency models: Modern consumers increasingly prefer sustainable brands. Transparency becomes a competitive advantage when companies publicly disclose environmental impacts, circular achievements, and material choices. Certifications, digital labels, and sustainability dashboards strengthen trust and differentiate brands in competitive markets.

Experience-based value delivery: Instead of merely selling products, circular businesses offer experiences repair cafés, upcycling workshops, product customization, and sustainability education. These experiences promote responsible consumption and foster deeper emotional connections with customers. Experiential value also supports premium pricing and customer loyalty.

Refurbished premiumization model: Refurbished products are no longer seen as inferior alternatives. Companies now position them as premium, eco-friendly, and technologically enhanced options. Certifications, warranties, and high-quality finishing strengthen the perception of refurbished goods. This model expands markets for items such as electronics, appliances, furniture, and automobiles.

6.2.4 Stakeholder collaboration layer

Circularity cannot be achieved in isolation. The stakeholder collaboration layer highlights the need for coordinated action among businesses, governments, consumers, NGOs, and local communities.

Industry symbiosis networks: Industry symbiosis allows companies to exchange materials, energy, and by-products. For example, waste heat from one factory may power another facility. Such networks reduce costs, minimize waste, and create regional circular hubs. Government-led industrial clusters further reinforce these networks.

Producer–consumer partnerships: Producers and consumers must collaborate to ensure that products are returned, repaired, or recycled effectively. Co-creation initiatives allow consumers to participate in product design, leading to longer-lasting and user-friendly goods. Return deposit schemes and take-back programs strengthen product recovery ecosystems.

Government–business model alignment: Government policies play a crucial role in enabling circular transitions. Extended Producer Responsibility (EPR), tax incentives, repair mandates, green public procurement, and waste regulations guide business behavior. Collaboration between businesses and policymakers ensures alignment with national sustainability goals.

NGO and community integration: NGOs and communities support circular models through awareness campaigns, local repair hubs, educational programs, and social recycling initiatives. Including informal waste workers in formal recycling chains promotes social equity and improves waste recovery systems.

6.2.5 Financial and economic enablement layer

Circular models require robust financial support to scale effectively. This layer includes green financing mechanisms, secondary material markets, incentives, and cost-saving approaches.

Circular investment and green financing: Green bonds, sustainability-linked loans, impact funds, and ESG financing support circular infrastructure, renewable sourcing, recycling facilities, and low-waste technologies. Investors increasingly prefer companies demonstrating circular business readiness, reducing financial risks and enabling long-term competitiveness.

Secondary material market platforms: Digital marketplaces for recycled materials enable companies to trade secondary raw materials with guaranteed quality and traceability. These platforms improve resource availability, reduce costs, and enhance circularity at the industrial level.

Incentive-based circular monetization: Governments and international bodies offer financial incentives through carbon credits, resource recovery credits, waste management subsidies, and circular procurement schemes. Such mechanisms accelerate business adoption of circular practices.

Cost-saving mechanisms: Circular models reduce raw material usage, energy consumption, waste disposal costs, and manufacturing expenses. Remanufactured and refurbished products often command high margins, creating profitable new markets.

6.2.6 Regenerative impact layer (long-term outcomes)

The final layer represents the long-term benefits created by circular business models across environmental, social, economic, and innovation dimensions.

Environmental regeneration: Circularity reduces pollution, minimizes waste, curtails carbon emissions, and encourages biodiversity restoration. By shifting to renewable materials and recycling systems, businesses significantly reduce their ecological footprint.

Social value creation: Circular initiatives generate green jobs, support marginalized communities, and expand skill development in repair, refurbishing, and recycling industries. Formalizing informal waste sectors contributes to social justice and economic inclusion.

Economic resilience: Circular economies are more resilient to resource shocks, supply chain disruptions, and market volatility. Localized production systems, diversified material sources, and reduced dependence on virgin resources contribute to long-term economic stability.

Innovation spillovers: Circularity fosters technological breakthroughs, new industries, cross-sector partnerships, and a culture of continuous improvement. Innovations in biomaterials, renewable energy, recycling technologies, and digital tools spill over into other sectors, accelerating sustainable development.

The innovative framework presented offers a holistic perspective on how businesses can shift toward circular models that regenerate value, leverage digital intelligence, enhance customer relationships, strengthen stakeholder collaboration, enable financial growth, and create long-term regenerative impacts. As global challenges intensify climate change, resource scarcity, pollution, and economic instability circular business models emerge as a viable pathway toward sustainable, resilient, and competitive futures.

6.3 Economic implications

The transition to a circular economy marks one of the most transformative economic shifts of the 21st century, reshaping how nations, industries, and organizations perceive value creation. Unlike the linear model based on extraction, production, consumption, and disposal the circular economy focuses on extending the lifecycle of products, optimizing resources, and minimizing waste through reuse, repair, refurbishment, recycling, and regenerative practices. These changes bring far-reaching economic implications that influence micro-level business decisions, national economic strategies, and global market dynamics. As countries widen their sustainability commitments, circular frameworks are becoming central drivers of innovation, competitiveness, and long-term economic resilience.

One of the most prominent economic benefits of the circular economy is the significant cost efficiency and resource optimization it enables. Industries traditionally dependent on virgin materials face rising extraction costs, supply chain disruptions, and volatile global commodity prices. Circular strategies such as using recycled materials, designing for durability, and recovering valuable components reduce the dependency on raw inputs and lower production expenses. Companies can realize long-term operational savings by prolonging product lifecycles, decreasing waste management fees, and reducing energy consumption associated with the production of new goods. For businesses operating in sectors like electronics, automotive, construction, and textiles, circular practices can cut costs by creating internal material loops, strengthening resource security, and minimizing exposure to global market instability. Fig. 6.2 illustrate economic implication in the circular economy.

Economic implications also arise from the emergence of entirely new business models and revenue streams. As the demand for sustainable consumption rises, companies are shifting from product ownership to service-based delivery systems. Subscription models, pay-per-use systems, leasing arrangements, and rental platforms redefine how consumers access goods. These circular models generate recurring revenue, improve customer retention, and reduce the financial risks associated with one-time sales. Additionally, remanufacturing, refurbishment, and second-life product industries offer new market opportunities, especially in electronics and machinery. The rise of reverse logistics, waste-to-resource enterprises, and recycling startups further expands the economic ecosystem, creating new pathways for entrepreneurship and innovation.

Innovation serves as a powerful economic engine within the circular transition. The need for environmentally responsible solutions pushes companies to invest in research and development related to eco-design, biomaterials, biodegradable packaging, modular components, and digital platforms for tracking materials. The integration of AI, IoT sensors, blockchain, and robotics into circular systems enhances efficiency, transparency, and material recovery rates. As a result, the innovation landscape expands, leading to increased intellectual property assets and new technological frontiers. This innovation-driven growth not only strengthens industrial competitiveness but also positions countries as leaders in the emerging global green economy.

Employment and skill development represent another substantial economic implication. The circular economy creates a variety of new "green jobs" across recycling, repair, reprocessing, remanufacturing, biomaterial production, and resource recovery sectors. These fields require specialized skills such as sustainable design, material innovation, advanced manufacturing, and supply chain analytics leading to the expansion of technical training and educational programs. The circular model also opens doors for micro-entrepreneurship in fields like repair services, upcycling, local recycling hubs, and community-based sharing initiatives. Moreover, countries with large informal waste sectors can formalize these roles, improving livelihoods and integrating marginalized workers into structured economic systems.

Market competitiveness is equally influenced by circular transitions. Companies adopting circular principles often enjoy improved brand reputation, attracting environmentally conscious consumers. The efficiency gains achieved through reduced waste and optimized material flows directly

FIGURE 6.2 Economic implication in the circular economy.

contribute to better financial performance. Businesses that integrate circularity across their value chains can outperform competitors by offering sustainable, durable, and economically attractive alternatives. In addition, circular strategies enhance supply chain resilience by lowering dependency on finite resources and mitigating the risk of supply disruptions caused by climate events, geopolitical tensions, or international trade fluctuations.

Financial markets increasingly recognize the economic value of circularity. Investors and financial institutions are promoting green finance tools such as sustainability-linked loans, green bonds, ESG-focused investment funds, and impact investment instruments to support circular projects. These financial incentives reduce the initial capital burden associated with establishing circular infrastructure and technologies. While the transition may require significant upfront investment particularly in redesigning production processes and developing reverse logistics systems the long-term returns include reduced operational costs, enhanced asset value, and decreased regulatory risks. As global financial markets shift toward ESG compliance, circular businesses benefit from better access to capital and lower investment risks.

The circular economy also produces powerful environmental–economic co-benefits. By reducing greenhouse gas emissions, energy consumption, and resource extraction, circular practices contribute to climate mitigation while generating economic returns. Countries integrating circular strategies can monetize environmental improvements through carbon credits, renewable energy certificates, and ecosystem service payments. Reductions in pollution and waste further decrease public expenditure on environmental remediation and healthcare. In this way, circular systems strengthen both environmental integrity and economic well-being, forming a mutually reinforcing cycle of sustainability.

At the macroeconomic level, circularity contributes to GDP growth by stimulating new industries, reducing dependence on imported resources, and expanding domestic resource recovery markets. Circular trade centered on refurbished products, recycled materials, biomaterials, and green technologies creates new export opportunities. Nations embracing circular models experience enhanced economic stability, as they are less affected by global resource price shocks and supply shortages. The circular economy also supports national strategies for sustainable development, aligning with global policy commitments such as the Paris Agreement and the United Nations Sustainable Development Goals.

Risk reduction and economic resilience are among the most compelling implications of circular transformation. Circular systems reduce vulnerability by diversifying material sources, decreasing reliance on finite resources, and enabling adaptive, closed-loop industry structures. This resilience is especially crucial in times of global crises such as pandemics, economic recessions, or disruptions in global trade when supply chains become uncertain. Circular models enhance business continuity by creating self-sustaining production cycles that depend less on external disruptions. Table 6.2 illustrate economic implications in the circular economy.

Finally, circular economies encourage social and economic inclusion. By enabling affordable refurbished products and creating accessible green jobs,

TABLE 6.2 Economic implications in the circular economy.

Dimension	Key components	Economic impacts
1. Cost efficiency and resource optimization	- Reduced use of virgin materials - Reuse, repair, recycling loops - Waste minimization	- Lower production costs - Reduced waste management expenses - Increased operational efficiency
2. New circular business models	- Product-as-a-Service (PaaS) - Leasing, sharing, renting - Refurbishment and remanufacturing industries	- New revenue streams - Higher customer retention - Growth of repair and second-life markets
3. Innovation and technology development	- Eco-design and biomaterials - Digital product passports - AI, IoT, blockchain in resource tracking	- Increased R&D investments - Competitive market advantage - Rise in patents and tech innovations
4. Employment and skill development	- Green jobs in recycling, repair, remanufacturing - Reverse logistics workforce - Technical and sustainability skills	- Boost in job creation - Entrepreneurship opportunities - Up-skilling and re-skilling of workforce
5. Industrial competitiveness	- Circular supply chains - Efficient material flows - Sustainable product design	- Stronger brand value - Increased resilience to global supply risks - Long-term market competitiveness
6. Financial and investment implications	- Green bonds and sustainability-linked loans - ESG and impact investing - Initial circular infrastructure costs	- Better access to green finance - Lower long-term financial risks - Increased investor confidence
7. Environmental–Economic Co-benefits	- Reduced GHG emissions - Less energy and resource consumption - Carbon credit generation	- Lower environmental remediation costs - New revenue from carbon markets - Strengthened resource security

8. Macroeconomic growth	- Expansion of circular industries - Circular exports and secondary material markets - Reduced resource import dependency	- GDP growth through green industries - National economic stability - Growth of domestic manufacturing
9. Risk reduction and resilience	- Diversified material sources - Localized circular production cycles - Crisis-resistant supply chains	- Lower exposure to global shocks - Improved business continuity - Enhanced long-term resilience
10. Social and economic inclusion	- Formalization of informal waste workers - Community repair centers - Affordable refurbished goods	- Income generation for marginalized groups - Reduced inequality - Strengthened local eccnomies

circular models uplift communities and reduce inequality. The formalization of informal waste workers, community repair centers, and local manufacturing cooperatives strengthens local economies and fosters shared prosperity. Through equitable access to sustainable goods and services, the circular economy promotes inclusive economic growth.

In summary, the economic implications of the circular economy are vast and transformative. By fostering cost efficiency, innovation, job creation, financial sustainability, environmental stewardship, and inclusive growth, the circular model offers a compelling pathway for building resilient and prosperous economies. As nations and industries accelerate circular transitions, they lay the foundation for a regenerative economic future driven by sustainability, opportunity, and long-term value.

6.4 Financing mechanism for circular economy initiatives

The transition from a linear economy to a circular economy (CE) represents a fundamental shift in the way we design, produce, consume, and dispose of products and services. Unlike the traditional "take-make-dispose" model, a circular economy focuses on resource efficiency, waste reduction, reuse, refurbishment, recycling, and regeneration. However, scaling circular economy solutions requires substantial investment and a multi-dimensional financing approach. The financial ecosystem supporting CE must align public policies, private sector investment, and innovative financing models to enable the systemic changes required. Financing mechanisms for circular economy initiatives include a diverse mix of public and private funding sources, green and sustainability-linked financial instruments, incentives, tax mechanisms, and innovative financing structures such as blended finance and impact investing. Effective financing enables the development of circular technologies, infrastructure, business models, and awareness programs. These financial tools not only reduce economic risk and improve the bankability of CE projects but also ensure long-term environmental, social, and economic returns. Fig. 6.3 illustrate integrated circular economy financing architecture.

6.4.1 Public finance and government support

Public funding plays a critical role in seeding and catalysing circular economy initiatives. Governmental financial mechanisms provide crucial support for early-stage projects, which often struggle to attract private capital due to perceived risks and uncertain returns.

Subsidies and grants: Subsidies and grants are among the most direct and powerful tools that governments use to support the transition toward a circular economy (CE). These financial aids reduce the burden of initial investment costs and encourage innovation and entrepreneurship in sustainable practices. For example, composting facilities, recycling plants, waste-to-energy units,

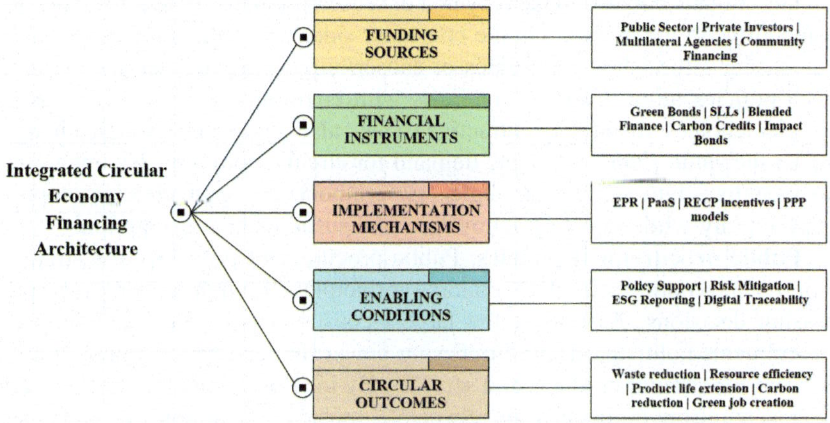

FIGURE 6.3 Integrated circular economy financing architecture.

and green startups often require significant capital to establish operations, purchase technology, and train staff. Such projects may not always be attractive to traditional investors due to long payback periods or untested business models. Here, government subsidies and grants can bridge the financial gap, de-risk projects, and provide the momentum needed to launch and sustain them. Subsidies can be in the form of reduced tariffs for sustainable equipment, land or utility subsidies, or operational incentives. Grants may be allocated for research and development, pilot projects, or community-led circular initiatives. Low-interest loans provided through public-sector banks or development financial institutions also enable long-term infrastructure development without causing a financial strain on small or emerging companies. These support mechanisms not only foster cleaner technologies and processes but also generate employment, build technical capacity, and strengthen the supply chain of circular industries. By prioritizing funding for CE initiatives, governments send a strong policy signal to private investors and society that sustainability is a national priority.

Tax incentives: Tax incentives serve as critical levers to nudge businesses and consumers toward more sustainable practices. In the context of the circular economy, these fiscal tools reward eco-friendly behavior and penalize unsustainable activities. For instance, companies that adopt eco-design where products are designed for durability, disassembly, and recyclability can benefit from reduced corporate tax or accelerated depreciation benefits. Likewise, firms engaging in reverse logistics, material recovery, or the use of secondary raw materials can receive tax deductions or exemptions. Governments may also offer VAT (Value Added Tax) reductions on goods made from recycled content or impose lower import duties on circular technologies. These incentives not only reduce operational costs for companies transitioning to circular models but also make recycled or reused products more affordable and competitive in the market. On the consumer side, tax credits may be given for purchasing refurbished electronics or subscribing to product-service systems. Such policies align economic interests with environmental goals by making circular choices financially attractive. They also drive behavioural changes across the value chain and create demand for circular solutions. By integrating tax incentives into national strategies, governments can scale circular adoption more rapidly while boosting innovation and sustainable consumption.

Public procurement policies: Public procurement policies are a strategic tool for promoting circular economy principles through government purchasing decisions. As one of the largest consumers of goods and services, governments hold significant purchasing power that can be leveraged to drive demand for circular products and services. By incorporating circularity criteria such as requiring products to be recyclable, reusable, repairable, or made from recycled materials into public tenders, governments set a precedent for sustainable market behavior. For instance, policies may mandate the purchase of refurbished computers in public schools, biodegradable packaging in government offices, or low-carbon construction materials in infrastructure projects. These requirements stimulate demand for circular products, thus encouraging businesses to invest in circular innovation and supply chain redesign. Furthermore, public procurement policies ensure a consistent market for circular goods, providing businesses with long-term certainty and scale. This, in turn, reduces costs and promotes economies of scale. In addition, these policies can include life-cycle costing to assess the total environmental and financial impact of products throughout their lifespan, rather than focusing solely on upfront prices. As a result, governments not only reduce their environmental footprint but also lead by example, encouraging private players to follow suit. In summary, public procurement policies act as a market-shaping mechanism, transforming purchasing habits across sectors and embedding sustainability into institutional frameworks. Table 6.3 illustrate financing mechanisms for circular economy initiatives.

Infrastructure investment: Infrastructure investment is foundational for enabling the systems and physical facilities required for the circular economy to function effectively. Circular practices such as recycling, composting, upcycling, and product remanufacturing cannot reach scale without appropriate infrastructure. This includes recycling hubs, urban mining centers (where valuable materials are extracted from electronic and industrial waste), waste segregation plants, repair stations, and smart logistics systems that facilitate material flow in reverse supply chains. Investment in such infrastructure often spearheaded by the government or in public-private partnerships creates the backbone for material circulation, resource efficiency, and waste minimization. For instance, smart collection and sorting centers powered by AI and robotics can improve recycling accuracy, reduce contamination, and lower operational costs. Similarly, digital platforms for asset sharing

TABLE 6.3 Financing mechanisms for circular economy initiatives.

Financing mechanism	Key property	Explanation	Relevance to circular economy (CE)
1. Green bonds	Use-of-proceeds transparency	Funds must be allocated only to certified green/circular projects, with clear documentation.	Ensures money flows into recycling plants, waste-to-energy, renewable inputs, and regenerative systems.
	Third-party verification	Independent auditors assess environmental impact reports.	Builds trust and attracts institutional investors for CE infrastructure.
	Long-term capital access	Typically 5–20 years maturity.	Suitable for large-scale circular facilities like material recovery centers, repair hubs, and green industrial parks.
2. Sustainability-linked loans (SLLs)	Performance-based interest rates	Loan interest decreases if sustainability KPIs (reuse %, recycling %, emission cuts) are achieved.	Incentivizes companies to shift from linear processes to circular KPIs.
	Flexible KPI selection	KPIs may include waste reduction, % recycled inputs, product lifespan increase.	Allows tailored finance aligned with the company's specific circular transition.
	Ongoing monitoring	Borrowers report progress periodically.	Embeds continuous circularity improvements in operations.
3. Carbon credit revenue financing	Credit-based cash flow	Projects earn revenue from carbon sequestration or avoided emissions.	Circular projects like biomass reuse, composting, and industrial symbiosis generate additional income.
	Market price volatility	Revenue depends on carbon price movements.	Encourages adoption of high-integrity circular solutions that attract premium carbon prices.
	MRV (measurement, reporting, verification) dependence	Accurate quantification is essential.	Ensures transparent carbon benefits of circular processes.

Continued

TABLE 6.3 Financing mechanisms for circular economy initiatives.—cont'd

Financing mechanism	Key property	Explanation	Relevance to circular economy (CE)
4. Extended producer responsibility (EPR) funds	Producer-pays principle	Manufacturers finance collection, reverse logistics, and recycling.	Drives product redesign, modularity, and long-life engineering.
	Compliance-based funding	Funds are mandatory in regulated sectors (e-waste, plastics, batteries).	Guarantees steady flow of money into circular recovery systems.
	Closed-loop investment	Funds reinvested into recycling centers, repair hubs, and waste-management innovation.	Strengthens the circular loop across supply chains.
5. Pay-as-you-save (PAYS) models	Customer-centric financing	Users pay for the circular product/service via savings achieved (energy, materials).	Encourages adoption of energy-efficient and resource-efficient products.
	Low upfront cost	Consumers avoid large capex.	Expands market for circular products such as energy-efficient appliances or reusable packaging systems.
	Performance guarantee	Payments depend on actual realized savings.	Ensures companies design truly efficient circular solutions.
6. Public–Private partnerships (PPP)	Risk sharing	Government and private entities share risks of large CE infrastructure.	Enables projects like smart waste parks, decentralized composting, and industrial symbiosis zones.
	Long-term concession models	10–25 year agreements.	Stabilizes funding for circular infrastructure requiring longer payback.
	Blended financing	Combination of grants, loans, and equity.	Makes large-scale circular transition financially feasible.

7. Venture capital (VC) and impact investors	Scalability focus	Invest in high-growth circular startups.	Supports innovative CE domains: 3D printing reuse, refill-tech, bio-based materials, digital circular platforms.
	High-risk, high-return appetite	Funds disruptive technologies with uncertain early returns.	Accelerates breakthrough circular innovations.
	Impact measurement requirements	Investors ask for quantified environmental/social impact KPIs.	Encourages transparency in circular impact.
8. Green public procurement (GPP) Budgets	Demand-creation mechanism	Government purchases circular products (recyclables, remanufactured items).	Creates guaranteed markets for CE companies.
	Policy-mandated funding	Budget allocated through national/state sustainability goals.	Pushes industries to shift from linear to circular supply chains.
	Lifecycle costing approach	Focus on total cost of ownership, not upfront cost.	Rewards durable, repairable, recyclable products.
9. Circular leasing/ Product-as-a-service models (PaaS)	Service-based revenue	Revenue generated from usage, not product sales.	Encourages companies to design long-life, repairable products with minimal waste.
	Ownership retention	Company retains ownership of the asset.	Enables product take-back, refurbishment, and reuse loops.
	Circular performance contracting	Payment tied to performance (uptime, energy savings).	Ensures circularity embedded in business model.
10. Blended finance for Circular transition	De-risking through grants	Grants lower risk for private investors entering new CE markets.	Helps launch innovative circular technologies (biochar, CCUS, reverse logistics).
	Stacked capital structure	Mix of concessionary and commercial capital.	Supports early-stage circular technologies with uncertain viability.
	Outcome-based disbursement	Funding linked to verified outcomes (material recovery %, CO_2 saved).	Encourages measurable circular progress.

or product-as-a-service (PaaS) models require robust IT and communication infrastructure. Urban areas particularly benefit from infrastructure investments in decentralised composting units, textile recycling clusters, and circular innovation parks. Rural areas can leverage such investments for bio-based energy projects and agricultural waste valorisation. These infrastructures are capital-intensive, and without government support or favourable financing mechanisms, private investment remains limited. By prioritizing circular infrastructure in national development plans and urban master plans, policymakers can ensure seamless implementation of CE across regions. Such investments also have multiplier effects creating jobs, supporting local economies, and enabling environmental stewardship on a wide scale.

Public financing often sets the direction and reduces risk for private investors, creating a ripple effect of co-investment. For instance, the European Union's Horizon Europe and LIFE programmes provide major CE-related R&D and deployment grants.

6.4.2 Private investment and venture capital

Private sector financing is increasingly flowing into circular ventures, particularly those involving technological innovations, materials science, or sustainable consumer products. These investments are crucial for scaling CE business models that demonstrate market viability. Private investment plays an increasingly pivotal role in advancing the circular economy (CE). As businesses and societies transition from linear "take-make-dispose" models to circular systems that emphasize reuse, recycling, and regeneration, private capital is becoming essential in driving innovation, scaling operations, and making circular models financially viable. Private sector financing flows into diverse areas such as green technologies, sustainable materials, renewable energy applications, and innovative business models that reduce resource dependency. Investors are recognizing not only the environmental and social value of CE but also the long-term financial opportunities associated with it—ranging from cost savings and regulatory incentives to brand differentiation and customer loyalty. Technological startups, mid-sized innovators, and large firms alike are benefiting from this growing interest. Particularly in the domains of waste-to-value technologies, biodegradable packaging, circular fashion, and electronics remanufacturing, private investments are fueling the development of new industries. These ventures demonstrate strong market potential, especially in a world where consumer preferences are shifting toward ethical and eco-conscious consumption. For investors, circular economy models offer dual benefits: they align with sustainability goals while also promising high returns on investment. In many cases, financial institutions and investors also integrate Environmental, Social, and Governance (ESG) metrics into their decision-making processes, further incentivizing support for CE projects. Whether through direct equity investment, venture capital funding, private equity participation, or structured lending, private capital is becoming one of the most critical enablers in scaling circular innovation and transitioning from pilot programs to commercially viable, mainstream enterprises.

Venture capital (VC): Startups focusing on circular design, reverse supply chains, biodegradable packaging, or upcycling often secure VC funding to scale up. Investors are drawn by high growth potential and sustainable branding value. Venture capital (VC) has emerged as a vital financing stream for early-stage startups working on circular economy innovations. These startups often explore cutting-edge solutions such as biodegradable materials, closed-loop supply chains, product-as-a-service (PaaS) models, and upcycling platforms. VC investors are drawn to these ventures not just for their high growth potential but also for their strong alignment with environmental and social goals. Circular economy startups typically face challenges in the early phases, including limited revenue generation and high upfront R&D costs, which traditional lenders are hesitant to support. This is where VC fills the gap providing essential risk capital in exchange for equity. Investors see value in the ability of these startups to disrupt traditional industries and generate scalable impact. For instance, companies creating plant-based packaging to replace plastics or digital platforms enabling reuse and repair systems attract VCs focused on both profitability and purpose. The attractiveness is also enhanced by global trends such as green consumerism, government regulation around sustainability, and pressure on corporations to decarbonize their supply chains. Moreover, VCs often bring more than just capital they offer mentorship, market access, strategic partnerships, and credibility to help these startups grow and attract additional funding rounds. Several specialized VC funds now focus solely on climate tech, clean tech, and circular innovations, recognizing the urgency and business opportunity in sustainability. The success of venture-funded CE start-ups contributes to the larger ecosystem by pushing technological boundaries, changing market norms, and inspiring a new generation of entrepreneurs to innovate for a regenerative economy.

Private equity (PE): More mature circular enterprises may attract PE investments, particularly in sectors like renewable materials, sustainable fashion, or industrial waste reprocessing. Private equity (PE) is increasingly playing a strategic role in scaling circular economy businesses that have moved past the startup phase and demonstrate solid operational and financial performance. These are typically companies that operate in sectors such as renewable materials, industrial waste management, sustainable construction, and ethical fashion. PE firms invest substantial capital to accelerate growth, restructure operations, expand into new markets, or integrate sustainable practices into legacy businesses. Unlike venture capital, which focuses on high-risk early-stage ventures, PE targets established enterprises with a clear path to profitability and scalability. Circular economy businesses attract PE investment because of their resilience, growing consumer demand, and potential for regulatory alignment especially in an era of stricter environmental policies and carbon accounting. For example, a PE firm might acquire a profitable recycling business or a company producing sustainable textiles, and then improve efficiencies, expand product lines, and introduce digital tools to optimize the supply chain. Circular enterprises often benefit from PE expertise in governance, performance tracking, and operational restructuring areas that can significantly improve their competitive positioning. Furthermore, PE firms are now under increasing pressure from institutional investors to integrate ESG criteria into their portfolios. This push has led many funds to actively seek companies that contribute to climate goals, reduce resource extraction, or promote material circularity. In this context, PE funding becomes a transformative force—not just injecting capital but also guiding businesses to operate more sustainably and strategically in the long run. These partnerships elevate circular businesses from niche players to mainstream industry leaders.

Corporate investments: Large corporations invest directly in CE projects as part of ESG (Environmental, Social, and Governance) goals, green branding, or supply chain resilience. For instance, manufacturers invest in closed-loop production systems or in-house recycling units. Corporate investments in the circular economy are driven by a mix of sustainability objectives, regulatory requirements, supply chain resilience, and branding strategies. Large corporations are increasingly allocating capital toward CE projects as part of their Environmental, Social, and Governance (ESG) mandates. These investments may be in the form of direct funding, corporate venture arms, strategic partnerships, or acquisitions. Common areas of investment include closed-loop manufacturing systems, in-house recycling units, circular product redesign, and take-back programs. For instance, electronics manufacturers may invest in systems to recover rare earth materials from discarded devices, while fashion brands may fund textile recycling initiatives or clothing rental platforms. Such efforts help reduce waste, secure access to scarce resources, and cater to a growing segment of environmentally conscious consumers. Beyond compliance and risk mitigation, corporate investments in CE can also generate cost efficiencies by reducing raw material dependency, energy consumption, and waste disposal expenses. Additionally, companies enhance their brand equity by aligning with sustainability values and demonstrating commitment to planetary health. These investments often form part of larger sustainability roadmaps or net-zero targets, positioning companies as future-ready and responsible players in global markets. Some corporations also establish internal innovation labs or collaborate with startups to co-develop circular technologies, bringing agility and external creativity into their operations. Ultimately, corporate investment in CE is not only a means of environmental stewardship but also a business strategy that unlocks innovation, drives customer loyalty, and ensures long-term competitiveness in a rapidly evolving global economy.

Bank loans and credit lines: Traditional banks are increasingly offering green credit lines or sustainability-linked loans that reward CE performance with lower interest rates. Banks and financial institutions are adapting to the growing demand for sustainable finance by offering green credit lines and sustainability-linked loans tailored for circular economy projects. These loans are often characterized by lower interest rates, flexible terms, and performance-linked conditions that reward environmental improvements. For example, a business that reduces landfill waste, increases recycling rates, or implements closed-loop production systems may qualify for interest rate reductions or fee waivers. Such financial products provide much-needed working capital and investment support to businesses pursuing circular initiatives, especially in capital-intensive sectors like manufacturing, logistics, and construction. Green loans can finance a range of CE-related activities such as installing renewable energy systems, setting up material recovery facilities, or investing in eco-efficient machinery. In many cases, banks assess circularity criteria and environmental impact metrics as part of their due diligence, signaling a shift from traditional credit evaluation models to more sustainability-aware frameworks. Furthermore, multilateral development banks and international financial institutions often partner with local banks to co-finance such green projects, offering partial guarantees or concessional lending terms. This approach reduces financial risk and encourages banks to expand their green finance portfolios. By aligning financial products with environmental performance, banks not only support the circular transition but also future-proof their lending operations against regulatory, reputational, and climate-related risks. This growing trend underscores the evolution of finance from a linear, short-term focus toward a longer-term, regenerative approach that supports economic resilience and environmental health.

6.4.3 Green bonds and sustainability-linked instruments

Green bonds and sustainability-linked financial instruments are transformative tools in mobilizing capital toward the development of a circular economy (CE). These financial vehicles not only provide returns to investors but also ensure that the funds are directed toward projects with tangible environmental benefits. As concerns over resource depletion, climate change, and waste management intensify, investors are increasingly aligning their portfolios with sustainability goals. In this context, green bonds and sustainability-linked loans (SLLs) have emerged as strategic financing solutions to support CE infrastructure, technology, and innovation across industries. These instruments help companies transition to sustainable business models by lowering capital costs and signalling their environmental commitment to stakeholders. Additionally, they offer transparency and accountability, as most require impact reporting and third-party verification, making them attractive to ESG-focused investors.Green and sustainability-linked financial instruments have emerged as powerful tools to channel capital towards CE infrastructure, technology, and innovation. They appeal to investors looking for both financial returns and environmental impact.

Green bonds: Green bonds are fixed-income securities issued specifically to finance projects that have a positive environmental impact. Within the circular economy domain, they are commonly used to support initiatives such as waste management systems, bio-refineries that convert organic waste into energy or chemicals, eco-industrial parks where industries collaborate to reuse byproducts, and facilities that produce circular or biodegradable packaging materials. The capital raised through green bonds enables large-scale infrastructure development while assuring investors that their funds are contributing to environmental sustainability. These bonds often follow strict frameworks, such as the Green Bond Principles (GBP), which ensure transparency, proper use of proceeds, and regular impact reporting. By attracting climate-conscious institutional investors, green bonds play a crucial role in driving systemic change and scaling up circular projects that would otherwise struggle to find affordable financing. These fixed-income securities are issued to finance environmentally beneficial projects, including waste management systems, bio-refineries, circular packaging facilities, or eco-industrial parks. Table 6.4 illustrate properties of green bonds and sustainability-linked instruments.

Sustainability-linked loans (SLLs): These are performance-based loans where interest rates are tied to the achievement of circularity goals, such as increased recycling rates or reduction in virgin resource use. Sustainability-linked loans (SLLs) are innovative financial instruments where the loan terms typically interest rates—are tied to a borrower's performance on sustainability targets. In the context of CE, these may include goals such as reducing virgin material consumption, increasing recycling rates, or implementing circular supply chains. If the borrower achieves or surpasses these key performance indicators (KPIs), they benefit from reduced interest rates or favorable repayment terms. Conversely, failure to meet targets may result in penalties. Unlike green bonds, SLLs are not limited to specific projects but instead reward company-wide circular transformation. This flexibility makes them attractive to businesses aiming for broader sustainability goals. SLLs foster a results-oriented approach, encouraging firms to integrate circularity into their operations while simultaneously improving access to capital.

Transition bonds: Used to fund companies shifting from linear to circular business models such as fossil-based plastic producers transitioning to bio-based polymers. Transition bonds serve as a bridge for companies moving from traditional, linear models to more sustainable and circular operations. These bonds finance projects that support environmental improvement but may not yet qualify under standard green definitions. For example, a company producing fossil-based plastics may use transition bonds to invest in facilities for producing bio-based or recyclable polymers. Similarly, a cement manufacturer could use them to fund carbon capture technologies or alternative raw material sourcing. Transition bonds recognize the need for industries to evolve gradually and provide capital to facilitate this process. They are particularly

TABLE 6.4 Properties of green bonds and sustainability-linked instruments.

Property	Green bonds	Sustainability-linked instruments (SLIs)
Nature of instrument	Project-based financing instrument	Performance-based financing instrument
Use of proceeds	Mandatory allocation to eligible green projects only	No restriction; proceeds can be used for general corporate purposes
Core mechanism	Financing specific environmental projects	Linking financial terms to sustainability KPIs and targets
Eligibility criteria	Must fit within recognized green taxonomies (e.g., CBI, EU taxonomy)	Must have material, ambitious, and measurable sustainability KPIs
Target structure	Not target-driven; project must be green	KPI-driven; coupon/interest depends on achieving sustainability targets
Financial incentives	Indirect (market reputation, investor demand)	Direct (step-up/step-down interest or coupon)
Reporting requirements	Allocation and impact reporting annually	Annual KPI performance reporting
Verification requirements	External review/certification of green project use	External verification/assurance of KPIs and target achievement
Environmental impact measurement	Required through measurable project metrics (e.g., CO_2 saved)	Required for KPI progress; high emphasis on accurate data
Flexibility across sectors	More suitable for green sectors (renewable energy, buildings, transport)	Highly flexible; applicable to heavy industries and hard-to-abate sectors
Issuer type	Governments, corporates, financial institutions, municipalities	Corporates across all sectors, including traditional and carbon-intensive
Risk perception	Often seen as lower risk due to transparency and impact focus	Risk varies depending on KPI ambition; performance risk exists
Standards/Guidelines	Green bond principles (GBP) by ICMA	Sustainability-linked bond principles (SLBP)/Loan principles (SLLP) by ICMA and LMA
Greenwashing risk	Potential risk if reporting is weak; mitigated by strict use-of-proceeds rules	Higher risk if KPIs are poorly selected or targets are easy

Continued

TABLE 6.4 Properties of green bonds and sustainability-linked instruments.—cont'd

Property	Green bonds	Sustainability-linked instruments (SLIs)
Market demand	Strong demand from ESG and climate-focused investors	Growing demand due to flexibility and performance-based structure
Impact on cost of capital	May benefit from "greenium" (lower yield)	Step-up penalty if targets not achieved increases cost
Transparency level	High transparency on project spending and environmental benefits	High transparency on KPI selection, measurement, and progress
Contribution to SDGs	Direct contribution to environmental SDGs (e.g., SDG 7, 13)	Broader contribution—environmental, social, governance-related
Ideal use case	When a specific green project needs financing	When an organization aims to embed sustainability into overall strategy

relevant for high-emission sectors that require substantial investment to overhaul legacy systems. By financing this shift, transition bonds enable a phased approach toward circularity, supporting long-term sustainability while offering measurable progress to investors.

6.4.4 Blended finance and multilateral development funding

Blended finance mechanisms leverage public or philanthropic funds to mobilize larger private investments in CE. They are especially useful in emerging economies or high-risk markets. Blended finance and multilateral development funding are critical enablers of the circular economy (CE), particularly in developing countries and high-risk markets where traditional financing is scarce or expensive. These mechanisms are designed to mobilize private sector investment by using public or philanthropic capital to absorb a portion of the risk. In essence, they de-risk circular projects that are often perceived as unproven or unprofitable by commercial investors. Blended finance helps build investor confidence in circular business models by sharing financial responsibility and providing flexible funding structures. It supports initiatives such as waste-to-energy plants, sustainable agriculture systems, recycling infrastructure, and circular manufacturing hubs. Additionally, multilateral development banks (MDBs) play a central role by offering technical expertise, concessional loans, grants, and capacity-building support.

Their involvement brings credibility and financial security to circular initiatives, helping governments and businesses bridge investment gaps and meet sustainability goals. Together, blended finance and multilateral funding provide a robust framework to scale CE innovations and promote sustainable development globally.

Blended finance structures: Blended finance structures are innovative financial arrangements that combine concessional capital typically sourced from governments, development finance institutions (DFIs), or philanthropic organizations—with commercial capital from private investors. The concessional component serves to absorb higher risk, reduce the cost of capital, or provide first-loss protection, making projects more attractive to risk-averse investors. In the context of the circular economy, blended finance can be used to support early-stage recycling technologies, bio-based product development, sustainable farming, or CE-focused SMEs. These structures are particularly effective in frontier markets, where infrastructure gaps and regulatory uncertainties deter commercial investors. By aligning incentives and redistributing risk, blended finance enables projects with high impact potential to move forward, paving the way for systemic circular transformation. These combine concessional capital (from governments or development agencies) with commercial capital (from private investors), reducing the risk for private financiers and improving project feasibility.

Multilateral development banks (MDBs): Multilateral Development Banks (MDBs), such as the World Bank, ADB, and EIB, are instrumental in advancing circular economy initiatives across the globe. These institutions provide not just financing but also policy advice, capacity-building programs, and technical assistance. MDBs fund large-scale CE projects like integrated waste management systems, circular water and sanitation programs, sustainable industrial zones, and green transportation networks. Their involvement enhances project viability by mitigating political and financial risks and attracting co-investment from other financiers. MDBs also work with governments to create enabling policy environments, develop national CE roadmaps, and strengthen institutional capacity. Their long-term, concessional lending terms and ability to mobilize global partnerships make them vital to mainstreaming CE in emerging and developing economies. Institutions like the World Bank, Asian Development Bank (ADB), and European Investment Bank (EIB) support CE through project financing, capacity-building, and technical assistance.

Guarantees and risk mitigation instruments: Guarantees and risk mitigation instruments are essential tools in facilitating private investment in circular economy projects. These include partial risk guarantees, political risk insurance, and first-loss capital arrangements that protect investors from specific financial or political uncertainties. Such instruments improve the risk-return profile of CE investments, particularly in markets where risks related to regulatory change, currency fluctuations, or political instability are high. For example, a partial guarantee might cover a percentage of loan repayment if a project fails, thereby encouraging banks to lend to CE initiatives that would otherwise be deemed too risky. Political risk insurance protects against expropriation, contract breaches, or political violence, offering assurance to international investors. First-loss capital, provided by public or philanthropic actors, absorbs initial losses in a blended finance structure, further de-risking the investment for senior lenders. These instruments not only build investor confidence but also enable the mobilization of large-scale private capital necessary for long-term circular infrastructure development and technological innovation. They act as financial cushions, encouraging more stakeholders to participate in the circular transition and making complex projects bankable in diverse economic settings. Partial risk guarantees, political risk insurance, and first-loss capital can improve the risk-return profile of CE investments.

6.4.5 Impact investing and ESG integration

Impact investors seek social and environmental outcomes along with financial returns, making them ideal partners for CE projects. Environmental, Social, and Governance (ESG) criteria are also becoming integral to mainstream investing. Fig. 6.4 illustrate impact investing and ESG integration framework.

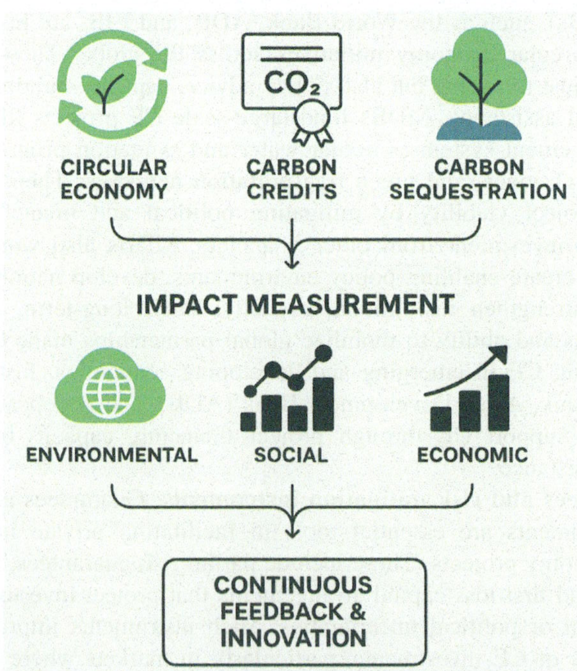

FIGURE 6.4 Impact investing and ESG integration framework.

Impact funds: Impact funds are specialized investment vehicles designed to generate measurable environmental or social outcomes alongside financial returns. In the context of the circular economy (CE), impact funds focus on ventures that promote resource efficiency, reduce environmental degradation, and contribute to sustainability goals. These funds typically invest in companies or projects that demonstrate tangible circular outcomes such as reduced landfill waste, increased material recovery rates, lower carbon emissions, and innovative product reuse systems. For instance, an impact fund might support a startup that transforms plastic waste into construction materials, or a company using artificial intelligence to improve recycling efficiency. The key feature of impact funds is accountability they require transparent metrics and regular reporting to track the impact of their investments. This aligns investor interest with long-term sustainability outcomes, creating pressure for businesses to operate responsibly and transparently. Impact funds are ideal for early-stage and mid-size circular ventures that may not yet be mainstream but have high potential for positive systemic change. As the demand for responsible investing grows, impact funds are becoming a powerful tool in scaling up CE business models while addressing global challenges like waste, pollution, and resource scarcity. These target ventures that produce measurable circular outcomes such as reduced landfill waste, increased material recovery, or carbon footprint reduction.

Social impact bonds: Social Impact Bonds (SIBs), also known as pay-for-success instruments, are innovative financing models that tie returns to the achievement of specific social or environmental outcomes. Though traditionally used in healthcare, education, and poverty reduction, they are increasingly being adapted for circular economy initiatives. In such cases, governments or philanthropic entities agree to repay investors only if the CE project meets predefined impact targets. A practical application could be funding inclusive waste management systems that integrate informal waste workers into formal recycling operations. Investors provide upfront capital to implement the program, and if targets like increased recycling volumes or worker welfare improvements are met, the government repays the investment with interest. SIBs transfer performance risk from public agencies to private investors, encouraging efficient, outcome-driven solutions. They are particularly useful in funding community-based circular projects that have strong social and environmental impact but limited access to traditional finance. Although more common in social sectors, such bonds can also be adapted for circular initiatives like inclusive waste management programs involving informal workers.

ESG-integrated portfolios: ESG-integrated portfolios are investment strategies that incorporate Environmental, Social, and Governance (ESG) criteria into the selection and management of assets. Increasingly, institutional investors and asset managers are prioritizing companies that comply with circular economy principles as part of their ESG mandates. This includes businesses that demonstrate resource efficiency, use recycled or renewable inputs, implement take-back schemes, and minimize environmental harm. Companies with circular business models are seen as forward-looking, resilient, and better positioned to adapt to regulatory and market shifts making them attractive investment options. ESG integration encourages firms to adopt more sustainable practices in order to remain competitive in capital markets. For example, a company reducing its carbon footprint by redesigning its packaging to be fully recyclable may be favored by ESG-focused funds. Investors in ESG-integrated portfolios use performance indicators like waste diversion rates, lifecycle assessments, and supply chain transparency to assess circularity. This integration supports a market-driven transition to the circular economy by aligning financial incentives with sustainability performance. It also signals to corporations that responsible resource management is no longer optional—it's a criterion for accessing mainstream capital. As a result, ESG-integrated portfolios are becoming a key driver in accelerating circular transformation across sectors and geographies. Asset managers and institutional investors now prioritize CE-compliant companies within ESG-focused portfolios, encouraging circular transformation across sectors.

6.4.6 Crowdfunding and community-based financing

For small-scale or community-based CE initiatives, crowdfunding provides an alternative financing method. It can also help validate the market potential of a circular product.

Reward-based crowdfunding: Reward-based crowdfunding is a popular method through which entrepreneurs and innovators can raise funds for circular economy (CE) initiatives by offering backers non-financial rewards. Platforms like Kickstarter and Indiegogo are commonly used for this purpose. In a reward-based model, individuals contribute small amounts of money to support a project in exchange for early access to a product, limited-edition versions, or other perks. This approach is especially effective for consumer-facing CE products such as reusable household goods, biodegradable packaging, or compostable items. It allows circular startups to validate their ideas, build customer loyalty, and secure funding without giving up equity or taking on debt. Moreover, crowdfunding campaigns can also serve as marketing tools, generating public interest and media attention around sustainable innovations. For example, a company developing zero-waste personal care products might offer contributors early delivery of product kits as a reward. This model democratizes access to funding by enabling anyone to support sustainable ventures, not just institutional investors. It's particularly valuable for first-time entrepreneurs and community projects that may lack traditional funding avenues. Ultimately, reward-based crowdfunding empowers innovators to bring their CE ideas to life while building an engaged, sustainability-conscious community of supporters.

Equity crowdfunding: Equity crowdfunding is a financing model that allows individuals to invest in early-stage startups in exchange for ownership shares. In the circular economy context, this approach enables circular-focused startups such as those working on upcycling platforms, renewable material development, or reverse logistics solutions to raise capital from a large pool of retail investors. Platforms like Crowdcube, Seedrs, and StartEngine facilitate such funding campaigns. Unlike reward based models, equity crowdfunding gives investors a stake in the company's future success. This model is particularly attractive to startups that want to scale quickly and engage with a broader investor base beyond venture capital. It also appeals to individuals who want to make impactful investments aligned with their sustainability values. For circular ventures, equity crowdfunding not only provides capital but also builds brand advocacy and long-term stakeholder relationships. This inclusive model supports entrepreneurial growth, fosters transparency, and helps democratize investment in the circular economy.

Community bonds: Community bonds are a unique financing tool that allows local residents, cooperatives, or citizen groups to invest directly in community-driven circular economy projects. These bonds are typically issued by non-profit organizations, municipalities, or community enterprises to raise capital for specific initiatives like community recycling centers, solar composters, tool libraries, or repair cafés. Investors earn a modest return while knowing their money supports local environmental and social goals. Community bonds strengthen local ownership and involvement, ensuring that circular infrastructure is designed with the needs of the community in mind. For instance, a town may issue green municipal bonds to finance a local biogas plant, with residents as bondholders. The transparency and localized nature of these bonds foster trust, engagement, and resilience. They are particularly useful in regions where commercial financing is inaccessible or not aligned with community priorities. Community bonds represent a powerful blend of civic participation and sustainable finance, reinforcing the idea that everyone can contribute to building a circular future.

6.5 Global scenario of carbon credit

The global carbon credit landscape has evolved into a vital mechanism to combat climate change by placing a monetary value on greenhouse gas (GHG) emissions. Carbon credits each representing the reduction or removal of one metric ton of CO_2-equivalent are traded in compliance and voluntary markets to help countries and companies meet emission reduction goals. Compliance Markets, regulated by governments, include major systems like the EU Emissions Trading System (EU ETS), China's National ETS, South Korea's K-ETS, and the UK ETS. These systems impose emission caps and allow trading of allowances to encourage efficiency and innovation. The EU ETS remains the world's largest and most mature, while China's ETS, launched in

2021, is now the largest by volume of emissions covered. Voluntary Carbon Markets (VCM), used by corporations and individuals to offset emissions, are rapidly growing driven by net-zero commitments and ESG (Environmental, Social, Governance) investing. Countries like Brazil, Kenya, and India are key suppliers of nature-based credits, especially through reforestation, renewable energy, and waste management projects. Following are the status of carbon credit of different countries:

6.5.1 European Union (EU)

System: Compliance (EU ETS—Emissions Trading System) Status: Most developed and largest carbon market (by value) Highlights:

- Covers power generation, manufacturing industries, and aviation.
- Operates on a cap-and-trade mechanism.
- The carbon price has reached over €90 per ton in recent years.
- EU Green Deal and Fit for 55 package are driving emissions reduction and expanding market sectors.
- Initiatives like CBAM (Carbon Border Adjustment Mechanism) aim to prevent carbon leakage from imports.

6.5.2 United States

System: Regional compliance markets + strong voluntary market Highlights:

- No national cap-and-trade system, but notable state-level schemes.
- California Cap-and-Trade Program: One of the most comprehensive in North America.
- Regional Greenhouse Gas Initiative (RGGI): Covers power sector in Northeastern states.
- Thriving voluntary carbon market driven by corporate climate pledges (Amazon, Microsoft, etc.).
- Emerging focus on carbon removal technologies (e.g., Direct Air Capture) supported by government incentives like the Inflation Reduction Act.

6.5.3 China

System: National Compliance Market (Launched in 2021) Status: World's largest carbon market by emissions volume Highlights:

- Initially covers the power generation sector (~40% of emissions).
- Gradual expansion expected to include steel, cement, and other heavy industries.

- China Certified Emissions Reduction (CCER) program is being revived for the voluntary carbon market.
- National ETS complements China's 2060 carbon neutrality goal.

6.5.4 India

System: In-development Compliance Market + Active Voluntary Participation Highlights:

- Large host of CDM (Clean Development Mechanism) projects during the Kyoto Protocol era.
- Government is now creating a domestic carbon market under the Carbon Credit Trading Scheme (CCTS), with sectoral pilot trading expected by 2025.
- Focus on renewable energy, afforestation, energy efficiency, and agriculture.
- Strong potential for nature-based and technology-based offsets in voluntary markets.

6.5.5 Japan

System: National Voluntary Carbon Offset Scheme + International Bilateral Offsets Highlights:

- Japan runs the J-Credit Scheme (voluntary), which certifies emission reductions and carbon removals.
- It also leads the Joint Crediting Mechanism (JCM) a bilateral carbon offset program with 17+ countries (like Vietnam, Kenya, and India).
- Strong focus on technology transfer and international cooperation for emission reduction.

6.5.6 South Korea

System: Korea ETS (K-ETS)—National Compliance Market Highlights:

- First nationwide cap-and-trade system in Asia (since 2015).
- Covers 70% of national emissions across industry, energy, and buildings.
- Includes offset mechanisms like Korea Offset Credit (KOC) and Korea Credit Unit (KCU).
- Strong linkage potential with international markets under Article 6 of the Paris Agreement.

6.5.7 Brazil

System: In-development Compliance Market + Strong Voluntary Market Highlights:

- Key player in nature-based solutions (REDD+, forest conservation).
- Draft legislation under debate to establish a regulated carbon market.
- Already home to large-scale voluntary carbon projects, particularly in the Amazon.
- International buyers seek Brazilian credits for biodiversity and social co-benefits.

6.5.8 Canada

System: Federal Carbon Pricing + Provincial Programs Highlights:

- Canada has a federal carbon pricing system for fuels and emissions.
- Provinces like British Columbia and Quebec have their own programs; Quebec is linked with California's ETS.
- Canada's Clean Fuel Standard (CFS) also creates credit-generating opportunities.
- Voluntary market engagement is rising with a focus on afforestation and carbon removal.

6.5.9 Australia

System: Voluntary Carbon Market (Australian Carbon Credit Units—ACCUs) Highlights:

- Safeguard Mechanism regulates large emitters with carbon baselines.
- Australian Carbon Credit Units (ACCUs) are issued for reforestation, methane capture, soil carbon, etc.
- Corporates use ACCUs for internal offsetting or compliance under the safeguard mechanism.
- Government reforms in 2023 have enhanced market transparency and integrity.

6.5.10 African countries (e.g., Kenya, Ghana, Senegal)

System: Active Voluntary Market Participation + Article 6 Readiness Highlights:

- Major contributors to nature-based offsets (REDD+, agroforestry).
- Kenya is a hub for voluntary projects in renewable energy and conservation.

- Ghana became the first country to sign a bilateral carbon trading agreement under Article 6.2 of the Paris Agreement (with Switzerland).
- African Carbon Markets Initiative (ACMI) launched to scale the market to 300 million credits/year by 2030.

6.5.11 Indonesia

System: ETS Pilots + Strong Voluntary Market Highlights:

- ETS pilot started in 2021 for the power sector.
- Key focus on REDD + forest conservation, peatland restoration, and sustainable land-use credits.
- Regulation No. 21/2022 formalized carbon trading mechanisms.
- Expected to scale under Indonesia's 2060 net-zero strategy. Table 6.5 illustrate global scenario of carbon credit. Table 6.6 illustrate major carbon-credit initiatives worldwide.

TABLE 6.5 Global scenario of carbon credit.

Country/ Region	Market type	Key mechanism	Status
EU	Compliance	EU ETS	Mature, global leader
USA	Regional + voluntary	California ETS, RGGI, voluntary markets	Strong in both areas
China	Compliance	National ETS (power sector)	Largest by volume
India	Emerging compliance	Carbon credit trading scheme (CCTS)	Under development
Japan	Voluntary + bilateral	J-credit, JCM	International cooperation
South Korea	Compliance	Korea ETS	Advanced, sectoral coverage
Brazil	In transition	Forest offsets, national market debate	Nature-based leader
Canada	Compliance + voluntary	Fed + provincial systems	Diverse mechanisms
Australia	Voluntary	ACCUs, safeguard mechanism	Well-established
Africa (multiple)	Voluntary + article 6	REDD+, ACMI	Emerging hotspot

TABLE 6.6 Major carbon-credit initiatives worldwide.

Initiative/program	Type and scope	What makes it important/Key features
EU ETS (European union emissions trading system)	Compliance (ETS)	The EU's flagship carbon trading system. Covers power, industry, and aviation within EU. One of the biggest and most mature ETS; a benchmark for carbon pricing.
China national ETS	Compliance (ETS)	The largest in terms of emissions coverage. Initially focused on power sector, with plans for expansion.
California cap-and-trade program	Compliance (State-level ETS)	Covers industrial, electricity, and fuel emissions. Well-established in North America.
Regional greenhouse gas initiative (RGGI)	Compliance (Regional ETS)	Multi-state cap-and-trade program in the U.S. (Northeast states) focused on power plants.
Korea emissions trading scheme (KETS)	Compliance (National ETS)	One of Asia's major ETS programs.
UK ETS	Compliance (Post-Brexit ETS)	United Kingdom-only emissions trading system after its exit from the EU ETS.
Joint crediting mechanism (JCM)	Bilateral/ Voluntary-compliance hybrid	A Japanese government–led initiative. Supports emission-reduction projects in partner developing countries.
Verified carbon standard (VCS)/Verra	Voluntary carbon market (VCM)	Widely used standard for certifying carbon credits. Very large number of projects (forestry, energy, waste, etc.).
Gold standard (GS)	Voluntary carbon market	High-quality standard focusing both on emissions reductions *and* sustainable development benefits.
Global carbon council (GCC)	Voluntary/Offset program	Based in MENA region but globally oriented. Has registry for carbon credits; recognized under some aviation carbon regimes (CORSIA).
Integrity council for voluntary carbon markets (ICVCM)	Governance/ Quality standard	Not a registry, but sets *Core Carbon Principles* (CCP) to define high-quality credits. Recently approved several major standards.
Coalition for rainforest nations (CfRN)/REDD+	Nature-based crediting/REDD + Initiative	Intergovernmental body focusing on deforestation avoidance (REDD+), bringing together rainforest nations.

6.6 Conclusion

Chapter a holistic and multi-dimensional understanding of the transition toward a circular economy, emphasizing the pivotal role of policy, economics, and business models in shaping a regenerative and sustainable future. The chapter concludes that achieving circularity requires more than isolated initiatives; it demands synchronized actions across regulatory institutions, industrial value chains, financial systems, and technological ecosystems. Policy frameworks form the structural foundation of this transition. The ten-pillar policy architecture spanning regulatory mechanisms, economic incentives, technological enablers, infrastructure development, and global cooperation ensures that circular practices are embedded within national priorities and industrial strategies. Instruments such as EPR mandates, eco-design standards, waste hierarchy regulations, and sector-specific guidelines enforce responsible production and consumption patterns across manufacturing, construction, electronics, textiles, and agriculture. Meanwhile, fiscal incentives, green procurement, and concessional financing reward circular innovation, reduce transition costs, and create strong market signals that accelerate adoption. Economically, the chapter demonstrates that circularity is not merely an environmental necessity but a strategic economic opportunity. Countries and industries that adopt circular models gain significant advantages through cost savings, improved supply chain resilience, reduced material dependency, and enhanced global competitiveness. Innovation thrives in circular ecosystems, leading to growth in digital solutions, green technology, biomaterials, and advanced recycling systems. Job creation expands across repair, maintenance, recycling, reverse logistics, and sustainability services, while new markets emerge for refurbished products, secondary raw materials, and shared service platforms. Circular investments through green bonds, ESG-linked portfolios, venture capital, blended finance, and community-based bonds further integrate sustainability into the financial mainstream. The chapter's business model framework reinforces the idea that circularity begins at the design stage and extends throughout the product lifecycle. Product life extension, resource recovery, sharing platforms, and PaaS models ensure that materials circulate efficiently, limiting waste and maximizing value. The digital intelligence layer powered by IoT, blockchain, and AI plays a transformative role in enabling traceability, predictive maintenance, automated sorting, and transparency within circular ecosystems. Stakeholder collaboration strengthens circular transitions by aligning the efforts of governments, businesses, consumers, NGOs, and communities, while financial enablement ensures scalability and long-term viability. Ultimately, Chapter concludes that the circular economy is not simply an environmental model but a developmental paradigm that addresses economic competitiveness, social equity, and ecological regeneration. It calls for sustained commitment, cross-sectoral collaboration, capacity building, and global policy harmonization to fully unlock the potential of circular systems. The chapter positions the circular economy as a pathway toward resilient, inclusive, and future-ready societies.

6.7 Case study

6.7.1 Case study 1: Circular electronics through extended producer responsibility (EPR)

Many countries have implemented EPR frameworks to manage electronic waste more effectively. Under these regulations, producers are responsible for collecting, recycling, and safely disposing of electronic products at the end of their lifecycle. Companies redesign products for durability, modularity, and ease of disassembly to reduce waste management costs and comply with recycling targets.

A leading electronics manufacturer created a nationwide take-back scheme, establishing collection centers and partnering with certified recyclers. The company introduced modular smartphones with easily replaceable batteries and screens, increasing product life and reducing electronic waste by nearly 40%. The firm also recovered precious metals through advanced recycling and reintroduced them into new devices, reducing dependency on virgin material and stabilizing production costs amid material price volatility. This case demonstrates how regulatory mechanisms, resource recovery models, and eco-design standards converge to enable efficient circularity.

6.7.2 Case study 2: Product-as-a-service in the furniture industry

A global furniture company transitioned from selling products to offering furniture-as-a-service for offices, schools, and co-working spaces. Instead of one-time sales, customers pay monthly fees for usage, while the company retains ownership of all furniture assets. This shift incentivized the company to design modular, long-life furniture using recyclable materials, enabling easy repair, refurbishment, and remanufacturing. Digital product passports track the material composition, repair history, and lifespan of each furniture unit, enabling optimized reverse logistics and reducing waste generation. Through circular leasing models, the company reduced raw material consumption by 65%, created new revenue streams from recurring payments, and expanded its refurbishment business.

6.8 Exercise questions

1. Explain the significance of Extended Producer Responsibility (EPR) in accelerating circular transitions.
2. Describe the role of eco-design regulations in reducing waste generation.

3. How does Green Public Procurement (GPP) create market demand for circular products?
4. Discuss the importance of digital product passports (DPPs) in material traceability.
5. Explain how infrastructure investments enable large-scale circular economy adoption.
6. Illustrate the value regeneration layer using examples from the chapter.
7. Discuss the economic implications of circular supply chains.
8. Explain how AI enhances circular material flow optimization.
9. Describe the role of international linkages in circular economy policy harmonization.
10. Discuss the benefits of shifting from product ownership to product-as-a-service.
11. How does stakeholder collaboration contribute to circular transformation?
12. Explain the function of monitoring and compliance systems in circular governance.
13. Describe the financial mechanisms supporting circular economy infrastructure.
14. Explain how refurbished premiumization supports consumer acceptance.
15. Discuss how ESG integration drives investment in circular businesses.

6.9 Objective type questions

1. EPR primarily places responsibility on:
   a) Consumers
   b) Local governments
   c) Producers
   d) Retailers

2. Eco-design focuses on:
   a) Product aesthetics
   b) Durability and reparability
   c) Advertising strategies
   d) Market forecasting

3. DPP stands for:
   a) Digital Pricing Protocol
   b) Direct Product Processing
   c) Digital Product Passport
   d) Data Provenance Policy

4. Green Public Procurement encourages purchase of:
   a) Single-use goods
   b) High-carbon materials
   c) Circular products
   d) Luxury consumer goods

5. Blockchain in circularity improves:
a) Production speed
b) Material traceability
c) Packaging design
d) Employee training

6. Circular supply chains prioritize:
a) Linear resource extraction
b) Renewable and recyclable materials
c) Single-use manufacturing
d) Centralized waste dumping

7. A key benefit of PaaS is:
a) Higher waste generation
b) Recurring revenue
c) Reduced customer engagement
d) Limited product lifespan

8. AI supports circularity through:
a) Predictive maintenance
b) Increasing landfill capacity
c) Promoting waste burning
d) Reducing data accuracy

9. Green bonds finance:
a) Fossil-fuel projects
b) Circular infrastructure
c) Single-use plastics
d) None

10. Reverse logistics involves:
a) Forward product delivery
b) Product returns for recovery
c) Employee transportation
d) Market promotions

11. Circular innovation labs support:
a) Fast product disposal
b) Experimentation and R&D
c) Landfill management
d) Import of virgin materials

12. Secondary material markets trade:
a) Virgin raw materials
b) Recycled materials
c) Luxury goods
d) Disposable items

13. Sharing economy platforms promote:
a) Under-utilization
b) Individual ownership
c) Shared access
d) Linear consumption

14. ESG-integrated portfolios prioritize:
a) Non compliant companies
b) Circular and sustainable firms
c) High-carbon emitters
d) Single-use manufacturers

15. Industrial symbiosis involves:
a) Waste serving as input for another industry
b) Exporting all waste
c) Importing new raw materials
d) Building more landfills

Chapter 7

Future directions and conclusion

7.1 Introduction

The circular economy has emerged as one of the most transformative frameworks for addressing the intertwined challenges of environmental degradation, resource scarcity, climate change, and socio-economic inequality. Moving beyond the traditional linear model of "take–make–dispose," the circular economy redefines growth by decoupling economic development from finite resource consumption and environmental harm. As global systems face increasing pressure from population growth, urbanization, and accelerating consumption, the future directions of the circular economy are no longer aspirational concepts but essential pathways for sustainable development. This chapter introduces the evolving landscape of the circular economy, focusing on how future-oriented strategies, technologies, policies, and behavioral shifts will shape resilient and regenerative economic systems. In recent years, the circular economy has progressed from a conceptual model to a strategic priority across governments, industries, and international organizations. Early efforts largely concentrated on waste management and recycling; however, the future of the circular economy extends far beyond end-of-life solutions. It emphasizes systemic change across entire value chains starting from product design and raw material extraction to production, distribution, consumption, and recovery. Future circular systems prioritize durability, modularity, repairability, and recyclability at the design stage, ensuring that materials remain in use for as long as possible while retaining their value. This forward-looking approach signals a shift from reactive environmental management to proactive system redesign.

Technological innovation will play a pivotal role in defining the future trajectory of the circular economy. Advances in digital technologies such as artificial intelligence, the Internet of Things (IoT), blockchain, and big data analytics are enabling smarter resource management, enhanced traceability, and predictive maintenance. These technologies allow organizations to optimize material flows, reduce waste, and improve product life cycles through

The Circular Economy and their Management. https://doi.org/10.1016/B978-0-443-44617-7.00010-3

real-time monitoring and data-driven decision-making. As Industry 4.0 converges with circular principles, the future economy is expected to become increasingly intelligent, interconnected, and adaptive, supporting circularity at scale across industries and regions. Another critical direction for the circular economy lies in the transformation of business models. Traditional ownership-based models are gradually giving way to service-oriented and performance-based models, such as product-as-a-service, sharing platforms, and leasing systems. These models encourage producers to retain ownership of products, thereby incentivizing long-term durability, easy repair, and efficient resource use. In the future, value creation will be increasingly defined by access, functionality, and outcomes rather than volume of production. This transition not only reduces material consumption but also opens new avenues for innovation, competitiveness, and inclusive economic participation. Fig. 7.1 illustrate future directions of circular economy.

Policy frameworks and governance mechanisms will significantly influence the pace and effectiveness of circular economy adoption in the coming decades. Governments are expected to play a more proactive role by integrating circular economy principles into national development plans, industrial policies, and climate strategies. Future regulatory approaches are likely to emphasize extended producer responsibility, eco-design standards, green public procurement, and incentives for circular innovation. International

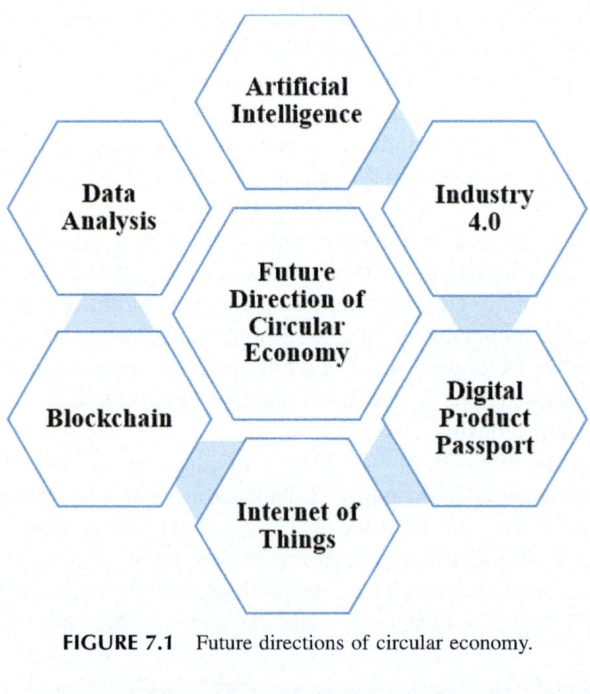

FIGURE 7.1 Future directions of circular economy.

cooperation will also become crucial, as global supply chains demand harmonized standards, shared knowledge, and coordinated action to address transboundary environmental challenges. The future of the circular economy is deeply intertwined with the global climate agenda and clean energy transition. Circular strategies such as material efficiency, product life extension, and resource recovery have the potential to significantly reduce greenhouse gas emissions across sectors including construction, manufacturing, energy, and transportation. By minimizing raw material extraction and energy-intensive production processes, circular systems contribute directly to climate mitigation and adaptation goals. As nations strive to meet net-zero targets, the integration of circular economy practices with renewable energy systems will become a defining feature of sustainable development pathways.

Social and behavioral dimensions will also shape the future directions of the circular economy. While technological and policy innovations are essential, the success of circular systems ultimately depends on changes in consumer behavior, cultural values, and societal norms. Future circular economies will require greater awareness, education, and participation from individuals and communities. Concepts such as responsible consumption, repair culture, and collaborative use models will need to be embedded into everyday life. Education systems, research institutions, and media platforms will play a vital role in fostering a circular mind-set and empowering citizens to act as active participants in sustainable transitions. Equity and inclusivity are increasingly recognized as central to the future of the circular economy. As circular strategies expand globally, it is essential to ensure that benefits are distributed fairly across regions, industries, and social groups. Developing and emerging economies, in particular, present both significant opportunities and unique challenges for circular implementation. Informal sectors, local innovation, and indigenous knowledge systems can contribute meaningfully to circular solutions if supported through inclusive policies and investments. The future circular economy must therefore be designed not only to be environmentally sustainable but also socially just and economically inclusive.

Finally, the future directions of the circular economy demand a shift from fragmented initiatives to integrated, system-wide transformation. Collaboration among industry, academia, government, and civil society will be critical for scaling circular solutions and overcoming structural barriers. Measuring progress through robust indicators and performance metrics will help track impacts, guide decision-making, and ensure accountability. As this chapter sets the stage for deeper exploration, it underscores that the circular economy is not a fixed endpoint but an evolving journey one that requires continuous innovation, learning, and collective commitment. Embracing these future directions offers a pathway toward resilient economies, healthier ecosystems, and a sustainable future for generations to come.

7.2 Innovation in materials, design, and sustainable production

Innovation in materials, design, and sustainable production forms the backbone of the circular economy, as it determines how resources are extracted, transformed, used, and reintegrated into economic systems. Unlike the linear model where materials are often designed for short lifespans and single-use applications the circular economy emphasizes regenerative, restorative, and value-retaining approaches. At the core of this transformation lies a fundamental rethinking of materials, product design philosophies, and production systems to minimize waste, reduce environmental impact, and maximize resource efficiency across the entire life cycle.

Innovation in materials for circularity: Material innovation is a critical enabler of circular systems, focusing on the development and adoption of materials that are renewable, recyclable, biodegradable, or capable of multiple life cycles without quality degradation. Advances in bio-based materials, such as bioplastics derived from agricultural waste, algae, or cellulose, are reducing dependence on fossil-based resources while offering lower carbon footprints. Similarly, the emergence of advanced composites and lightweight materials enables the production of durable goods that require fewer raw materials and less energy during manufacturing and transportation. In addition, innovations in material chemistry are improving recyclability and recovery rates. Monomaterial solutions, for example, replace complex multi-layer structures that are difficult to recycle, particularly in packaging and consumer goods. Smart materials that can self-heal, change properties, or signal degradation are also gaining attention, as they extend product life and improve maintenance efficiency. By prioritizing materials that can circulate safely and continuously within technical or biological cycles, the circular economy moves closer to closed-loop resource systems.

Circular design principles and product innovation: Design plays a decisive role in determining whether products can be reused, repaired, remanufactured, or recycled. Circular design shifts the focus from aesthetics and cost efficiency alone to life-cycle performance and environmental impact. Key principles such as design for durability, modularity, disassembly, and standardization enable products to be easily repaired, upgraded, or dismantled at the end of their use phase. This approach contrasts sharply with planned obsolescence, which has historically driven consumption and waste. Innovative design strategies also include cradle-to-cradle design, which ensures that products are created with the intention of becoming inputs for new products, either through technical recycling or biological decomposition. Digital design tools and simulation technologies further support circularity by allowing designers to evaluate material flows, energy use, and environmental impacts during the early stages of product development. As a result, future products are expected to be smarter, longer-lasting, and more adaptable to changing user needs.

Sustainable and circular production processes: Sustainable production in the circular economy focuses on minimizing resource input, energy consumption, and emissions while maximizing output value and operational efficiency. Cleaner production techniques, such as closed-loop manufacturing, industrial symbiosis, and zero-waste factories, are redefining how goods are produced. In closed-loop systems, waste and by-products from one process become valuable inputs for another, reducing the need for virgin resources and lowering environmental burdens. Industry 4.0 technologies are accelerating this transformation by enabling real-time monitoring, predictive maintenance, and adaptive manufacturing. Automation, robotics, and additive manufacturing (3D printing) allow for precise material usage, reduced waste generation, and localized production. Additive manufacturing, in particular, supports circularity by enabling on-demand production, customization, and the reuse of recycled feedstock materials. These innovations not only enhance efficiency but also reduce environmental impact across the production lifecycle.

Integration of renewable energy and resource efficiency: Innovation in sustainable production is closely linked to the use of renewable energy and energy-efficient technologies. Circular production systems increasingly rely on solar, wind, and other renewable energy sources to reduce carbon emissions and dependency on fossil fuels. Energy-efficient machinery, heat recovery systems, and water recycling technologies further enhance the sustainability of production processes. By integrating energy and material efficiency, circular production contributes simultaneously to resource conservation and climate mitigation goals. Moreover, life cycle assessment (LCA) tools are being widely adopted to measure environmental impacts and guide decision-making in material selection, design choices, and production methods. These analytical tools support transparency and accountability, ensuring that innovations deliver genuine environmental benefits rather than shifting impacts from one stage of the life cycle to another.

Economic and social impacts of innovation: Innovation in materials, design, and sustainable production also generates significant economic and social value. Circular innovations reduce production costs through resource efficiency, create new markets for secondary materials, and foster job creation in areas such as repair, remanufacturing, and recycling. For industries, adopting circular practices enhances resilience against resource price volatility and supply chain disruptions. For society, it promotes sustainable consumption patterns and supports long-term economic stability. At the same time, these innovations encourage collaboration across value chains, involving material scientists, designers, manufacturers, policymakers, and consumers. Such collaborative ecosystems are essential for scaling circular solutions and overcoming technical, economic, and regulatory barriers. In the context of the circular economy, innovation in materials, design, and sustainable production represents a paradigm shift toward regenerative and resilient systems. By rethinking what materials are used, how products are designed, and how goods are produced, the circular economy transforms waste into value and challenges conventional notions of growth and consumption. As technologies evolve and awareness increases, these innovations will play a central role in shaping sustainable industrial systems and driving the transition toward a truly circular and low-carbon future.

7.3 Role of digital technologies in circular transitions

The transition from a linear economic model to a circular economy represents a fundamental shift in how societies produce, consume, and manage resources. While circular principles such as reduce, reuse, repair, and recycle provide the conceptual foundation for this transformation, their large-scale implementation would be difficult without the enabling power of digital technologies. In recent years, digitalization has emerged as a critical catalyst for accelerating circular transitions by improving visibility, efficiency, coordination, and decision-making across complex economic systems. This chapter introduces the pivotal role that digital technologies play in supporting and scaling circular economy practices across industries and societies. Digital technologies enable the collection, analysis, and exchange of vast amounts of data throughout product life cycles, from raw material extraction and manufacturing to consumption, recovery, and reintegration. Technologies such as the Internet of Things (IoT), artificial intelligence (AI), big data analytics, and cloud computing allow real-time monitoring of material flows, asset conditions, and environmental performance. This enhanced transparency helps organizations identify inefficiencies, predict maintenance needs, and optimize resource utilization. By transforming physical products into data-rich assets, digital tools make circular strategies more measurable, manageable, and economically viable. One of the most significant contributions of digital technologies to circular transitions lies in traceability and accountability. Blockchain and digital product passports, for instance, enable secure and transparent tracking of materials and components across supply chains. This traceability supports responsible sourcing, facilitates compliance with environmental regulations, and improves the quality of recycling and remanufacturing processes. As products carry digital identities, stakeholders across the value chain can make informed decisions that preserve material value and reduce waste generation. Such digital infrastructures are increasingly essential for building trust and collaboration in circular ecosystems. Table 7.1 illustrate comparison of different emerging technology with circular economy.

7.3.1 Data analysis in circular economy

Data Analysis in the Circular Economy plays a foundational role in transforming circular principles into measurable, scalable, and actionable practices.

TABLE 7.1 Comparison of different emerging technology with circular economy.

Dimension	Data analysis	Artificial intelligence (AI)	Internet of things (IoT)	Blockchain
Primary role in circular economy	Converts circular data into insights for monitoring and decision-making	Adds intelligence, prediction, and optimization to circular systems	Acts as the sensory layer by collecting real-time life-cycle data	Acts as a trust and transparency layer for circular value chains
Core function	Measurement, evaluation, and performance assessment	Learning, prediction, optimization, and autonomous decision-making	Real-time sensing, monitoring, and connectivity	Secure, immutable recording and verification of transactions
Key objective	Enable evidence-based circular strategies	Optimize resource use and system performance	Provide continuous visibility of products and materials	Ensure traceability, accountability, and trust
Type of data handled	Structured and semi-structured life-cycle data	Large, heterogeneous, and high-dimensional data	Real-time sensor-generated data	Transactional and provenance data
Position in circular system	Analytical and evaluation layer	Intelligence and optimization layer	Data generation and communication layer	Governance and verification layer
Life-cycle stages supported	Design, production, use, end-of-life (analysis across stages)	Design optimization, production, usage, recovery	Production, usage, logistics, waste management	Material sourcing, production, use, recycling, compliance
Contribution to resource efficiency	Identifies inefficiencies and material losses	Predicts failures and optimizes resource allocation	Monitors actual resource usage in real time	Verifies recycled content and responsible sourcing
Role in product life extension	Measures usage and maintenance patterns	Predictive maintenance and failure forecasting	Condition monitoring of products and assets	Stores verified maintenance and ownership history
Support for closed-loop supply chains	Analyzes forward and reverse flows	Optimizes reverse logistics and recovery decisions	Tracks product movement and returns	Enables transparent take-back and recovery records
Decision-making capability	Descriptive, diagnostic, predictive, prescriptive insights	Autonomous and adaptive decision support	Supports operational decisions through live data	Supports compliance and governance decisions

Continued

TABLE 7.1 Comparison of different emerging technology with circular economy.—cont'd

Dimension	Data analysis	Artificial intelligence (AI)	Internet of things (IoT)	Blockchain
Key circular applications	Material flow analysis, circular indicators, dashboards	Waste sorting, eco-design, predictive maintenance	Smart manufacturing, asset tracking, smart waste systems	Material traceability, EPR compliance, circular marketplaces
Role in circular business models	Evaluates economic viability and performance	Enables PaaS, sharing, and leasing optimization	Tracks usage for pay-per-use models	Enables secure transactions and smart contracts
Contribution to transparency	Improves internal visibility and reporting	Improves predictive and strategic transparency	Improves operational transparency	Provides end-to-end supply chain transparency
Environmental impact	Quantifies emission and resource reductions	Reduces waste and energy use through optimization	Prevents losses through monitoring and control	Prevents greenwashing via verified data
Social and governance impact	Supports policy evaluation and reporting	Supports strategic and policy simulations	Improves service reliability and safety	Builds stakeholder trust and accountability
Technological dependency	Depends on quality and integration of data	Depends on high-quality data and computing power	Depends on sensors, connectivity, and infrastructure	Depends on accurate data inputs and network participation
Major challenges	Data silos, data quality, standardization	Data bias, high energy and computational costs	Security, scalability, energy consumption	Scalability, energy use, regulatory uncertainty
Key strength	Converts data into measurable circular insights	Enables intelligent and adaptive circular systems	Provides real-time, life-cycle visibility	Ensures trust, traceability, and verification
Key limitation	Cannot act autonomously	Requires human oversight and ethical governance	Generates large volumes of data requiring analysis	Cannot verify physical reality without reliable inputs
Overall contribution	Makes circular economy measurable	Makes circular economy intelligent	Makes circular economy observable	Makes circular economy trustworthy

As circular systems are inherently complex spanning multiple life-cycle stages, stakeholders, and material flows data-driven approaches are essential for understanding resource movements, identifying inefficiencies, and guiding strategic decision-making. By systematically collecting, processing, and analyzing data across the value chain, organizations and policymakers can move from intuition-based sustainability efforts to evidence-based circular transitions.

The first pillar of data analysis in the circular economy lies in the diversity of data sources that capture activities across the entire life cycle of products and materials. Material-related data includes information on raw material extraction, secondary materials, recycled content, and material quality. Product life-cycle data spans design specifications, manufacturing parameters, usage patterns, maintenance history, and end-of-life treatment. Operational data further provides insights into energy consumption, water usage, emissions, and waste generation, while supply chain data captures logistics, inventory management, and reverse flows. In addition, consumer behavior data and policy-related datasets enrich circular analysis by linking technical performance with social participation and regulatory compliance. Once data sources are identified, effective data collection and integration mechanisms become critical. Digital technologies such as Internet of Things sensors enable real-time monitoring of products, equipment, and processes, creating continuous data streams throughout the life cycle. Digital product passports enhance traceability by storing material composition and sustainability attributes, while enterprise systems such as ERP and supply chain platforms consolidate operational data. Blockchain-based solutions further support secure and transparent data sharing across stakeholders, and external databases provide benchmarking and contextual insights. Together, these mechanisms ensure that fragmented circular data is integrated into a unified analytical environment. Data processing and management form the backbone of reliable circular analysis. Raw data must be cleaned, validated, and standardized to ensure accuracy, consistency, and interoperability across systems. Cloud-based storage solutions and data lakes allow organizations to manage large volumes of heterogeneous data, while edge computing supports real-time processing in dynamic environments. Strong data governance frameworks are essential to define data ownership, protect privacy, and ensure ethical data use. Without robust data management practices, circular insights risk being incomplete or misleading.

Analytical methods transform managed data into actionable intelligence. Descriptive analytics provides visibility into material flows, life-cycle inventories, and circular performance through tools such as material flow analysis and dashboards. Diagnostic analytics helps identify the root causes of waste, losses, and inefficiencies within processes and supply chains. Predictive analytics leverages historical and real-time data to forecast demand for secondary materials, predict equipment failure, and estimate product return rates at end of life. Prescriptive analytics goes a step further by recommending optimal decisions, such as reverse logistics routes, resource allocation strategies, and scenario-based interventions to improve circular performance.

The outcomes of these analytical processes are captured through circular performance indicators that translate data into measurable impacts. Material circularity indicators assess recycled content and recovery rates, while environmental indicators measure reductions in carbon emissions, energy use, and water consumption. Economic indicators evaluate cost savings, value retention, and revenue generation from circular activities, and operational indicators track product life extension and remanufacturing efficiency. Social indicators, such as job creation and consumer engagement, provide a broader understanding of circular economy impacts beyond technical and financial metrics. Fig. 7.2 illustrate process of data analysis in circular economy.

Data-driven insights directly support decision-making and strategic applications across organizations and systems. In product development, analytics informs design choices that enhance durability, modularity, and recyclability. In operations, it enables continuous process improvement and waste minimization. At the business model level, data analysis supports the evaluation of service-based and sharing models, while in supply chain management it enables the design of closed-loop and resilient networks. For policymakers, data analytics provides evidence for regulatory compliance, policy evaluation, and long-term planning.

A defining feature of data analysis in the circular economy is the presence of a continuous feedback loop. Performance outcomes are monitored, lessons are learned, and analytical models are refined over time to improve accuracy and effectiveness. This iterative process allows circular systems to adapt to changing technologies, market conditions, and regulatory requirements. By embedding learning and adaptation into decision-making, organizations can progressively enhance circular performance.

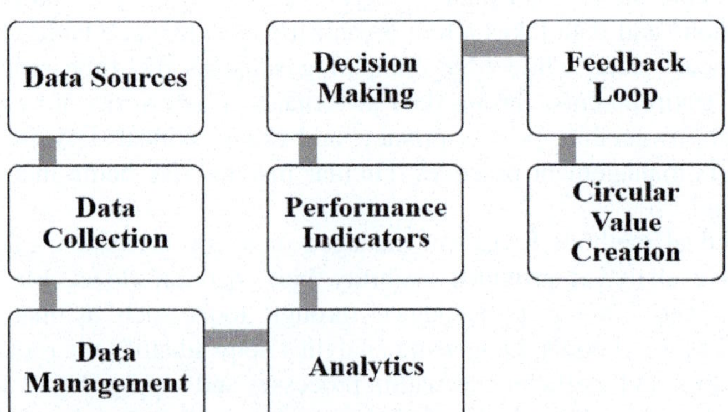

FIGURE 7.2 Process of data analysis in circular economy.

7.3.1.1 Framework: Data analysis in the circular economy

1. **Data sources in the circular economy**
   This layer identifies where data originates across the circular value chain.
   - **Material data**
   - Raw material extraction
   - Secondary materials and recyclates
   - Material composition and quality
   - **Product life-cycle data**
   - Design specifications
   - Manufacturing data
   - Usage and maintenance records
   - End-of-life and recovery data
   - **Operational and process data**
   - Energy and water consumption
   - Emissions and waste generation
   - Production efficiency metrics
   - **Supply chain and logistics data**
   - Transportation flows
   - Inventory and reverse logistics
   - Supplier sustainability data
   - **Consumer and behavioral data**
   - Usage patterns
   - Repair, reuse, and return behavior
   - Consumption preferences
   - **Policy and regulatory data**
   - Compliance indicators
   - Reporting standards
   - Circular economy benchmarks
2. **Data collection and integration mechanisms**
   This stage focuses on how circular data is captured and unified.
   - **Internet of things (IoT) sensors**
   - Smart products and equipment
   - Real-time monitoring of assets
   - **Digital product passports**
   - Material traceability
   - Lifecycle transparency
   - **Enterprise systems**
   - ERP, MES, SCM platforms
   - **Blockchain-based data records**
   - Secure and immutable tracking
   - **External data integration**
   - Market data
   - Environmental databases
   - Government and NGO datasets

3. **Data processing and management**

   This layer ensures data quality, accessibility, and usability.

   - **Data cleaning and validation**
   - Removing inconsistencies
   - Ensuring accuracy and completeness
   - **Data standardization**
   - Harmonized units and formats
   - Interoperability across systems
   - **Data storage**
   - Cloud-based data lakes
   - Edge computing for real-time data
   - **Data governance**
   - Data ownership
   - Privacy and security
   - Ethical data use

4. **Analytical methods and tools**

   This is the core analytical engine of the framework.

   1. **Descriptive analytics**
      - Material flow analysis (MFA)
      - Life cycle inventory analysis
      - Circularity indicators and dashboards
   2. **Diagnostic analytics**
      - Root cause analysis of waste and losses
      - Process bottleneck identification
      - Inefficiency mapping
   3. **Predictive analytics**
      - Demand forecasting for secondary materials
      - Predictive maintenance for asset life extension
      - End-of-life return prediction
   4. **Prescriptive analytics**
      - Optimization of reverse logistics
      - Resource allocation and recovery decisions
      - Scenario-based decision support

5. **Circular performance indicators**

   This layer translates analytics into measurable outcomes.

   - **Material circularity indicators**
   - Recycled content ratio
   - Material recovery rate
   - **Environmental indicators**
   - Carbon footprint reduction
   - Energy and water savings
   - **Economic indicators**
   - Cost savings from circular practices
   - Value retention and resale revenue

- **Operational indicators**
- Product lifetime extension
- Repair and remanufacturing rates
- **Social indicators**
- Job creation in circular activities
- Consumer participation levels

6. **Decision-making and strategic applications**
   Analytics-driven insights support action.
   - **Product design optimization**
   - Design for disassembly and reuse
   - **Process improvement**
   - Waste minimization strategies
   - **Business model innovation**
   - Product-as-a-service feasibility
   - **Supply chain optimization**
   - Closed-loop supply chain design
   - **Policy and compliance decisions**
   - Regulatory reporting
   - Circular policy impact assessment
7. **Feedback loop and continuous improvement**
   This ensures adaptability and long-term circularity.
   - Performance monitoring
   - Learning from outcomes
   - Model refinement and recalibration
   - Scaling successful circular interventions
8. **Enablers and challenges**
   Cross-cutting elements affecting the framework.
   Enablers
   - Digital technologies (AI, IoT, Blockchain)
   - Standardized circular metrics
   - Cross-sector collaboration

   Challenges
   - Data silos and interoperability issues
   - Data quality and availability
   - High implementation costs
   - Privacy and governance concerns
9. **Outcome: data-driven circular economy**
   The final outcome of the framework.
   - Evidence-based decision-making
   - Optimized resource loops
   - Reduced environmental impact
   - Resilient and sustainable economic systems

Data analysis serves as a critical enabler of a data-driven circular economy, linking information, intelligence, and action. While digital technologies and analytical tools provide powerful enablers, challenges such as data silos, quality issues, high implementation costs, and governance concerns must be addressed. When effectively implemented, data analysis supports optimized resource loops, reduced environmental impacts, and resilient economic systems, ensuring that circular economy principles translate into tangible and sustained value creation.

7.3.2 Artificial intelligence in circular economy

Artificial Intelligence (AI) is increasingly recognized as a powerful enabler of the circular economy, offering advanced capabilities to manage complexity, enhance efficiency, and support intelligent decision-making across circular systems. The circular economy seeks to optimize resource use, extend product lifecycles, and minimize waste and environmental impacts. However, achieving these objectives at scale requires the ability to analyse large volumes of heterogeneous data, anticipate system behavior, and dynamically optimize processes capabilities that traditional analytical methods often struggle to provide. AI addresses these challenges by embedding intelligence into circular economy processes, enabling more adaptive, predictive, and resilient systems.

At the foundation of AI-driven circularity lies a robust data ecosystem that captures information across the entire product and material life cycle. This ecosystem includes material composition data, product design and usage data, operational data related to energy and emissions, supply chain and logistics data, and consumer behavior data reflecting usage, repair, and return patterns. Environmental indicators and policy-related datasets further contextualize circular performance within regulatory and sustainability frameworks. By integrating these diverse data sources, AI systems gain a holistic view of circular value chains, allowing them to identify hidden patterns, correlations, and opportunities for improvement.

AI techniques form the analytical core that transforms circular data into actionable intelligence. Machine learning algorithms are widely used to recognize patterns in material flows, forecast demand for secondary materials, and improve waste classification and sorting processes. Deep learning models enhance image- and sensor-based applications, such as automated waste identification, quality inspection of recycled materials, and detection of material contamination. Natural language processing techniques support the analysis of unstructured data, including sustainability reports, regulatory documents, and supplier disclosures, enabling organizations to assess compliance and sustainability performance more effectively. Reinforcement learning and optimization algorithms further allow systems to adaptively optimize processes such as production scheduling, resource allocation, and reverse logistics routing under dynamic and uncertain conditions.

These AI capabilities are translated into tangible circular applications across all phases of the product life cycle. During the design phase, AI supports eco-design by evaluating material alternatives, simulating environmental impacts, and recommending design choices that improve durability, modularity, and recyclability. In the production phase, AI enables predictive maintenance, energy optimization, and process control, reducing resource consumption and preventing waste before it occurs. During the use phase, intelligent monitoring systems help extend product lifetimes by predicting failures, optimizing usage patterns, and guiding maintenance interventions. At the end-of-life stage, AI enhances waste sorting, disassembly, and recycling efficiency by identifying materials and guiding automated recovery processes. In parallel, AI-driven reverse logistics systems optimize collection routes, recovery decisions, and inventory management in closed-loop supply chains.

The impact of AI in the circular economy is measured through enhanced circular performance metrics. AI improves the accuracy and timeliness of indicators related to material circularity, recycling rates, energy efficiency, and carbon emission reductions. It also supports economic performance measurement by tracking cost savings, value recovery, and productivity gains from circular practices. By continuously monitoring these metrics, AI systems enable organizations to assess progress, compare scenarios, and prioritize interventions that deliver the greatest environmental and economic benefits. Beyond operational improvements, AI serves as a critical decision-support tool for strategic planning and policy development. AI-driven insights help organizations evaluate circular business models, such as product-as-a-service or sharing platforms, by analyzing demand patterns, asset utilization, and financial viability. At the policy level, AI can simulate the impacts of regulatory measures, assess compliance risks, and support evidence-based policymaking. These capabilities enhance resilience and long-term planning in the face of resource constraints, market volatility, and climate-related risks. Fig. 7.3 illustrate artificial intelligence with circular economy.

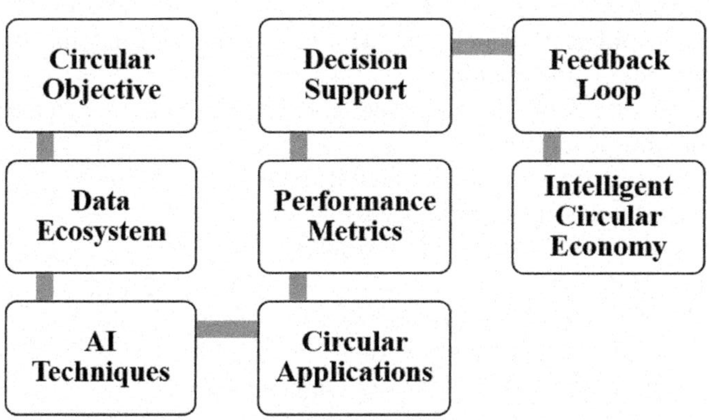

FIGURE 7.3 Artificial intelligence with circular economy.

Human AI collaboration is an essential dimension of successful AI integration in the circular economy. While AI systems provide powerful analytical capabilities, human expertise is needed to interpret results, set objectives, and ensure ethical and responsible use. Explainable AI approaches enhance transparency and trust by making AI-driven decisions understandable to stakeholders. Building skills and capacity among workers, managers, and policymakers ensures that AI complements human judgment rather than replacing it, fostering inclusive and responsible circular transitions.

A defining feature of AI-enabled circular systems is continuous learning through feedback loops. As new data becomes available from circular operations, AI models are retrained and refined to improve accuracy and adaptability. This iterative learning process allows circular systems to evolve in response to changing technologies, consumer behavior, and regulatory environments. Over time, this adaptability strengthens the effectiveness and scalability of circular solutions.

7.3.2.1 Framework of artificial intelligence in the circular economy

1. **Circular economy objectives**
   This layer defines what the circular system aims to achieve through AI.
   - Resource efficiency and material value retention
   - Waste minimization and pollution reduction
   - Product life extension and durability
   - Closed-loop supply chain optimization
   - Climate impact reduction and sustainability performance
2. **AI data ecosystem**
   AI functions as an intelligence layer built on diverse circular data sources.
   - **Material data:** composition, quality, recyclability
   - **Product life-cycle data:** design, usage, maintenance, end-of-life
   - **Operational data:** energy, emissions, waste streams
   - **Supply chain data:** logistics, reverse flows, inventory
   - **Consumer behavior data:** usage, repair, return patterns
   - **Environmental & policy data:** regulations, sustainability metrics
3. **AI techniques and models**
   This layer represents the analytical core enabling circular intelligence.
   1. **Machine learning**
      - Pattern recognition in material flows
      - Waste classification and sorting
      - Demand forecasting for secondary materials
   2. **Deep learning**
      - Image-based waste identification
      - Automated quality inspection
      - Material composition recognition

3. **Natural language processing (NLP)**
   - Analysis of sustainability reports
   - Regulatory and policy interpretation
   - Supplier compliance assessment
4. **Reinforcement learning**
   - Adaptive process optimization
   - Dynamic reverse logistics routing
   - Resource allocation under uncertainty
5. **Optimization algorithms**
   - Closed-loop supply chain design
   - Production scheduling with minimal waste

4. **AI-enabled circular applications**

   This layer translates AI intelligence into real-world circular actions.
   - **Design phase**
   - AI-driven eco-design recommendations
   - Material substitution analysis
   - **Production phase**
   - Predictive maintenance
   - Energy and resource optimization
   - **Use phase**
   - Product life extension through condition monitoring
   - Usage optimization and user behavior insights
   - **End-of-life phase**
   - Intelligent waste sorting
   - Automated disassembly and recycling guidance
   - **Reverse logistics**
   - AI-optimized collection, routing, and recovery

5. **Circular performance metrics**

   AI enhances measurement, monitoring, and forecasting.
   - Material circularity indicators
   - Carbon and energy reduction metrics
   - Cost efficiency and value recovery
   - Product lifetime extension rates
   - Recycling and remanufacturing efficiency

6. **Decision support and strategy layer**

   AI outputs support strategic and operational decision-making.
   - Circular investment prioritization
   - Business model selection (PaaS, sharing models)
   - Policy impact simulation and compliance forecasting
   - Risk assessment and resilience planning

7. **Human–AI collaboration**

   Ensures responsible and effective AI adoption.
   - Expert-in-the-loop decision-making
   - Explainable AI for transparency

- Ethical and responsible AI governance
- Skill development and workforce adaptation

8. **Feedback loop and continuous learning**
   AI systems continuously evolve with new data.
   - Model retraining with real-time circular data
   - Adaptive learning across life cycles
   - Performance-based refinement
9. **Enablers and challenges**
   Cross-cutting factors influencing AI adoption.
   Enablers
   - Digital infrastructure and IoT integration
   - High-quality, standardized circular data
   - Cross-sector collaboration

   Challenges
   - Data availability and bias
   - High computational and energy costs
   - Regulatory and ethical concerns
   - Interoperability and scalability issues
10. **Outcome: Intelligent circular economy**
    Final system-level impact.
    - Data-driven and AI-optimized resource loops
    - Reduced environmental footprint
    - Resilient and adaptive economic systems
    - Scalable and measurable circular transitions

Artificial intelligence acts as an intelligence layer that connects data, decision-making, and action within the circular economy. While challenges such as data quality, energy consumption of AI systems, ethical concerns, and regulatory uncertainties remain, the potential benefits are substantial. When responsibly deployed, AI enables optimized resource loops, reduced environmental impacts, and resilient economic systems. As circular economy initiatives continue to expand, AI will play an increasingly central role in shaping intelligent, sustainable, and future-ready economies.

7.3.3 Internet of things in circular economy

The Internet of Things (IoT) has emerged as a foundational digital technology for enabling the transition toward a circular economy by providing real-time visibility into how products, materials, and resources move through economic systems. The circular economy seeks to extend product lifetimes, reduce waste, and maintain the value of materials through reuse, repair, remanufacturing, and recycling. Achieving these objectives requires continuous and accurate information across the entire life cycle of products and infrastructure. IoT addresses this need by acting as the sensory and communication layer of the circular economy, transforming physical assets into connected, data-generating entities.

At the core of IoT-enabled circularity is the generation of data through smart products, machines, and infrastructure. Sensors embedded in products capture information on usage patterns, wear, and performance, offering valuable insights into how products function in real-world conditions. In manufacturing environments, IoT enabled machines generate data on equipment health, process efficiency, and resource consumption, supporting cleaner and more efficient production. Smart infrastructure systems monitor energy, water, and waste flows in buildings and cities, while smart logistics solutions track assets, shipments, and reverse flows. In waste management systems, IoT sensors measure fill levels, material types, and contamination, improving collection efficiency and recovery outcomes. Connectivity and communication form a critical layer that enables IoT data to flow seamlessly across circular systems. Through wireless networks such as 5G, low-power wide-area networks, and industrial Wi-Fi, sensor data is transmitted to edge devices, gateways, and cloud platforms. This connectivity enables both real-time and periodic data transfer, ensuring timely insights for operational control and strategic planning. Reliable communication infrastructure is essential for maintaining system responsiveness, especially in dynamic environments such as logistics networks and urban waste management.

Once collected, IoT data must be effectively managed and integrated to support circular decision-making. Data management systems aggregate and store large volumes of sensor data using cloud and edge computing architectures. Standardization and interoperability are crucial at this stage, allowing data from different devices, platforms, and organizations to be combined and analyzed. Integration with enterprise systems such as enterprise resource planning, supply chain management platforms, and advanced analytics tools ensures that IoT data contributes meaningfully to organizational workflows and sustainability strategies. IoT-enabled data streams support a wide range of circular applications across all phases of the product life cycle. In the design and production stages, feedback from product usage data informs design improvements that enhance durability, modularity, and re-pairability. Predictive maintenance applications use sensor data to anticipate equipment failures and reduce downtime, extending the life of machines and reducing material waste. During the use phase, condition monitoring allows products to be maintained, upgraded, or reused more effectively, delaying end-of-life and maximizing value retention. At the end-of-life stage, IoT plays a crucial role in enabling efficient recovery and recycling processes. Smart identification technologies help track products and materials, guide disassembly, and improve sorting accuracy. In reverse logistics, IoT data supports optimized collection routes, inventory visibility, and asset recovery decisions, reducing transportation emissions and improving economic efficiency. These applications help close material loops and strengthen circular supply chains.

The impact of IoT in the circular economy is reflected through improved circular performance indicators. By providing accurate and timely data, IoT enables organizations to measure material efficiency, product lifetime extension, energy and water savings, waste reduction, and emission reductions more effectively. Continuous monitoring ensures that performance deviations are detected early and corrective actions are implemented promptly.

IoT also enhances decision support and automation within circular systems. Real-time dashboards, automated alerts, and sensor-driven control systems enable faster and more informed decision-making. When integrated with artificial intelligence and advanced analytics, IoT data supports predictive and prescriptive insights, enabling adaptive resource management and continuous optimization of circular processes. A defining feature of IoT-enabled circular systems is the presence of feedback loops that support continuous improvement. Data collected during production, use, and recovery phases feeds back into design, operational, and strategic decisions, enabling systems to evolve over time. This iterative learning process strengthens circular performance and supports long-term sustainability goals. Fig. 7.4 illustrate internet of things with circular economy.

The Internet of Things acts as the connective tissue of the circular economy, linking physical assets with digital intelligence. While challenges related to data security, privacy, interoperability, energy consumption, and deployment costs remain, the benefits of IoT for circular transitions are substantial. By enabling real-time visibility, informed decision-making, and system-wide optimization, IoT plays a central role in building connected, resilient, and sustainable circular economies.

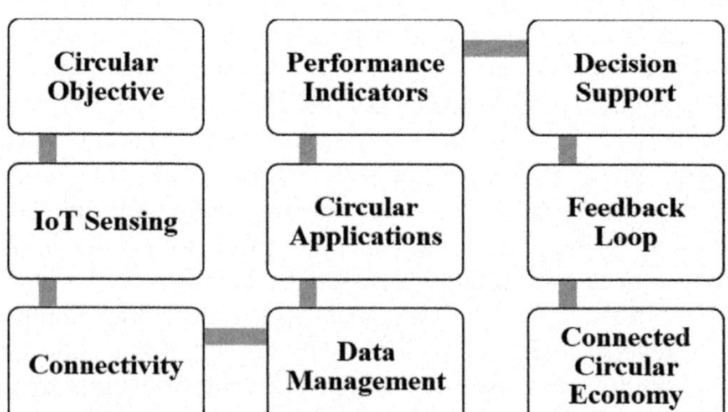

FIGURE 7.4 Internet of things with circular economy.

7.3.3.1 Framework of internet of things (IoT) in the circular economy

1. **Circular economy objectives**
   This layer defines how IoT supports circular outcomes.
   - Efficient resource utilization
   - Product life extension and durability
   - Waste reduction and recovery
   - Closed-loop material flows
   - Environmental impact minimization
2. **IoT data generation layer**
   IoT acts as the sensory layer of the circular economy.
   - **Smart products**
   - Usage, wear, and performance data
   - **Smart manufacturing systems**
   - Machine health and process efficiency
   - **Smart infrastructure**
   - Energy, water, and waste monitoring
   - **Smart logistics**
   - Asset tracking and reverse logistics
   - **Smart waste systems**
   - Fill-level sensing and material detection
3. **Connectivity and communication layer**
   Ensures seamless data transmission.
   - Sensors and actuators
   - Wireless networks (5G, LPWAN, Wi-Fi)
   - Edge devices and gateways
   - Real-time and batch data transfer
4. **Data management and integration layer**
   Transforms raw IoT data into usable information.
   - Data collection and aggregation
   - Cloud and edge data storage
   - Data standardization and interoperability
   - Integration with ERP, SCM, and AI platforms
5. **IoT-enabled circular applications**
   IoT translates data into circular actions.
   - **Design and production**
   - Design feedback from usage data
   - Predictive maintenance
   - **Use phase**
   - Condition monitoring
   - Product life extension

- **End-of-life**
- Intelligent sorting and tracking
- Disassembly guidance
- **Reverse logistics**
- Collection optimization
- Asset recovery management

6. **Circular performance indicators**
   Measures circular outcomes enabled by IoT.
   - Material usage efficiency
   - Product lifetime extension
   - Energy and water savings
   - Waste reduction rates
   - Emission reductions
7. **Decision support and automation**
   IoT supports intelligent operations.
   - Real-time operational dashboards
   - Automated alerts and controls
   - AI-driven optimization and forecasting
   - Adaptive resource management
8. **Feedback loop and continuous improvement**
   IoT enables learning systems.
   - Continuous performance monitoring
   - Process optimization based on data
   - Design and system refinement
9. **Enablers and challenges**
   Cross-cutting factors for IoT adoption.
   Enablers
   - Digital infrastructure
   - Interoperable standards
   - Integration with AI and analytics

   Challenges
   - Data security and privacy
   - Scalability and interoperability
   - Energy consumption of devices
   - High deployment costs
10. **Outcome: connected circular economy**
    Final system-level impact.
    - Real-time visibility of resource flows
    - Optimized closed-loop systems
    - Reduced environmental footprint
    - Resilient and adaptive circular systems

7.3.4 Blockchain in circular economy

Blockchain technology has emerged as a powerful digital enabler for advancing the circular economy by addressing critical challenges related to transparency, trust, and coordination across complex value chains. The circular economy depends on the continuous circulation of materials and products through multiple life cycles, involving diverse stakeholders such as manufacturers, consumers, recyclers, and regulators. Ensuring reliable information exchange and accountability among these actors is essential for closing material loops. Blockchain supports this objective by providing a decentralized, secure, and immutable system for recording and sharing data, thereby strengthening the foundations of circular systems.

At the core of Blockchain enabled circularity lies the technological infrastructure of distributed ledger technology. In a blockchain network, data is stored across multiple nodes operated by participating stakeholders, reducing dependence on centralized intermediaries. Public, private, and consortium blockchain's offer different levels of accessibility and control, allowing circular economy applications to be tailored to specific industry and governance needs. Cryptographic mechanisms ensure data integrity and security, making blockchain particularly suitable for tracking materials, products, and transactions throughout their life cycles.

The effectiveness of blockchain in the circular economy depends heavily on accurate and reliable data inputs. Blockchain platforms integrate data from various sources, including Internet of Things sensors, digital product passports, and enterprise systems such as enterprise resource planning and supply chain management platforms. IoT devices automatically capture real-time information on product usage, location, and condition, while digital product passports store detailed data on material composition and sustainability attributes. Data validation mechanisms, both automated and manual, are essential to ensure that information recorded on the blockchain accurately reflects physical reality, as the immutability of blockchain makes errors difficult to correct once data is recorded.

Smart contracts represent a key functional layer that enables automation and efficiency within blockchain-based circular systems. These self-executing contracts operate according to predefined rules and conditions, automatically triggering actions when specified criteria are met. In the context of the circular economy, smart contracts can enforce extended producer responsibility obligations, verify regulatory compliance, and automate incentive mechanisms for recycling and reuse. They also support innovative business models such as pay-per-use and product-as-a-service by enabling transparent and secure transactions based on actual product usage.

Blockchain-enabled circular applications span the entire product and material life cycle. In material traceability, blockchain records the origin, transformation, and movement of raw, secondary, and recycled materials, ensuring transparency and ethical sourcing. For product life-cycle management, blockchain securely stores records of ownership, maintenance, repair, and refurbishment, enabling informed decisions about reuse and remanufacturing. In waste management, blockchain provides verified records of collection, sorting, recycling, and recovery, reducing fraud and improving accountability. Reverse logistics systems benefit from transparent take-back and return processes, while circular marketplaces leverage blockchain to enable trusted trading of secondary materials and refurbished products.

A significant contribution of blockchain to the circular economy lies in performance verification and measurement. By ensuring data integrity and traceability, blockchain enables reliable calculation and verification of circular performance indicators, such as recycling rates, recycled content, and material recovery levels. Environmental metrics, including carbon emissions reductions and environmental credits, can also be securely tracked and audited. This verification capability supports credible sustainability reporting, reduces greenwashing, and enhances confidence among regulators, investors, and consumers. Governance and trust mechanisms are central to the success of blockchain-based circular systems. Blockchain supports decentralized governance models that distribute decision-making authority among stakeholders while maintaining transparency and accountability. Role-based access controls allow participants to manage data visibility and ownership, addressing privacy and commercial sensitivity concerns. These governance features facilitate collaboration across organizations and borders, which is essential for global circular supply chains. Beyond operational transparency, blockchain plays an important role in decision support and policy enablement. Policymakers can use blockchain-based data to assess compliance, evaluate the effectiveness of circular economy regulations, and design evidence-based interventions. Fig. 7.5 illustrate blockchain with circular economy.

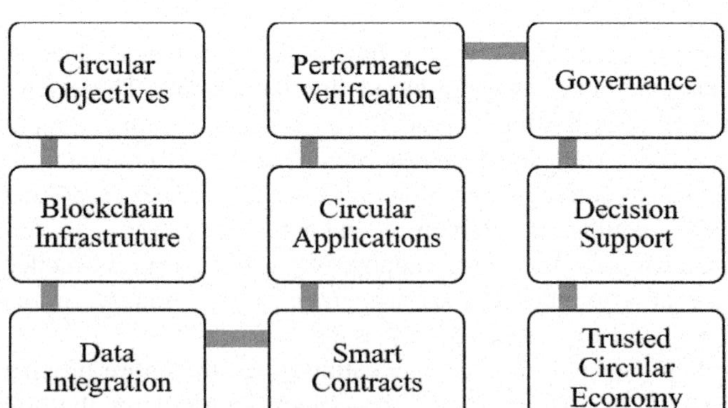

FIGURE 7.5 Blockchain with circular economy.

For businesses, blockchain enhances supply chain risk management and sustainability disclosures, while for consumers it builds trust by providing verifiable information on product origins, environmental impacts, and end-of-life pathways. Despite its significant potential, the adoption of blockchain in the circular economy faces several challenges. Scalability, transaction costs, and energy consumption particularly in energy-intensive consensus mechanisms remain concerns. Data accuracy at the point of entry is critical, as blockchain can only ensure the integrity of recorded data, not its initial correctness. Regulatory uncertainty and the need for standardized data frameworks further influence adoption. Addressing these challenges through technological innovation, policy alignment, and stakeholder collaboration is essential for realizing blockchain's full potential.

7.3.4.1 Framework: Blockchain in the circular economy

1. **Circular economy objectives**
   This layer defines how blockchain supports circularity.
   - Transparency and traceability of material flows
   - Trust and accountability among stakeholders
   - Efficient resource recovery and reuse
   - Reduction of fraud and information asymmetry
   - Support for sustainable and ethical supply chains
2. **Blockchain infrastructure layer**
   This forms the technological foundation.
   - Distributed ledger technology (DLT)
   - Public, private, and consortium blockchains
   - Nodes operated by supply chain stakeholders
   - Cryptographic security mechanisms
3. **Data input and integration layer**
   Ensures accurate and reliable data entry.
   - IoT and sensor data integration
   - Digital product passports
   - Enterprise systems (ERP, SCM)
   - Manual and automated data validation
4. **Smart contracts and automation**
   Blockchain-enabled logic for circular actions.
   - Automated compliance verification
   - Extended Producer Responsibility (EPR) execution
   - Incentive and reward mechanisms
   - Pay-per-use and product-as-a-service models
5. **Blockchain-enabled circular applications**
   This layer translates blockchain capabilities into practical use cases.
   - **Material traceability**
   - Tracking raw, secondary, and recycled materials

- **Product life-cycle management**
- Secure recording of usage, repair, and ownership history
- **Waste management**
- Verified recycling and recovery records
- **Reverse logistics**
- Transparent return and take-back systems
- **Circular marketplaces**
- Trading of secondary materials and refurbished products

6. **Circular performance and verification**
   Blockchain ensures data integrity and verification.
   - Verified circularity indicators
   - Auditable sustainability metrics
   - Carbon and environmental credit tracking
   - Regulatory compliance reporting
7. **Governance and trust mechanisms**
   Supports collaboration across stakeholders.
   - Decentralized governance models
   - Role-based access control
   - Data ownership and permission management
   - Cross-border collaboration
8. **Decision support and policy enablement**
   Blockchain-informed insights support strategic decisions.
   - Evidence-based policymaking
   - Supply chain risk management
   - Sustainability reporting and disclosures
   - Consumer trust and engagement
9. **Enablers and challenges**
   Cross-cutting factors affecting adoption.
   Enablers
   - Integration with IoT and AI
   - Standardized circular data frameworks
   - Multi-stakeholder collaboration

   Challenges
   - Scalability and transaction costs
   - Energy consumption of blockchains
   - Data accuracy at entry points
   - Regulatory and legal uncertainty
10. **Outcome: Trusted circular economy**
   Final system-level impact.
   - Transparent and accountable circular systems
   - Reduced fraud and greenwashing
   - Enhanced collaboration and trust
   - Scalable and verifiable circular transitions

Blockchain acts as a trust-enabling infrastructure for the circular economy, supporting transparent, accountable, and verifiable circular systems. By enhancing traceability, automating compliance, and strengthening governance, blockchain reduces information asymmetry and builds confidence among stakeholders. When integrated with complementary technologies such as IoT and artificial intelligence, blockchain can significantly accelerate scalable and trustworthy circular transitions, contributing to more resilient, ethical, and sustainable economic systems.

7.4 Industry 4.0 as an enabler of the circular economy

Industry 4.0 represents a paradigm shift from conventional industrial systems to digitally connected, intelligent, and autonomous production environments. Its core technologies cyber-physical systems, IoT, AI, big data, cloud computing, robotics, and advanced automation enable unprecedented levels of transparency, responsiveness, and optimization. In the context of the circular economy, Industry 4.0 acts as a critical enabler by providing the technological infrastructure required to manage circular flows of materials, products, and information across complex value chains. Traditional linear production systems are limited by fragmented data, static processes, and minimal visibility beyond factory boundaries. Industry 4.0 overcomes these limitations by integrating digital intelligence throughout the product life cycle. Real-time monitoring of materials, processes, and assets allows organizations to identify inefficiencies, reduce waste, and make informed decisions that support reuse, remanufacturing, and recycling. As a result, circular strategies that were previously difficult to implement at scale become economically feasible and operationally efficient. Fig. 7.6 illustrate circular economy with industry 4.0.

Smart design and digital twins for circularity: The design phase determines up to 80% of a product's environmental impact, making it a critical leverage point for circularity. Industry 4.0 introduces advanced digital design tools and digital twin's virtual representations of physical products and systems that allow designers to simulate product performance, durability, energy use, and environmental impacts before production begins. Digital twins enable life-cycle thinking by integrating data from material selection, manufacturing processes, usage conditions, and end-of-life scenarios. Designers can test modular configurations, evaluate ease of disassembly, and assess recyclability under different conditions. This capability supports design-for-circularity principles, such as durability, reparability, upgradability, and material recovery. Furthermore, continuous feedback from real-world usage data allows products to be redesigned and improved over time, ensuring long-term value retention and reduced resource consumption.

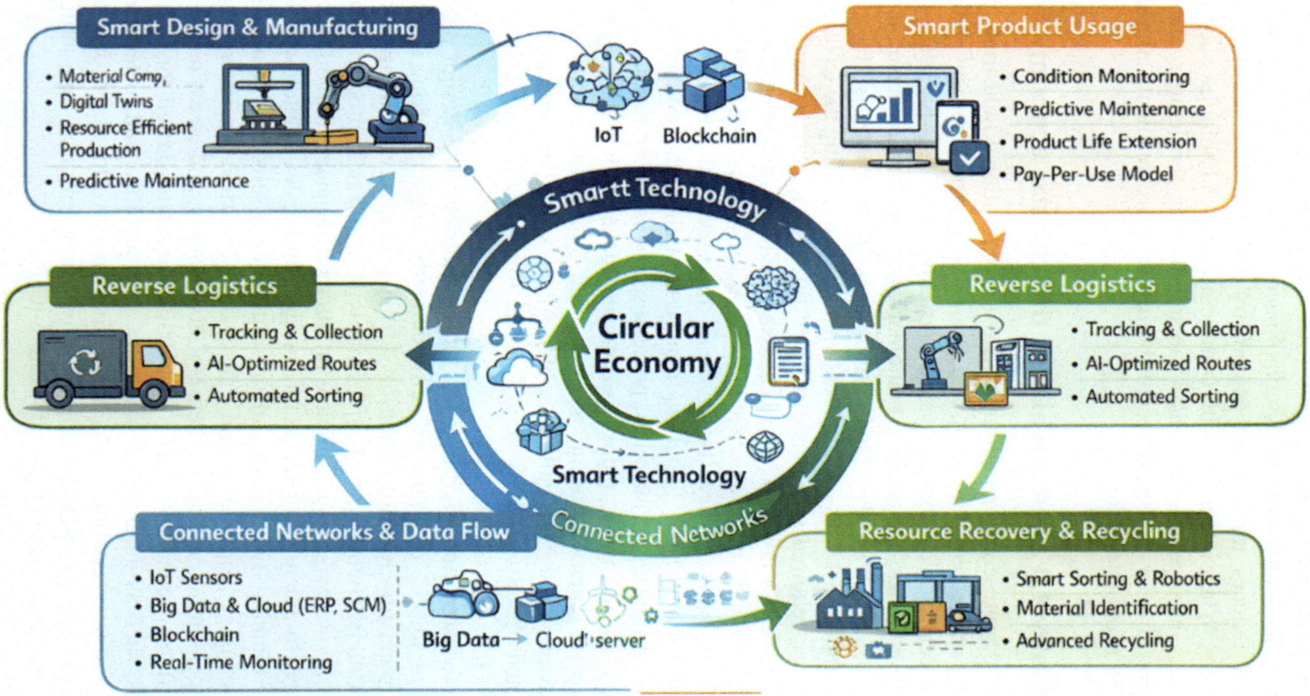

FIGURE 7.6 Circular economy with Industry 4.0.

Circular and intelligent manufacturing systems: Manufacturing lies at the heart of the circular economy–Industry 4.0 integration. Smart factories equipped with IoT sensors, automation systems, and AI-driven analytics continuously monitor production parameters such as material usage, energy consumption, emissions, and machine performance. This data-driven visibility enables manufacturers to minimize resource losses, reduce scrap, and optimize production efficiency. Predictive maintenance powered by AI extends equipment lifetimes by identifying early signs of wear or failure, preventing breakdowns and reducing the need for replacement. Additive manufacturing (3D printing) further enhances circularity by enabling near-zero material waste, lightweight product designs, and the use of recycled or bio-based materials. Collectively, these technologies transform manufacturing into a closed-loop, low-waste, and adaptive system aligned with circular economy objectives.

Product life extension through smart usage and maintenance: Extending the use phase of products is one of the most effective strategies for reducing environmental impacts. Industry 4.0 enables this through smart, connected products that continuously generate data on usage, condition, and performance. Sensors embedded in products allow manufacturers and users to monitor wear, detect anomalies, and schedule maintenance based on actual conditions rather than fixed intervals. AI-driven analytics predict failures and recommend timely repairs, upgrades, or component replacements, significantly extending product lifetimes. This capability supports circular practices such as refurbishment, remanufacturing, and reuse. Moreover, digital platforms enable manufacturers to remain connected with products throughout their life cycle, fostering long-term service relationships and reducing incentives for premature replacement.

Closed-loop supply chains and reverse logistics: A defining characteristic of the circular economy is the establishment of closed-loop supply chains in which products, components, and materials flow back into production systems after use. Industry 4.0 technologies enable this by providing end-to-end visibility and traceability across forward and reverse supply chains. Technologies such as RFID, GPS, blockchain, and IoT sensors track products throughout their life cycles, enabling efficient collection, sorting, and recovery at end-of-life. AI-powered optimization models determine the most sustainable recovery pathways reuse, repair, remanufacturing, or recycling based on economic and environmental criteria. These intelligent reverse logistics systems reduce transportation emissions, lower recovery costs, and increase the overall efficiency of circular supply chains.

Intelligent waste management and advanced recycling: Waste management and recycling are critical components of circular systems, and Industry 4.0 significantly enhances their effectiveness. Smart waste collection systems use sensors to monitor fill levels and material composition, optimizing collection schedules and reducing unnecessary transportation. This improves operational efficiency while lowering emissions.

Advanced robotics and AI-based computer vision systems enable high-precision sorting of waste streams, addressing one of the key challenges in recycling material contamination. By improving the quality of recovered materials, Industry 4.0 technologies enable higher-value recycling and reduce dependence on virgin resources. These intelligent recycling systems help close material loops more effectively and support the transition toward zero-waste industrial ecosystems.

Data-driven circular performance measurement: Measuring circular performance is essential for monitoring progress, ensuring accountability, and guiding strategic decisions. Industry 4.0 enables the collection and integration of large volumes of real-time data across production, usage, and recovery stages. Big data analytics and cloud-based platforms transform this data into actionable insights through dashboards, indicators, and predictive models. Key circular performance metrics include material circularity, product lifetime extension, recycling efficiency, energy and water savings, and carbon emission reductions. Continuous monitoring allows organizations to identify deviations, evaluate interventions, and drive continuous improvement. This data-driven approach enhances transparency and supports compliance with sustainability regulations and reporting frameworks.

New circular business models enabled by industry 4.0: Industry 4.0 plays a central role in enabling innovative circular business models that shift value creation from product ownership to performance and service delivery. Digital platforms, sensors, and analytics enable models such as product-as-a-service, sharing platforms, leasing, and pay-per-use systems. By retaining ownership of products, manufacturers are incentivized to design durable, repairable, and upgradeable products. Real-time usage data ensures accurate billing, asset management, and customer engagement. These models reduce material consumption, strengthen customer relationships, and create stable revenue streams while reinforcing circular economy principles.

Integration with sustainability and climate objectives: The convergence of circular economy and Industry 4.0 directly contributes to climate change mitigation and broader sustainability goals. Improved material efficiency reduces the need for energy-intensive extraction and processing of virgin resources. Smart energy management systems optimize energy use and integrate renewable energy sources into industrial operations. Digital technologies also support life cycle assessment and carbon footprint analysis, enabling organizations to quantify environmental impacts and align strategies with net-zero targets. As climate policies increasingly emphasize systemic solutions, Industry 4.0-enabled circular strategies are becoming essential components of national and corporate sustainability agendas.

Despite its transformative potential, the integration of circular economy and Industry 4.0 faces several challenges. High investment costs, lack of data interoperability, cybersecurity risks, and skills shortages can hinder adoption. Small and medium-sized enterprises, in particular, may face barriers due to limited resources and technical expertise. Looking ahead, advances in AI autonomy, blockchain governance, and standardized digital infrastructures are expected to further strengthen circular systems. Policy support, workforce upskilling, and cross-sector collaboration will be critical for scaling these solutions. As these challenges are addressed, the synergy between circular economy and Industry 4.0 is likely to define the future of sustainable industrial development. Industry 4.0 provides the digital backbone that enables circular economy principles to be implemented effectively and at scale. Through intelligent design, smart manufacturing, closed-loop supply chains, and data-driven decision-making, Industry 4.0 transforms circularity into an operational reality. Together, these paradigms offer a powerful pathway toward resilient, competitive, and environmentally sustainable industrial systems, shaping the future of global industry in the digital and green transition era. Table 7.2 illustrate comparison of traditional industrial systems and circular economy in industry 4.0.

7.5 Digital product passports (DPPs) in the circular economy

Digital Product Passports (DPPs) are emerging as a key enabler of circular economy strategies, providing structured, standardized, and accessible information about products throughout their life cycle. In the context of a circular economy, the main objective is to maximize resource efficiency, extend product life, facilitate repair and reuse, and improve recycling outcomes. DPPs serve as a digital record that contains all relevant data about a product, creating transparency and traceability across supply chains, from design to end-of-life. Fig. 7.7 illustrate circular economy with digital product passport.

Concept and structure of DPPs: A digital product passport is essentially a digital record or ledger that stores detailed information about a product. This includes:

- Material composition: Types of raw materials, recycled content, and hazardous substances.
- Manufacturing information: Production processes, energy consumption, emissions, and location of manufacture.
- Usage data: Instructions, warranties, maintenance schedules, and usage history.
- End-of-life guidance: Recycling instructions, disassembly steps, component recoverability, and recommended disposal or reuse pathways.
- Ownership and certification data: Compliance with sustainability standards, certifications, and regulatory information.

DPPs can be hosted on centralized or decentralized platforms, and in some advanced implementations, integrated with blockchain for immutability and secure access. IoT sensors and connected devices can automatically update DPPs in real time, ensuring that information stays accurate throughout the product's life cycle.

TABLE 7.2 Comparison of traditional industrial systems and circular economy in Industry 4.0.

Dimension	Traditional linear industry	Circular economy enabled by industry 4.0
Economic model	Linear model based on take–make–dispose	Circular model based on reduce reuse–repair–remanufacture–recycle
Design philosophy	Design focused on cost and functionality	Design for circularity (durability, modularity, disassembly, recyclability)
Product life-cycle perspective	Limited to production and sales stage	Full life-cycle management (design to end-of-life recovery)
Material usage	High dependence on virgin raw materials	Optimized use of secondary and recycled materials
Manufacturing approach	Mass production with limited flexibility	Smart, adaptive, and resource-efficient manufacturing
Technology integration	Low digital integration	High integration of IoT, AI, robotics, digital twins, and analytics
Data availability	Fragmented and siloed data	Real-time, integrated, and data-driven systems
Process optimization	Reactive and manual optimization	Predictive and autonomous optimization using AI
Waste management	Waste treated as an unavoidable by-product	Waste treated as a valuable resource
Product lifetime	Short product life and planned obsolescence	Extended life through monitoring, maintenance, and upgrades
Maintenance strategy	Preventive or breakdown-based maintenance	Predictive and condition-based maintenance
Supply chain structure	One-way forward supply chains	Closed-loop supply chains with reverse logistics
Traceability	Limited visibility beyond manufacturing	End-to-end traceability using IoT, RFID, and blockchain
Reverse logistics	Inefficient and costly	Optimized through digital tracking and AI algorithms
Recycling efficiency	Low recovery rates due to poor sorting	High recovery through smart sorting and automation
Business models	Product ownership and one-time sales	Product-as-a-service, leasing, sharing, pay-per-use

TABLE 7.2 Comparison of traditional industrial systems and circular economy in Industry 4.0.—cont'd

Dimension	Traditional linear industry	Circular economy enabled by industry 4.0
Customer relationship	Transactional and short-term	Long-term, service-oriented relationships
Energy usage	High energy intensity and fossil fuel dependence	Energy-efficient systems integrated with renewables
Environmental impact	High emissions and pollution	Reduced carbon footprint and environmental impact
Performance measurement	Limited environmental performance metrics	Comprehensive circular and sustainability indicators
Decision-making	Experience-based and reactive	Data-driven, predictive, and prescriptive
Innovation focus	Incremental and cost-driven	System-level, sustainability-driven innovation
Regulatory compliance	Compliance as a constraint	Compliance as a strategic advantage
Workforce skills	Manual and repetitive skill sets	Digital, analytical, and interdisciplinary skills
Resilience to disruptions	Low resilience to supply chain shocks	High resilience through adaptive and flexible systems
Economic value creation	Value ends at point of sale	Continuous value creation across product life cycles

7.5.1 Role of DPPs in circular economy

Enhancing traceability and transparency: DPPs provide a complete and accessible record of a product's materials, components, and history. This transparency allows manufacturers, recyclers, and consumers to track materials and ensure compliance with environmental regulations. For example, a DPP can verify whether a product contains recycled materials or hazardous substances, supporting responsible sourcing and recycling.

Supporting product life extension: By providing detailed usage history, maintenance records, and repair instructions, DPPs help extend the product's operational life. Consumers and service providers can access accurate information for maintenance, refurbishment, or upgrading, reducing the need for premature disposal.

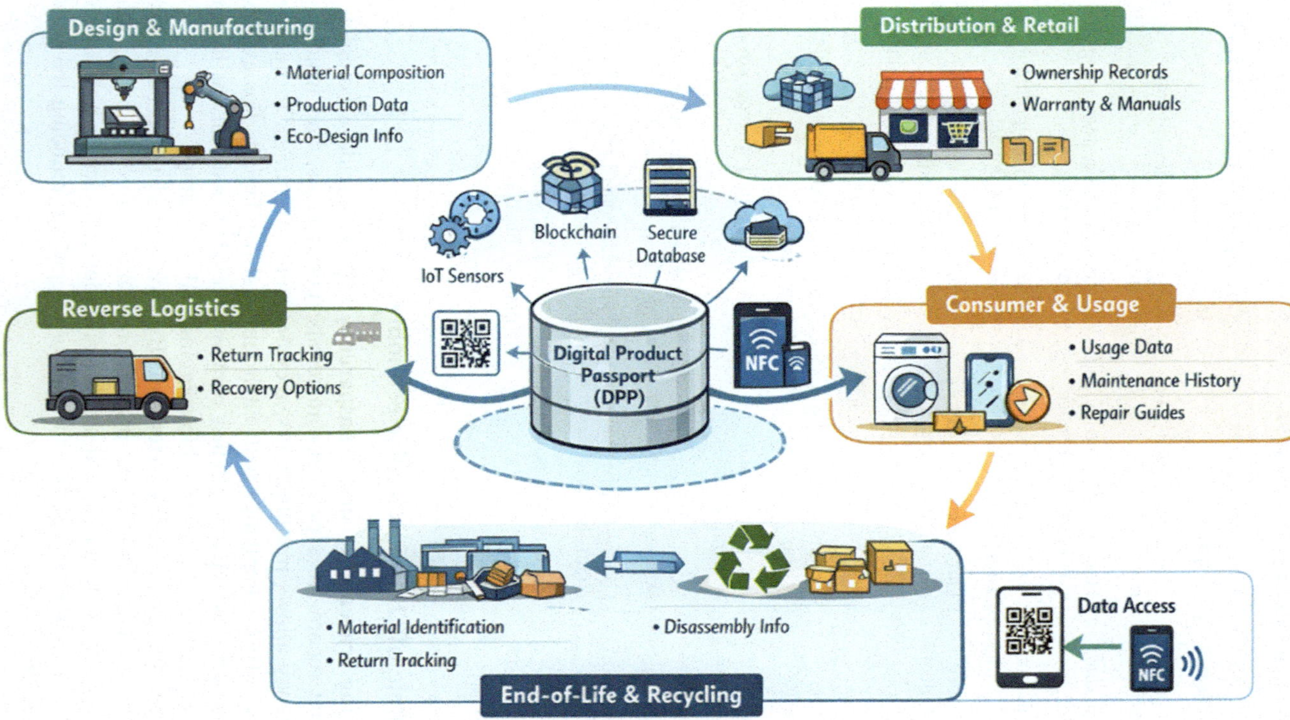

FIGURE 7.7 Circular economy with digital product passport.

Facilitating recycling and material recovery: Recycling processes often fail due to insufficient or inaccurate material information. DPPs enable recyclers to identify material composition, hazardous content, and optimal separation methods. This improves the quality and quantity of recovered materials and reduces contamination in recycling streams.

Enabling circular business models: DPPs underpin innovative business models such as product-as-a-service, leasing, or sharing platforms. By providing detailed product usage data, ownership history, and condition reports, companies can efficiently manage leased or shared assets, track returns, and optimize maintenance schedules.

Regulatory compliance and reporting: Governments and regulatory bodies are increasingly pushing for material transparency and sustainable production. DPPs allow companies to comply with regulations such as extended producer responsibility (EPR) schemes, eco-design requirements, and EU Sustainable Products Initiative mandates. They also facilitate environmental reporting, carbon footprint assessment, and auditing.

7.5.2 Technologies supporting DPPs

- Internet of things (IoT): Sensors embedded in products can update usage, condition, and location in real-time.
- Blockchain: Provides secure, immutable, and decentralized storage of product information, enhancing trust among stakeholders.
- Cloud computing and big data analytics: Enable large-scale storage, processing, and analysis of DPP data across supply chains.
- QR codes or NFC tags: Provide easy consumer access to product information and guidance.

7.5.3 Benefits of DPPs

- Improved resource efficiency and material circularity
- Higher quality and efficiency in recycling and recovery
- Support for eco-design and sustainable product innovation
- Enhanced consumer awareness and engagement
- Reduced environmental impacts and carbon footprint
- Facilitation of compliance with global sustainability standards

7.5.4 Challenges and considerations

- Standardization: Lack of global standards for data formats and interoperability
- Data Privacy: Protecting consumer and commercial data in DPPs
- Accuracy: Ensuring real-time and reliable updates throughout the product life cycle

- Cost and Implementation: Investment in digital infrastructure, IoT devices, and blockchain integration
- Stakeholder Collaboration: Requires coordination among manufacturers, retailers, service providers, and recyclers

Digital Product Passports are expected to become a cornerstone of the circular economy, especially in sectors like electronics, automotive, textiles, and packaging. With regulatory pressure increasing and technological integration advancing, DPPs will enable fully transparent, traceable, and circular supply chains, making sustainable production and consumption practices a practical reality. When combined with IoT, AI, and blockchain, DPPs can serve as dynamic, intelligent records that continuously guide circular actions from eco-design to end-of-life recovery.

7.6 Conclusion

The transition toward a circular economy represents a fundamental reorientation of economic systems, shifting from linear patterns of extraction, consumption, and disposal to regenerative, value-retaining, and resilient models of growth. As discussed throughout this chapter, the future directions of the circular economy are no longer theoretical aspirations but practical necessities driven by escalating environmental pressures, climate change imperatives, and resource constraints. Achieving circularity requires system-wide transformation that integrates technological innovation, sustainable design, supportive policy frameworks, and inclusive social participation.

Innovation in materials, product design, and sustainable production emerges as a cornerstone of circular systems. By prioritizing renewable, recyclable, and durable materials, along with design principles that support modularity, re-pairability, and disassembly, industries can significantly extend product lifecycles and reduce dependency on virgin resources. Sustainable and intelligent production processes, supported by cleaner technologies and renewable energy integration, further enable the reduction of waste, emissions, and resource inefficiencies across industrial operations.

Digital technologies play a decisive role in enabling and scaling circular economy transitions. The convergence of data analytics, artificial intelligence, the Internet of Things, blockchain, and Industry 4.0 technologies provides the digital backbone required for real-time visibility, predictive decision-making, and closed-loop supply chain management. Data-driven approaches transform circular principles into measurable and actionable strategies, while AI introduces adaptive intelligence that optimizes resource flows, extends asset lifetimes, and supports strategic planning. IoT acts as the sensory layer of circular systems, generating continuous life-cycle data, whereas blockchain establishes trust, transparency, and accountability across complex value chains. Digital Product Passports further strengthen these capabilities by ensuring standardized, accessible, and verifiable product information throughout the product life cycle.

The integration of circular economy principles with Industry 4.0 significantly enhances operational efficiency and economic viability. Smart manufacturing, digital twins, predictive maintenance, and intelligent waste management systems convert circularity into an operational reality rather than a conceptual ideal. These technologies enable new circular business models such as product-as-a-service, leasing, and sharing platforms, which redefine value creation around performance, access, and longevity rather than volume of production. As a result, organizations can achieve both environmental sustainability and long-term competitiveness.

Equally important are the social, behavioral, and governance dimensions of the circular economy. Policy alignment, regulatory incentives, and international cooperation are essential for creating enabling environments that support circular innovation and investment. At the same time, consumer awareness, education, and behavioral change are critical for embedding circular practices into everyday life. Ensuring equity and inclusivity, particularly for developing economies and informal sectors, remains a central challenge and opportunity for future circular systems.

The circular economy represents an evolving journey rather than a fixed destination, requiring continuous learning, collaboration, and innovation. The synergistic integration of technological advancements, sustainable design, intelligent production, and responsible governance offers a clear pathway toward resilient economies, healthier ecosystems, and climate-aligned development. As global systems confront unprecedented environmental and economic uncertainties, embracing the future directions of the circular economy will be essential for achieving sustainable prosperity for present and future generations.

7.7 Case study

7.7.1 Case study 1: AI- and IoT-enabled circular manufacturing in the automotive sector

7.7.1.1 Background

AutoNova Motors is a mid-sized automotive manufacturer operating in an emerging economy. The company faces increasing regulatory pressure to reduce carbon emissions, comply with extended producer responsibility (EPR) norms, and improve resource efficiency. At the same time, volatile raw material prices and supply chain disruptions are affecting profitability. Traditionally, AutoNova followed a linear production model with limited visibility beyond manufacturing and sales. To address these challenges, AutoNova initiated a transition toward a circular economy model enabled by Industry 4.0 technologies, focusing on AI, IoT, and data analytics.

7.7.1.2 Circular strategy implemented

1. Smart product design: AutoNova redesigned key vehicle components using modular architecture. Critical parts such as batteries, engines, and electronic modules were designed for easy disassembly, repair, and remanufacturing.
2. IoT-enabled manufacturing: Sensors were installed across production lines to monitor material usage, energy consumption, machine health, and emissions in real time. This enabled predictive maintenance and reduced production downtime.
3. AI-driven data analytics: AI models analyzed production and usage data to identify material inefficiencies, predict component failure, and optimize energy consumption. Deep learning algorithms were introduced for automated defect detection and waste reduction.
4. Closed-loop supply chain: AI-optimized reverse logistics systems were deployed to collect end-of-life vehicles and components. Recovered parts were sorted, remanufactured, or recycled based on condition and material quality.
5. Digital product passports (DPPs): Each vehicle was assigned a DPP containing material composition, maintenance history, usage data, and end-of-life recovery instructions, improving traceability and recycling efficiency.

7.7.1.3 Outcomes

- Reduction in raw material consumption by 18%
- Extension of component life by 25% through predictive maintenance
- 30% improvement in recycling and remanufacturing efficiency
- Enhanced compliance with EPR and sustainability regulations
- Improved profitability and resilience against supply chain disruptions

7.7.1.4 Key challenges

- High initial investment in digital infrastructure
- Integration of legacy systems with Industry 4.0 platforms
- Workforce upskilling and change management
- Data governance and cybersecurity concerns

7.7.2 Case study 2: Blockchain-enabled circular economy in the textile and fashion industry

7.7.2.1 Background

EcoThread Collective is a global apparel brand operating in the fast-fashion segment. The company faces growing criticism related to textile waste, lack of supply chain transparency, labor concerns, and greenwashing. Additionally, recycling rates for textiles remain low due to complex material blends and insufficient product information. To address these issues, EcoThread launched a blockchain-enabled circular economy initiative integrated with IoT and Digital Product Passports.

7.7.2.2 Circular strategy implemented

1. Blockchain-based material traceability: Blockchain was used to record the origin, processing, and movement of raw materials such as organic cotton, recycled polyester, and dyes. Each transaction was securely logged on a distributed ledger.
2. Digital product passports (DPPs): Every garment carried a QR code linked to a DPP that provided information on material composition, manufacturing location, environmental impact, care instructions, and recycling guidance.
3. IoT-enabled usage monitoring: Smart tags tracked garment usage patterns and condition (wash cycles, wear frequency), helping EcoThread design more durable products and offer repair or resale options.
4. Smart contracts for circular incentives: Blockchain-based smart contracts automatically rewarded consumers with store credits when garments were returned for reuse or recycling.
5. Circular business models: EcoThread introduced resale, rental, and repair services supported by verified product histories stored on blockchain platforms.

7.7.2.3 Outcomes

- 40% increase in garment return rates
- Significant reduction in textile waste sent to landfills
- Improved consumer trust and brand transparency
- Enhanced compliance with sustainability reporting standards
- New revenue streams through resale and rental platforms

7.7.2.4 Key challenges

- High cost of blockchain and IoT integration
- Data accuracy at the point of entry
- Interoperability with suppliers across different regions
- Consumer adoption and digital literacy

7.8 Exercise questions

1. Explain how the circular economy differs from the traditional linear economic model. Why is this shift critical for sustainable development?

2. Discuss the future directions of the circular economy in the context of global challenges such as climate change, resource scarcity, and socio-economic inequality.
3. Examine the role of material innovation in enabling circular economy systems. Provide examples of bio-based and advanced materials.
4. Analyze the importance of circular design principles such as modularity, reparability, and design for disassembly in extending product life cycles.
5. Explain how sustainable and circular production processes contribute to waste minimization and resource efficiency.
6. Discuss the role of data analysis in transforming circular economy principles into measurable and actionable strategies.
7. Describe the framework of data analysis in the circular economy, highlighting key layers from data sources to decision-making.
8. Explain how artificial intelligence (AI) enhances circular economy performance across the product life cycle.
9. Discuss the applications of AI in waste sorting, predictive maintenance, and reverse logistics within circular systems.
10. Evaluate the role of the Internet of Things (IoT) as the sensory and communication layer of the circular economy.
11. Explain how blockchain technology improves transparency, traceability, and trust in circular economy value chains.
12. Discuss the significance of Digital Product Passports (DPPs) in supporting product life extension and efficient recycling.
13. Analyze how Industry 4.0 technologies act as enablers of circular economy implementation at scale.
14. Compare traditional industrial systems with circular economy-enabled Industry 4.0 systems, highlighting key differences.
15. Critically examine the challenges and enablers associated with the large-scale adoption of digital technologies in the circular economy.

7.9 Objective type questions

1. The traditional linear economic model is best described as:
   a) Reduce–Reuse–Recycle
   b) Take–Make–Dispose
   c) Share–Repair–Reuse
   d) Design–Recover–Regenerate
2. Which design principle focuses on enabling easy repair and upgrading of products?
   a) Planned obsolescence
   b) Modularity
   c) Mass customization
   d) Cost minimization

3. Which technology enables real-time monitoring of product usage and condition?
a) Blockchain
b) Artificial Intelligence
c) Internet of Things
d) Cloud Computing

4. Digital Product Passports primarily support which circular economy objective?
a) Increased consumption
b) Product transparency and traceability
c) Faster disposal
d) Market monopolization

5. Which analytical method focuses on forecasting future outcomes such as material return rates?
a) Descriptive analytics
b) Diagnostic analytics
c) Predictive analytics
d) Prescriptive analytics

6. In the circular economy, waste is best described as:
a) An unavoidable by-product
b) A disposal problem
c) A valuable resource
d) A regulatory burden

7. Which AI technique is most suitable for automated waste classification using images?
a) Natural Language Processing
b) Reinforcement Learning
c) Deep Learning
d) Linear Regression

8. Blockchain contributes to circular economy mainly by enabling:
a) Faster manufacturing
b) Centralized data control
c) Transparent and immutable records
d) Higher energy consumption

9. Which business model encourages manufacturers to retain ownership of products?
a) One-time sales
b) Product-as-a-Service
c) Bulk manufacturing
d) Spot market sales

10. Industry 4.0 technologies primarily enable circular economy by:
   a) Increasing production volume
   b) Eliminating human involvement
   c) Providing real-time, data-driven decision-making
   d) Reducing product customization

11. Which indicator measures the efficiency of material reuse and recovery?
   a) GDP growth rate
   b) Material circularity indicator
   c) Inflation index
   d) Market share

12. Predictive maintenance in circular systems helps to:
   a) Shorten product life
   b) Increase breakdown frequency
   c) Extend equipment and product life
   d) Eliminate the need for repairs

13. Which technology combination most effectively enables closed-loop supply chains?
   a) ERP and manual audits
   b) IoT, AI, and Blockchain
   c) Traditional logistics and warehousing
   d) Paper-based documentation

14. Circular economy-enabled Industry 4.0 systems are characterized by:
   a) One-way material flows
   b) Fragmented data
   c) Closed-loop material flows
   d) High waste generation

15. The ultimate goal of a data-driven circular economy is to achieve:
   a) Increased consumption
   b) Optimized resource loops and reduced environmental impact
   c) Faster disposal cycles
   d) Higher dependence on virgin resources

Index

'*Note:* Page numbers followed by "f" indicate figures and "t" indicate tables.'

F

G

H

I

P

R

Printed and bound by CPI Group (UK) Ltd, Croydon, CR0 4YY
14/09/2026
02216408-0006